Crop Science: Growth and Development of Crops

Crop Science: Growth and Development of Crops

Edited by Harold Salazar

SYRAWOOD
PUBLISHING HOUSE

New York

Published by Syrawood Publishing House,
750 Third Avenue, 9th Floor,
New York, NY 10017, USA
www.syrawoodpublishinghouse.com

Crop Science: Growth and Development of Crops
Edited by Harold Salazar

International Standard Book Number: 978-1-68286-738-9 (Hardback)

Cataloging-in-Publication Data

Crop science : growth and development of crops / edited by Harold Salazar.
    p. cm.
Includes bibliographical references and index.
ISBN 978-1-68286-738-9
1. Crop science. 2. Crops. 3. Crops--Growth. 4. Crops--Development.
I. Salazar, Harold.
SB91 .C76 2019
631--dc23

TABLE OF CONTENTS

Permissions

List of Contributors

Index

PREFACE

Crop science is a sub-field of biology that studies the practice of agriculture. It is a multidisciplinary science that integrates the principles of several natural, economic and social sciences. It incorporates the study of plant breeding, genetics, pathology, entomology, soil science, etc. Improving agricultural productivity and minimizing the effects of pests are other areas of interest in this field. Modern crop science uses the tools of agricultural biotechnology to achieve these objectives. The various advancements in crop science are glanced at in this book and their applications as well as ramifications are looked in detail. It elucidates the concepts and innovative models around prospective developments with respect to crop science. This book is a vital tool for all researching or studying agricultural science and crop science as it gives incredible insights into emerging trends and their applications.

The information shared in this book is based on empirical researches made by veterans in this field of study. The elaborative information provided in this book will help the readers further their scope of knowledge leading to advancements in this field.

Finally, I would like to thank my fellow researchers who gave constructive feedback and my family members who supported me at every step of my research.

Editor

THE INFLUENCE OF ORGANIC AND MINERAL FERTILIZATION ON YIELD OF THE WHEAT GROWN ON REDDISH PRELUVOSOIL

Theodora BORUGĂ, Costică CIONTU, Iulian BORUGĂ, Dumitru-Ilie SĂNDOIU

University of Agronomic Sciences and Veterinary Medicine of Bucharest, 59 Mărăşti Blvd., District 1, Bucharest, Romania

Corresponding author email: theodora.boruga@yahoo.com

Abstract

The purpose of this paper is to present the results obtained on the reddish preluvosoil of the Moara Domnească experimental field belonging to the Department of Soil Sciences from Faculty of Agriculture, University of Agronomic Sciences and Veterinary Medicine of Bucharest, in the agricultural year 2014-2015. Wheat crop was part of a wheat-barley-sugar beet crop rotation. The experiment was a two-factor concept based on the split plot method, organised in three replications, with factor A - organic fertilization and factor B - mineral fertilization. Plant residues and manure were used as natural organic fertilizers. The variants of factor B were different rates of mineral nitrogen. Obtained results showed that organic and mineral fertilization increased wheat yield. Variant a_3 was the best for organic fertilization - 40 t/ha leaves and epicotyls of sugar beet+N_{50}, while the optimum solution for mineral fertilization was b_4 - N_{150}. For the organically unfertilized variants, the highest yield resulted from mineral fertilizer with N_{200} (b_5), but N_{100} (b_2) was the best economic solution because of the number of fall plants.

Key words: wheat, organic fertilization, mineral fertilization, nitrogen.

INTRODUCTION

Wheat is a multi-purposes cereal, which is rich in nutrients necessary for human consumption (carbohydrates, proteins, lipids), but which is also used for feeding animal. Wheat straw can be used in animal nutrition and bedding, and not least as fertilizer. Wheat stubble incorporated under the furrow increases organic matter in the soil (Săndoiu et al., 2012; Ştefanic et al., 2015). To obtain a high yield of wheat, it is recommended to include it in a rational rotation (Ciontu et al., 2010; Săndoiu et al., 2007). Wheat monoculture should be avoided as it leads to the expansion of wheat-specific diseases, pests and weed appearance, which results in the gradual reduction in the obtained yields (Săndoiu et al., 2008, 2012). As a basic element in plant nutrition, nitrogen is the main ingredient of wheat proteins and additional application helps increase its protein content (Borlan and Hera, 1994; Criste, 2012)

In their nutrition, plants accumulate nitrogen from the soil in ammonia and nitric form, and convert it through amination and transamination reactions in protein substances (Borlan and Hera, 1994).

To optimize the application of mineral nitrogen, wheat requirements must be taken into consideration, together with the conditions provided by the soil (Dincă et al., 2010; Gîdea et al., 2015). The exclusive use of chemical fertilizers does not reduce the importance of manure. The largest gains at harvest are obtained from the combined action of chemical fertilizers with manure (Obrişcă et al., 2010; Criste, 2012).

MATERIALS AND METHODS

Research on organic and mineral fertilization for growing winter wheat on reddish preluvosoil began in 1991, in the experimental field belonging to the Department of Soil Sciences from Faculty of Agriculture, University of Agronomic Sciences and Veterinary Medicine of Bucharest, which is located at Moara Domnească Farm (Ilfov County from South Romania). The present paper presents the results obtained in 2014-2015. Wheat crop was part of a wheat-barley-sugar beet rotation.

The two factors experience was organised by the split plot method, in three replications with factor A - organic fertilization and B - mineral fertilization.

Factor A - organic fertilization included three experimental graduations: a_1 - unfertilized; a_2 -

30 t/ha of manure applied at preceding crop (wheat benefited of the residual effect of the manure applied at the preceding crop, which was sugar beet); a_3 - 40 t/ha leaves and epicotyls of sugar beet + N_{50} (50 kg/ha of nitrogen applied in autumn).

Factor B - mineral fertilization included five graduations: b_1 - unfertilized; b_2 - N_{60}; b_3 - N_{100}; b_4 - N_{150}; b_5 - N_{200}.

A mineral background of 70 kg.ha^{-1} of phosphorous was provided for all experimental plots. In the field experiment, it was used Glosa wheat cultivar.

Each variant of organic and mineral fertilization was placed in three replications for a better accuracy of the results. So, the field experiment had in total 45 experimental variants.

The determinations carried out in the experimental variants were aimed at measuring the yield, the hectoliter mass and the thousand grains weight (TGW). The results were calculated and statistically interpreted by method of variance analysis.

RESULTS AND DISCUSSIONS

The influence of organic fertilization on the winter wheat yield is presented in Table 1.

Table 1. Influence of organic fertilization on winter wheat yield (Moara Domnească, 2014/2015)

Experimental variant	Yield		Difference		Significance
	q/ha	%	q/ha	%	
a_1 - unfertilized organically	52.17	100.00	Control	-	
a_2 - 30 t/ha of manure applied to the preceding crop	55.60	106.57	3.43	6.57	*
a_3 - 40 t/ha of leaves and epicotyls of sugar beet + N_{50}	75.35	144.43	23.18	44.43	***

LSD$_{5\%}$ = 2.235 q/ha; LSD$_{1\%}$ = 3.698 q/ha; LSD$_{0.1\%}$ = 6.922 q/ha

The analysis of the data presented in Table 1 shows the following outcome: the yield of winter wheat was influenced by organic fertilization, so it increased from 52.17 q/ha for a_1 - unfertilized organically variant, to 55.60 q/ha in a_2 - 30 t/ha of manure applied to the preceding crop, reaching 75.35 g/ha in a_3 - 40 t/ha of leaves and epicotyls of sugar beet + N_{50}.

The 30 t/ha of manure applied to the preceding crop caused an increase in yield by 3.43 q/ha, which represented an increase of 6.67%.

The application of 40 t/ha leaves and epicotyls of sugar beet + N_{50} resulted in an increased wheat yield by 23.18 q/ha, which represented a 44.43% growth in yield.

In terms of organic fertilization, optimal variant is the variant a_3, which ensures the highest yield increase.

Table 2 shows the influence of mineral nitrogen fertilization in the absence of organic fertilization on wheat yield.

Table 2. Influence of mineral nitrogen fertilization in the absence of organic fertilization on yield of winter wheat (Moara Domnească, 2014/2015)

Experimental variant	Yield		Difference		Significance
	q/ha	%	q/ha	%	
b_1 – N_0	52.17	100.00	Control	-	
b_2 – N_{60}	70.08	134.33	17.91	34.33	***
b_3 – N_{100}	81.99	157.16	29.82	57.16	***
b_4 – N_{150}	83.97	160.95	31.80	60.95	***
b_5 – N_{200}	95.31	182.69	43.14	82.69	***

LSD$_{5\%}$ = 1.526 q/ha; LSD$_{1\%}$ = 2.074 q/ha; LSD$_{0.1\%}$ = 2.779 q/ha

Winter wheat yield increased proportionally with the amount of applied mineral nitrogen, as follows: for the unfertilized variant 52.17 q/ha, for N_{60} - 70.80 q/ha, for N_{100} - 81.99 q/ha, for N_{150} - 83.97 q/ha, and for N_{200} - 95.31 q/ha. Compared to the control variant (unfertilized), the rate of N_{60} resulted in a yield increase by 17. 91 q/ha, meaning a growth of 34.33%.

The application of a rate of N_{100} generated an increased yield, compared to the control variant of 29.82 q/ha, representing 57.16%.

The rate of N_{150} led to an increase in yield by 31.8 q/ha, representing 60.95%. Increasing the rate of nitrogen to N_{200} generated a yield growth of 43.14 q/ha, compared to the control variant by 82.69%

N_{200} variant was not recommended for mechanized harvesting since the degree of fall plants was 42% on average, which would decrease the harvested yield.

Variant b_4 - fertilization N_{150}, was also not recommended from an economic point of view because, compared with the N_{100} variant, it led to an increase of yield of only 1.98 q/ha. The added yield value was 99 lei/ha while the amount of fertilizer applied in addition to the N_{100} cost about 181 lei/ha. Therefore, from the economic point of view, the optimal variant was b_3 - N_{100}.

The analysis of the data presented in Table 3 shows the following results: mineral nitrogen fertilization, without any organic fertilization, had the greatest influence on all experimental variants; the maximum increase in yield was 43.14 q/ha at b_5 - N_{200}.

Mineral nitrogen application rates on a background of 30 t/ha of manure applied to the preceding crop resulted in a progressive yield increase; however, the increase rates were much lower, compared with the variant unfertilized organically. In variant a_2 - 30 t/ha of manure applied to the preceding crop, the optimal economic choice was N_{150}.

The influence of mineral nitrogen fertilization on a background of 40 t/ha of leaves and epicotyls of sugar beet + N_{50} has a much lower effect, of only 29.46 q/ha, although it obtained the highest yield of winter wheat.

Variant b_5-N_{200} caused a decrease in yield with 1.08 q/ha, compared to version b_4 - N_{150}, plus a high degree of fall plants; therefore, the optimal variant was b_4 - N_{150}.

Table 3. Influence of organic and mineral fertilization on the yield of winter wheat (Moara Domnească, 2014/2015)

Mineral fertilization	a_1 - unfertilized organically			a_2 - 30 t/ha of manure applied to the preceding crop			a_3 - 40 t/ha of leaves and epicotyls of sugar beet + N_{50}		
	Yield (q/ha)	Difference (q/ha)	Significance	Yield (q/ha)	Difference (q/ha)	Significance	Yield (q/ha)	Difference (q/ha)	Significance
b_1 - N_0	52.17	Control		55.60	Control		75.35	Control	
b_2 - N_{60}	70.08	17.91	***	71.01	15.41	***	86.37	11.02	***
b_3 - N_{100}	81.99	29.82	***	76.07	20.47	***	100.38	25.03	***
b_4 - N_{150}	83.97	31.80	***	87.92	32.32	***	104.81	29.46	***
b_5 - N_{200}	95.31	43.14	***	89.23	33.63	***	102.73	27.38	***

LSD$_{5\%}$ = 2.645 q/ha; LSD$_{1\%}$ = 3.595 q/ha; LSD$_{0.1\%}$ = 4.815 q/ha

The analysis of the data presented in Table 4 shows the following: organic nitrogen fertilization recorded in 40 t/ha of leaves and epicotyls of sugar beet + N_{50}, determined the most significant yield increase.

Table 4. Influence of organic fertilization on the yield of winter wheat at the same levels of mineral fertilization (Moara Domnească, 2014/2015)

Mineral fertilization	Comparison between organic fertilization variants					
	a_2 - a_1		a_3 - a_1		a_3 - a_2	
	Difference (q/ha)	Significance	Difference (q/ha)	Significance	Difference (q/ha)	Significance
b_1 - N_0	3.43	*	23.18	***	19.75	***
b_2 - N_{60}	16.29	***	16.29	***	15.36	***
b_3 - N_{100}	5.92	**	18.39	***	24.31	***
b_4 - N_{150}	3.95	*	20.84	***	16.89	***
b_5 - N_{200}	-6.08	00	7.42	**	13.50	***

LSD$_{5\%}$ = 3.218 q/ha; LSD$_{1\%}$ = 4.752 q/ha; LSD$_{0.1\%}$ = 4.752 q/ha

Organic fertilization on a mineral background of N_{200} caused lower yield than mineral fertilization on a N_{150} background, therefore is not justified economically and the optimal variant for mineral fertilizer must be lower than N_{150}.

As we can see, the yield was lower than the control variant with 6.08 q/ha. The best option was organic fertilization variant of a_3 - 40 t/ha leaves and epicotyls of sugar beet + N_{50}, with a mineral fertilization background b_4 - N_{150}. The influence of 40 t/ha leaves and epicotyls of sugar beet + N_{50} is visible for all variants of mineral fertilizer.

In the analysis of the data presented in Table 5, there is shows that hectoliter mass was influenced by the increase of organic fertilization. The highest hectoliter mass was obtained in variant a_3 - 40 t/ha leaves and epicotyls of sugar beet + N_{50}, generating a variation (1.76%) between a_2 - 30 t/ha of manure applied to the preceding crop.

Table 5. Influence of organic nitrogen fertilization on hectoliter mass for winter wheat (Moara Domnească, 2014/2015)

Experimental variant	Hectoliter mass		Difference		Significance
	kg/hl	%	kg/hl	%	
a_1 - unfertilized organically	79.8	100.00	Mt	-	
a_2 - 30 t/ha of manure applied to the preceding crop	80.1	100.37	0.30	0.37	
a_3 - 40 t/ha of leaves and epicotyls of sugar beet + N_{50}	81.5	102.13	1.7	2.13	***

LSD$_{5\%}$ = 0.52 kg/hl; LSD$_{1\%}$ = 0.86 kg/hl; LSD$_{0.1\%}$ = 1.60 kg/hl

The hectoliter mass analysis is widely used in agriculture and cereal processing industry. Hectoliter mass is mainly used in determining the degree of extraction of flour and determining storage space.

The analysis of the data in Table 6 shows that hectoliter mass increased with the amount of total nitrogen and variant a_3b_5 ensured the highest hectoliter mass (82 kg/hl). For some organic fertilization variants, using b_5 - N_{200} was not the best option as the hectoliter mass decreased once the rate of mineral nitrogen was higher. For the variant a_1 - unfertilized organically, we can observe that the hectolitre mass increases until a certain dose of mineral nitrogen (b_4 - N_{150}), after that it declines. For organic variant a_2 - 30 t/ha of manure applied to the preceding crop, the maximum hectolitre mass is reached for mineral dose of b_3 - N_{100}, after which the decline starts. Nevertheless, analysing the values below we can see that the best results were obtained in the organic variant a_3 - 40 t/ha leaves and epicotyls of sugar beet + N_{50}, regardless of the mineral nitrogen used. The variation between the nitrogen rates used for variant a_3 was not significant, meaning that for the hectoliter mass the plant residue worked very well in combination with mineral nitrogen.

Table 6. Influence of mineral and organic fertilization on winter wheat hectoliter mass (Moara Domnească, 2014/2015)

Mineral fertilization	a_1 . unfertilized organically			a_2 - 30 t/ha of manure applied to the preceding crop			a_3 - 40 t/ha of leaves and epicotyls of sugar beet + N_{50}		
	H.M. (kg/hl)	Difference (kg/hl)	Significance	H.M. (kg/hl)	Difference (kg/hl)	Significance	H.M. (kg/hl)	Difference (kg/hl)	Significance
b_1 - N_0	79.8	Control		80.1	Control		81.5	Control	
b_2 - N_{60}	80.6	0.80		81.3	1.20	**	81.8	0.30	
b_3 - N_{100}	81.3	1.50	**	81.7	1.60	***	81.7	0.20	
b_4 - N_{150}	81.7	1.90	***	81.5	1.40	**	81.6	0.10	
b_5 - N_{200}	81.0	1.20	**	81.2	1.10	*	82.0	0.50	

LSD$_{5\%}$ = 0.86 kg/hl; LSD$_{1\%}$ = 1.17 kg/hl; LSD$_{0.1\%}$ = 1.56 kg/hl

Thousand grains weight (TGW) is also an important indicator in evaluating cereal yield. TGW is the mass of a thousand seeds, with the humidity at time of measurement and it is expressed in grams with two decimals. TGW is a good indicator in determining the protein content. In this respect, it was found that thousand grains weight is directly proportional to the protein content in cereals.

The data presented in Table 7 reveal a general trend of decreasing, with increasing the amount of total nitrogen (mineral and organic) for thousand grains weight. As we can see below, for variant a_1 - unfertilized organically and a_2 - 30 t/ha of manure applied to the preceding crop, the thousand grains weight reaches it's maximum value, for mineral variant b_3 - N_{100}, after that it begins to decline. For a_3 - 40 t/ha leaves and epicotyls of sugar beet + N_{50}, the maximum thousand grains weight it is found for variant b_2, and a dose of only N_{60}.

Table 7. Influence of mineral and organic fertilization on TGW in winter wheat (Moara Domnească, 2014/2015)

Mineral fertilization	a_1 - unfertilized organically			a_2 - 30 t/ha of manure applied to the preceding crop			a_3 - 40 t/ha leaves and epicotyls of sugar beet + N_{50}		
	TKW (g)	Difference (g)	Signification	TKW (g)	Difference (g)	Signification	TKW (g)	Difference (g)	Signification
b_1 - N_0	48.53	Control		48.89	Control		49.71	Control	
b_2 - N_{60}	50.02	1.49		49.45	0.56		49.99	0.28	
b_3 - N_{100}	50.46	1.93		50.07	1.18		48.68	-1.03	
b_4 - N_{150}	48.61	0.08		47.05	-1.82		47.82	-1.89	
b_5 - N_{200}	46.80	-1.73		46.01	-2.88	o	47.57	-2.14	

LSD$_{5\%}$ = 2.65 g; LSD$_{1\%}$ = 3.60 g; LSD$_{0.1\%}$ = 4.82 g

CONCLUSIONS

Organic and mineral fertilization resulted in the increased winter wheat yield. The best option for organic fertilization was considered to be variant a_3 - 40 t/ha of leaves and epicotyls of sugar beet.

The variants of mineral fertilization depended on organic fertilization.

Winter wheat yield increased proportionally with the amount of applied mineral nitrogen.

For the organically unfertilized variant, the highest yield was obtained with mineral fertilization of N_{200} (b_5); however, economically, the optimum solution was N_{100} (b_2) because of the number of fall plants.

For the organically fertilized variant a_2 - 30 t/ha of manure applied to the preceding crop, the best mineral fertilizer variant was N_{150} (b_4), which also represented the best economic choice.

For the organically fertilized variant a_3 - 40 t leaves and epicotyls of sugar beet + N_{50}, the optimum solution for mineral fertilizer, in terms of yield, was N_{150} - b_4, but variant b_3 with N_{100} was the most suitable economically.

Mineral nitrogen application rates on a background of 30 t/ha manure applied to the preceding crop resulted in a progressive yield increase; however, the increase rates were much lower, compared with the variant unfertilized organically.

The variation between the nitrogen rates used for variant a_3 was not significant, meaning that for the hectoliter mass the plant residue worked very well in combination with mineral nitrogen.

REFERENCES

Borlan Z., Hera C., 1994. Fertilitatea si fertilizarea solurilor (compediu de agrochimie). Ed. Ceres, Bucureşti.

Ciontu C., Săndoiu D.I., Penescu A., Gâdea M., Obrişcă M., 2010. Research concerning influence of crop rotation to winter wheat on the reddish preluvosoil from Moara Domnească. Lucrări Ştiinţifice USAMV Bucureşti, Seria A, LIII, p. 240 -245.

Criste D., 2012. Fertilizarea cu gunoi de grajd şi efectele utilizării lui. www.recolta.eu.

Dincă L., Săndoiu D.I., Obrişcă M., Răducioiu G.A., 2010. Influence of organic and mineral nitrogen fertilization on wheat yield on the reddish preluvosoil in Danube Plain. Lucrări Ştiinţifice USAMV Bucureşti, Seria A, LIII, p. 200-206.

Gîdea M., Ciontu C., Săndoiu D.I., Penescu A., 2015. The role of Rotation and Nitrogen Fertilization Level upon the Economic Indicators at Wheat and Crops in Condition of a Long Term Experience. Journal of Agriculture and Agricultural Science- Procedia, Elsevier, Vol. 6, p. 24-29.

Obrişcă M., Săndoiu D.I., Ştefanic Gh., 2010. Studies and methodical applications for estimating the fertility state of soils. Lucrări Ştiinţifice UŞAMV Bucureşti, Seria A, LIII, p. 234-239.

Săndoiu D.I., Boguslawski E., Săndoiu I.F., 1996. Effect des engrais azotée et organiques sur le rendement du blé, dans des différentes conditions de milieu. In vol. Proceeding of the international Scientific Conference Bucharest, 7-12 Oct., "The effect of Cropping systems on yield, farm, produce quality, profitability and environment protection in the main crop and pasture lands".

Săndoiu D.I., Dumitrescu N.C., Ştefanic Gh., Obrişcă M., Gheorghiţă N., Ciontu C., Săndoiu I.F., 2007. Influenţa rotaţiei culturilor şi a fertilizării cu azot asupra producţiilor de grâu de toamnă şi porumb boabe şi a fertilităţii solului brun-roşcat. In vol. Ameliorarea, conservarea şi valorificarea solurilor degradate prin intervenţii antropice, Ed. Ion Ionescu de la Brad Iaşi, p. 20-40.

Săndoiu D.I., Gheorghiţă N., Obrişcă M., Dumitrescu N., Săndoiu I.F., 2008. Tehnici neconvenţionale în agrotehnica diferenţiată a zonei preluvosolului roşcat din Câmpia Română - implicaţii în fertilitatea solului şi creşterea producţiei. Revista de scientometrie CNCSIS, Braşov.

Săndoiu D.I, Dincă L., Ştefanic Gh., Ciontu C., Penescu A., 2012. Effects of organic and mineral manuring on microbial processes of reddish preluvosoil in Romanian Field after 19 years of experimentation. Agriculture Research, Proceeding of Romanian Academy, Series B, 14(3), p. 245-249.

Ştefanic Gh., Săndoiu D.I., Oprea G., 2015. Necessity to ameliorate the soils with compost for simultaneous increase of fertility and productivity. Soil Science No. 1, Vol. XLIX, p. 69-75.

2

MOLECULAR IDENTIFICATION AND METABOLIC SCREENING OF SOME YEAST STRAINS FROM FOODS

**Ortansa CSUTAK, Emilia SABĂU, Diana PELINESCU, Viorica CORBU,
Ioana CÎRPICI, Tatiana VASSU**

University of Bucharest, Faculty of Biology, Department of Genetics, 1-3 Aleea Portocalelor,
060101 Bucharest, Romania, Email: cs_ortansa@hotmail.com

Corresponding author email: cs_ortansa@hotmail.com

Abstract

Yeast strains from Saccharomyces, Kluyveromyces and Candida genera are present in most foods representing the basis for various industrial and biotechnological processes. The strains CMGB79 and CMGB159 were identified using PCR-RFLP on the ITS1-5.8S-ITS2 region as belonging to Candida parapsilosis, respectively, to Kluyveromyces marxianus. The primer OPA03 yield the highest degree of intraspecific RAPD polymorphism for the strains Saccharomyces cerevisiae CMGB59, CMGB121and ATCC201583. Lipase production was observed in the presence of Tween 80 in concentration of 0.1 and 0.5% for Candida parapsilosis CMGB79, respectively, 0.1 to 0.8% for Candida parapsilosis CBS604. The oleic acid represented the best substrate for lipase induction and cell growth for Kluyveromyces marxianus CMGB159. All the yeast strains tested positive for lipase synthesis in the presence of tributyrin. The antagonistic interactions between the studied strains were assessed using killer assays against Candida parapsilosis CMGB79 and CBS604. The killer activity was high for Kluyveromyces marxianus CMGB159 and good for the Saccharomyces cerevisiae strains, the toxin representing a stress factor which determined modifications in the sensitive cells. The results obtained during the present work showed that the characterized yeast strains present an important potential for applications in food industry, in obtaining probiotic compounds or as therapeutic agents of biomedical interest.

Key words: *Candida, Kluyveromyces, Saccharomyces, killer activity, lipases.*

INTRODUCTION

The intensive development of numerous industry and biotechnology domains and the advances of knowledge on the molecular and metabolic characteristics of the microorganisms, revealed that many genera and species present strains with multiple and various abilities. The ubiquitous presence of yeasts in natural and industrial environments and their ability to assimilate a wide range of substrates and to produce proteins or specific metabolites, represent the basis for their important practical applications.

The lipases are a family of enzymes involved in catalyzing the hydrolysis of triglycerides producing mono- or diglycerides, glycerol and fatty acids. Yeast lipases produced by *Candida, Kluyveromyces* and *Saccharomyces* species are intensively studied and used in food industry (for taste and flavour improvement, preservation of food products, hydrolysis of milk fat), in production of cosmetics, detergents, leather, pharmaceutics, in biomedicine and in bioremediation of polluted environments (Saxena et al., 1999; Vakhlu and Kour, 2006; Lock et al., 2007; Karigar and Rao, 2011).

Strains belonging to *S. cerevisiae, K. lactis* and *K. marxianus* produce extracellular killer toxins which induce the death of cells from the same species or even belonging to different genera and species. While the killer systems and toxins are well studied for *S. cerevisiae* (Marquina et al., 2002) and *K. lactis* (Schraffrath and Breunig, 2000), the knowledge concerning *K. marxianus* is still poor (Abranches et al., 1997). Even though the killer toxins are of main interest in beer, wine or bakery products, they gain a growing importance in obtaining monoclonal killer toxin-like antibodies with candidacidal activity (Magliani et al., 1997).

In the present work yeast strains isolated from foods are identified and characterized at molecular level and screened for the lipolytic activity, with emphasis on the influence of carbon and nitrogen sources on lipase induction and cell growth. Also, the killer activity was used to evaluate the antagonistic relations between the strains and to assess the effect of the killer toxin on *Candida* sensitive cells.

MATERIALS AND METHODS

1. Yeast strains
The yeast strains CMGB79, CMGB159, *Saccharomyces cerevisiae* CMGB59 and *S. cerevisiae* CMGB121 were previously isolated from fermented food products (diary), food industry and fodder and maintained in the Collection of Microorganisms of the Department of Genetics, Faculty of Biology, University of Bucharest, Romania (CMGB). For the experiments were used fresh cultures grown for 20 hours on Yeast Peptone Glucose (YPG) medium (0.5% yeast extract, 1% peptone, 0.2% glucose).

2. Analysis of the ITS1-5.8S-ITS2 amplicons
The genomic DNA of the strains CMGB79, CMGB159, *S. cerevisiae* CMGB59, *S. cerevisiae* CMGB121 and *S. cerevisiae* ATCC201583 was isolated (Csutak et al., 2014) and then amplified using the PCR program: initial denaturation 5 min at 94°C, 40 cycles of 1 min at 94°C, 30 sec at 55°C, 2 min at 72°C, and a final extension 5 min at 72°C. The reaction mix comprised 1 µl genomic DNA, 25 µl Thermo Scientific DreamTaq Green PCR Master Mix (2X) and 1.2 µM ITS1 (5'TCCGTAGGTGAACCTGCGG) and ITS4 (5'TCCTCCGCTTATTGATATGC) primers. The amplicons were digested with 0.5 µl of each endonuclease *Cfo* I (5'-GCG/C-3'), *Hae* III (5'-GG/CC-3'), *Hinf* I (5'-G/ANTC-3') and *Msp* I (5'-C/CGG-3') (10U/µl, Promega) and vizualized by gel electrophoresis using 1.7% agarose and Tris-Borate-EDTA (TBE) 0.5X. The sizes of the amplicons and restriction fragments were determined using the program Quantity One (Bio-Rad).

3. PCR-RAPD
The PCR-RAPD analysis of the strains *S. cerevisiae* CMGB59, *S. cerevisiae* CMGB121 and *S. cerevisiae* ATCC201583 was performed in a total volume of 25 µl using GoTaq Green Master Mix 2X (Promega), 1 µM of OPA01 (5'CAGGCCCTTC-3'), OPA03 (5'-AGTCAGCCAC-3'), OPB17 (5'-AGGGAACGAG-3') or M13 (5'-AGGGTGGCGGTTCT-3') and the

amplification program: initial denaturation 5 min at 94°C, 45 cycles of 1 min at 94°C, 1 min at 36°C and 2 min at 72°C, and a final extension of 10 minutes at 72°C. The RAPD fragments were analysed in 1.5% agarose gels in TBE 0.5X.

4. Screening for lipase production
Lipase production was evaluated using two types of tests. In a first place, we used the Tween opacity test with six different media: S (1% BactoPeptone, 0.5% NaCl, 0.01% CaCl$_2$, 2% agar-agar, 0.5% Tween 80), S w/o Ca (S without CaCl$_2$), S-0.4Ca (S with 0.04 % CaCl$_2$), S-1T (S with 0.1% Tween 80), S-8T (S with 0.8% Tween 80) and S-10T (S with 1% Tween 80) (Slifkin, 2000). Lipase production resulted in a white halo of precipitate of calcium salts surrounding the colonies of the tested strains. In parallel, lipase production was also determined by growing the yeast strains for five days on YPTA medium (0.3% yeast extract, 0.5% peptone, 1% tributyrin, 2% agar) (Shirazi et al., 1998; Darvishi, 2012). The qualitative evaluation of the lipolytic activity was based on the ratio between the measured diameters of the tributyrin hydrolysis halo (TH) and those of the cell colonies (CC). Thus, if the TH/CC ratio is 1, the yeast strain is not able to hydrolyse the tributyrin and does not produce lipases. On the contrary, a high TH/CC ratio indicates active lipase secretion.

5. Influence of the growth media on the lipolytic activity
The studied strains (2 x 10^7 cells/ml) were grown on liquid media: T20 (0.7% YNB, 2.5% Tween 20), T80 (0.7% YNB, 2.5% Tween 80) and D (1% olive oil, 0.2% yeast extract, 0.05 % KH$_2$PO$_4$, 0.05% K$_2$HPO$_4$, 0.05% MgSO$_4$ x 7 H$_2$O, 0.01% CaCl$_2$, 0.01% NaCl) (Tsuboi et al., 1996; Darvishi et al., 2009). The ability of the yeast strains to produce lipase for hydrolyzing Tween 20, Tween 80 and olive oil was estimated by monitoring cell growth after 24, 48 and 72 hours with a Thoma counting chamber. The aspect of the cells was also microscopically observed.

6. Antagonistic activity
The antagonistic activity was assessed using the killer assay. Colonies from overnight YPG grown cultures of CMGB159, *S. cerevisiae* CMGB59, *S. cerevisiae* CMGB121 and *K. lodderae* CMGB64 were spotted onto Petri

dishes with killer medium (0.1 M phosphate citrate buffer pH 4.8, 2% glucose, 1% yeast extract, 2% agar, 0.03% methylene blue) inoculated with 10^7 cells/ml of potential sensitive yeast strains: CMGB79 and *C. parapsilosis* CBS604. The plates were checked daily during seven days of incubation at 22°C. A strain was considered as killer positive when an inhibition halo or a zone with reduced growth of the sensitive strain appeared surrounding the colonies (Vassu et al., 2001). The killer toxin effect was also microscopically analyzed (40X) by observing the modified sensitive yeast cells.

RESULTS AND DISCUSSIONS

Molecular identification

Accurate identification of yeasts using molecular approaches represents the basis for the development of new research strategies aimed to improve the metabolic abilities of yeast strains for various practical applications such as food industry. Besides sequencing, the PCR-RFLP analysis of the ITS1-5.8S rDNA-ITS2 region is one of the techniques extensively used in interspecific differentiation of yeast species from food industry (Clemente-Jimenez et al., 2004; Naumova et al., 2004).

The genomic DNA obtained from the strains CMGB79 and CMGB159 was amplified using the primers ITS1 and ITS4 and the amplicons were digested with four endonucleases. The PCR products had 540 bp for CMGB79 and 740 bp for CMGB159 (Table1).

The restriction profiles obtained with *Cfo* I and *Hinf* I for CMGB79 and CMGB159 (Table 1, Figure 1) were highly similar with those obtained for other strains from the species *C. parapsilosis* (Guillamon et al., 1998; Esteve-Zarzoso et al., 1999; Jeyaram et al., 2008), respectively, *K. marxianus* (Bockelmann et al., 2008; Pham et al., 2011; Verdugo Valdez et al., 2011).

No restriction sites were found for *Msp* I for both CMGB79 and CMGB159. Our results were identical with those obtained during previous studies for *C. parapsilosis* (Mirhendi et al., 2001; Mirhendi et al., 2006; Basilio et al., 2008; Ayatollahi Mousavi et al., 2012) and for *K. marxianus* for which, in present, only few data can be found (Pérez-Brito et al., 2007).

Table 1. The size of the amplicons and restriction fragments of ITS1-5.8S-ITS2 region for the strains CMGB79 and CMGB159

Yeast strain	Amplicon (bp)	Fragment size (bp)			
		Cfo I	*Hae* III	*Hinf* I	*Msp* I
CMGB79	540	270, 240	420, 110	260, 255	520
CMGB159	740	300, 210, 180, 90	660, 80	285, 190, 120, 70	720

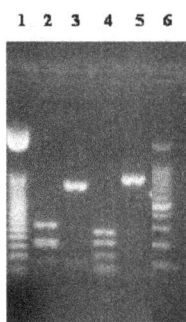

Figure 1. The restriction profile of the ITS1-5.8S-ITS2 amplicon of the strain CMGB159 obtained with: 1 -50 bp DNA Step ladder (Promega), 2- *Cfo* I, 3 - *Hae* III, 4 - *Hinf* I, 5 - *Msp* I, 6-Benchtop 100 bp DNA ladder (Promega)

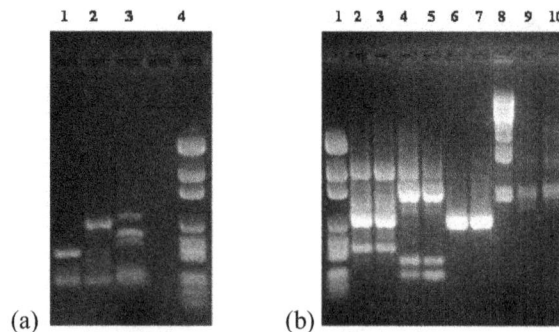

Figure 2. The PCR-RAPD amplicons obtained for: (a) *S. cerevisiae* ATCC201583 with the primers: 1 - OPA03, 2 - OPB17, 3 - M13; (b) *S. cerevisiae* CMGB59 (2, 4, 6, 9) and *S. cerevisiae* CMGB121 (3, 5, 7, 10) with the primers: 2, 3 - M13; 4, 5 - OPB17; 6, 7 - OPA01; 9, 10 - OPA03. Molecular markers used: (a) 4 and (b) 1 - Benchtop pGEM marker (Promega), 8 - 1 kb DNA ladder (Promega)

On the contrary, the variations observed in the restriction patterns obtained with *Hae* III for our strains compared to strains from the works mentioned above, are probably the result of the intraspecific variability related to the different environments from which the strains were isolated. The data from the analysis of the ITS1-5.8S rDNA-ITS2 amplicons allowed us to conclude that the strain CMGB79 belongs to *C. parapsilosis* and the strain CMGB159 to *K. marxianus*.

PCR-RAPD analyses were performed in order to determine intraspecific polymorphisms within the *Saccharomyces* specie using four primers: OPA01, OPA03, OPB17 and M13. Only two amplified fragments were obtained in the case of OPA01, while OPB17 and M13 yield the highest number of bands, followed by OPA03 (Figure 2b). The profiles were identical for *S. cerevisiae* CMGB59 and *S. cerevisiae* CMGB121 except when the OPA03 primer was used (Figure 2b). Therefore, in order to obtain a more accurate discriminatory characterization of the *S. cerevisiae* strains, we decided to also use a reference strain: *S. cerevisiae* ATCC201583. The compared analysis of the RAPD profiles (Figure 2a and b) showed that, although the OPA03 primer yield a total of only seven bands for all three strains, it provided a maximum degree of polymorphism - 100% (Table 2). Our results can be correlated with similar studies using the primers OPA03, OPB17 and M13 (Gomes et al., 2003; Andrade et al., 2006; Zhang et al., 2015).

Table 2. Polymorphic frequency of *S. cerevisiae* strains generated with primers M13, OPB17 and OPA03

Primer	Total amplified lanes	Common lanes	Polymorphic lanes	Polymorphic frequency (%)
M13	10	6	4	40
OPB17	10	3	7	70
OPA03	7	0	7	100

Lipolytic activity

The strains *C. parapsilosis* CMGB79, *C. parapsilosis* CBS604 and *K. marxianus* CMGB159 were tested for lipase production, since they belong to genera described as having lipolytic activity (Vakhlu and Kour, 2006). The strain *C. parapsilosis* CBS604 showed best results on S-1T medium with 0.1% Tween 80. Good results were also obtained for Tween 80 at 0.5% (S and S-0.4 Ca media) and 0.8% (S-8T medium) (Table 3).

Table 3. Lipase production on media with Tween 80 and CaCl$_2$ after seven days of incubation

Growth media	Opacity halos (mm)	
	C. parapsilosis CMGB79	*C. parapsilosis* CBS604
S	2	3
S w/o Ca	-	-
S-0.4Ca	2	3
S-1T	2	5
S-8T	-	3
S-10T	-	2

(-) no halo

C. parapsilosis CMGB79 was able to hydrolyze Tween 80 when added in small concentrations (0.1 and 0.5%). Moreover, it seems that lipase production was not influenced by the amount of Tween 80 from the growth media, or the variation of the activity was reduced and therefore could not be detected using the Tween opacity test even in the presence of higher concentrations of CaCl$_2$ (S-0.4Ca medium). In contrast, no lipase synthesis could be observed for the strain *K. marxianus* CMGB159 under the same conditions.

Since *C. parapsilosis* CBS604 showed the best screening results from the two *Candida* strains tested, we decided to investigate the factors that influence its lipolytic activity. We observed the ability of the yeast strain to use as sole carbon source for growth the oleic acid (C$_{18}$) and the lauric acid (C$_{12}$) liberated by the hydrolysis of Tween 80 (polyoxyethylene (20) sorbitan monooleate), respectively, Tween 20 (polyoxyethylene (20) sorbitan monolaurate), added in various concentrations.

In general, *C. parapsilosis* CBS604 preferred Tween 20 (E20-T20) as source for lipase production. The growth was reduced in the presence of 0.1% Tween. Best results with similar curve profiles were obtained using 2.5% Tween. The lipolytic activity was high within the first 24 hours, followed by a plateau phase at 48 hours and a slower ascending curve at 72 hours (Figure 3). This might indicate that, in the first day of incubation, only part of the Tween was hydrolyzed forming fatty acids. Once these carbon sources exhausted, the growth was reduced. In the next hours, a possible enhancement of the lipolytic activity could take place due to residual Tween from the environment. New quantities of fatty acids were liberated, providing thus a basis for the cell growth. It was reported previously that Tween 20 and 80 from the culture media activate lipase production in *Candida* and promote synthesis of multiple lipase isoforms (Aravidan et al., 2007). On the other hand, the ethylene oxide (oxyethylene) groups present in Tween, are successfully used in production of copolymers aimed to increase lipase activity (Park et al., 2006).

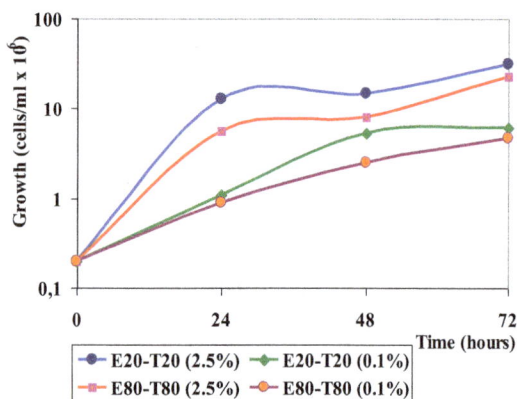

Figure 3. Effect of carbon source on cell growth of *C. parapsilosis* CBS604

The influence of carbon and nitrogen substrates on lipase synthesis of the strain *K. marxianus* CMGB159 was determined by monitoring the cell growth in liquid media. Cell multiplication was fast within the first 24 hours similar to *C. parapsilosis* CBS604, then it maintained a constant rate or decreased rapidly in the case of Tween 80 (E80), respectively, Tween 20 (E20) (Figure 4). The lipolytic activity was enhanced in the presence of olive oil (medium D). The olive oil and the oleic acid, component of the olive oil and Tween 80, were determined as best carbon source for lipase production also for other *K. marxianus* strains (Lock et al., 2007; Stergiou et al., 2012).

The nitrogen source played an important role in the process. In D medium, the yeast extract provided a complex nutritional basis for the cell growth, comparatively to the YNB (Yeast Nitrogen Base) used in E20 and E80 cultures. This could be observed also at microscopic level, *K. marxianus* CMGB159 presenting numerous cells and developed pseudohyphae after 48 hours of incubation on D medium (Figure 5).

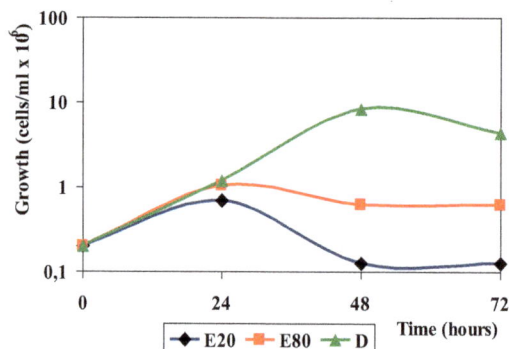

Figure 4. The influence of media on the cell growth of *K. marxianus* CMGB159

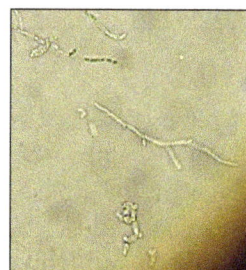

Figure 5. *K. marxianus* CMGB159 pseudohyphae on D media with olive oil

When tested on media with tributyrin, *K. marxianus* CMGB159 showed the highest lipolytic ability, followed closely by the *S. cerevisiae* strains CMGB59 and CMGB121 and, finally, by *C. parapsilosis* CMGB79 (Table 4). All three species can be found in fermented food, such as various dairy products, in which the butyric acid released by tributyrin hydrolysis is one of the main chemical compounds. Recent studies showed that the butyric acid inhibits the growth of potential pathogenic bacteria such as *Helicobacter pylori* (Yonezawa et al., 2012), has a major positive impact on the energy metabolism of the host when produced by the intestinal microbiota and is involved in gene regulation processes (Kasabuchi et al., 2015).

Therefore, the ability to hydrolyze trybutirin represents an important characteristic that recommend our strains for potential probiotic use.

Table 4. Tributyrin hydrolysis after seven days of incubation

Yeast strains	K. marxianus CMGB159	C. parapsilosis CMGB79	S. cerevisiae CMGB59	S. cerevisiae CMGB121
Lipolytic activity (HT/CC)	1.8	1.5	1.7	1.7

Antagonistic interactions

The antagonistic interactions between yeast strains are widespread not only in nature, but also in industry, having a major importance on the quality of the product and on the rate of success of the process.

S. cerevisiae and K. marxianus are species with antimicrobial activities related to the presence of killer toxins acting both intra and interspecific, including against Candida species (Marquina et al., 2002; Hernández et al., 2008). C. parapsilosis can be found in fermented food such as the fruit juice (Arias et al., 2002), but the most important aspect is represented by its pathogenicity, being one of the most known opportunistic pathogens which causes nosocomial infections (van Asbeck et al., 2009).

In fact, in present, the killer toxins present a growing importance not only for food industry, but also as therapeutic agents (Magliani et al., 2008).

Therefore, it was interesting to determine the potential interactions between Candida, Kluyveromyces and Saccharomyces strains. Besides the strains described during this work, we also tested the strains C. parapsilosis CMGB79 and K. lodderae CMGB64 from the pharmaceutical industry (Csutak et al., 2014). K. marxianus CMGB159 showed good killer activity against C. parapsilosis CMGB79 while for C. parapsilosis CBS604 the results were less important (Figure 6 a, b). S. cerevisiae CMGB59 and S. cerevisiae CMGB121 formed growth inhibition halos for both Candida strains tested (Figure 6 c, d). When observed at the microscope, the presence of the killer toxin from K. lodderae CMGB64 determined modifications in the C. parapsilosis CBS604 culture: the appearance of pseudohyphae with large

vacuoles and dead cells which absorbed the methylene blue (Figure 7).

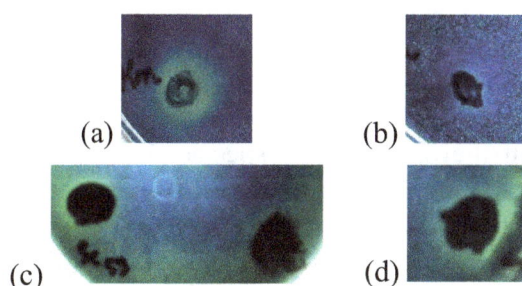

(a)   (b)

(c)   (d)

Figure 6. Killer activity of:
K. marxianus CMGB159 against C. parapsilosis CMGB79 (a) and C. parapsilosis CBS604 (b); S. cerevisiae CMGB59 and S. cerevisiae CMGB121 against C. parapsilosis CMGB79 (c); S. cerevisiae CMGB59 against C. parapsilosis CBS604 (d)

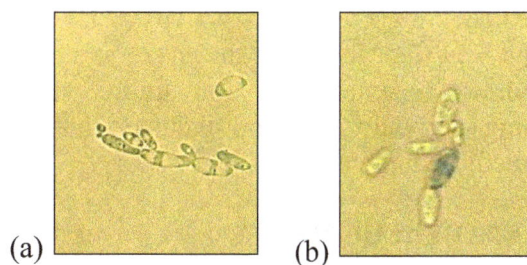

(a)   (b)

Figure 7. Modifications of C. parapsilosis CBS604 cells due to K. lodderae CMGB64 killer toxin: pseudohyphae with large vacuoles (a) and dead cells coloured with methylene blue (b)

Until present, the zymocin produced by K. lactis is the best known toxin of the Kluyveromyces genera. It acts in G1 before DNA replication causing cell-cycle arrest in the sensitive cells, similar to the K28 toxin from S. cerevisiae (Schaffrath and Breunig, 2000). It has been suggested that the presence of low concentrations of killer toxins in the environment determine caspase-like apoptosis and accumulation of reactive oxygen species in the susceptible cells (Schmitt and Breinig,

2006). The vacuoles play an important role in cell adaptation to these processes. Moreover, emergence of pseudohyphae at *Candida* is accompanied by appearance of numerous large vacuoles (Palmer et al., 2005). This might explain our observations regarding the mechanism of action observed for *K. lodderae* CMGB64 toxin on *C. parapsilosis* CBS604.

CONCLUSIONS

Yeast strains isolated from foods were characterized at molecular level and screened for specific metabolic abilities. The strains CMGB79 and CMGB159 were identified as belonging to *C. parapsilosis*, respectively, to *K. marxianus*. The primer OPA03 assured the highest degree of intraspecific polymorphisms within the *S. cerevisiae* speciae. When tested for the lipolytic activity, *C. parapsilosis* CBS604 hydrolyzed Tween 20 and Tween 80 in concentration up to 2.5%, while the presence of oleic acid and nitrogen substrates had an important impact on lipase induction and cell proliferation in *K. marxianus* CMGB159. All the strains showed good ability to hydrolyze tributyrin which might represent a basis for studies concerning a potential probiotic use. Moreover, the antagonistic interactions revealed high killer activity for the *Kluyveromyces* and *Saccharomyces* strains, an important characteristic for future studies aiming not only the food industry, but also biomedical applications.

ACKNOWLEDGMENTS

This work was supported by CNFIS - UEFISCDI project PCCA number 105/2012.

REFERENCES

Abranches J., Mendonça-Hagler L.C., Hagler A.N., Morais P.B., Rosa C.A., 1997. The incidence of killer activity and extracellular proteases in tropical yeast communities. Canandian Journal of Microbiology, 43(4): p. 328-336.

Andrade M.J., Rodriguez M., Sánchez B., Aranda E., Córdoba J.J., 2006. DNA typing methods for differentiation of yeasts related to dry-cured meat products. International Journal of Food Microbiology, 107: p. 48-58.

Aravindan R., Anbumathi P., Viruthagiri T., 2007. Lipase applications in food industry. Indian Journal of Biotechnology, 6: p. 141-158.

Arias C.R., Burns J.K., Friedrich L.M., Goodrich R.M., Parish M.E., 2002. Yeast species associated with orange juice: evaluation of different identification methods. Applied and Environmental Microbiology, 68(4): p. 1955-1961.

Ayatollahi Mousavi S.A., Salari S., Rezaie S., Shahabi Nejad N., Hadizadeh S., Kamyabi H., Aghasi H., 2012. Identification of *Candida* species isolated from oral colonization in Iranian HIV-positive patients, by PCR-RFLP method. Jundishapur Journal of Microbiology, 5 (1): p. 336-340.

Basílio A.C.M., de Araujo P.R.L., de Morais J.O.F., da Silva Filho E.A., de Morais Jr. M.A., Simoes D.A., 2008. Detection and identification of wild yeast contaminants of the industrial fuel ethanol fermentation process. Current Microbioogy, 56: p. 322-326.

Bockelmann W., Heller M., Heller K.J., 2008. Identification of yeasts of dairy origin by amplified ribosomal DNA restriction analysis (ARDRA). International Dairy Journal, 18: p. 1066-1071.

Clemente-Jimenez J.M., Mingorance-Cazorla L., Martinez-Rodriguez S., Las Heras-Vázquez F.J., Rodriguez-Vico F., 2004. Molecular characterization and oenological properties of wine yeasts isolated during spontaneous fermentation of six varieties of grape must. Food Microbiology, 21: p. 149-155.

Csutak O., Stoica I., Vassu T., 2014. Molecular identification and antimicrobial activity of two new *Kluyveromyces lodderae* and *Saccharomyces cerevisiae* strains. Biointerface Research in Applied Chemistry, 4 (6): p. 873-878.

Darvishi F., Nahvi I., Zarkesh-Esfahani H., Momenbeik F., 2009. Effect of plant oils upon lipase and citric acid production in *Yarrowia lipolytica* yeast. Journal of Biomedicine and Biotechnolology, article ID:562943.

Darvishi F., 2012. Expression of native and mutant extracellular lipases from *Yarrowia lipolytica* in *Saccharomyces cerevisiae*. Microbial Biotechnology, 5(5): p. 634-641.

Esteve-Zarzoso B., Belloch C., Uruburu F., Querol A., 1999. Identification of yeasts by RFLP analysis of the 5.8S rRNA gene and the two ribosomal internal transcribed spacers. International Journal of Systematic Bacteriology, 49: p. 329-337.

Gomes L.H., Echeverrigaray S., Conti J.H., Lourenço M.V.M., Duarte K.M.R., 2003. Presence of the yeast *Candida tropicalis* in figs infected by the fruit fly *Zaprionus indianus* (dip.: Drosophilidae). Brazilian Journal of Microbiology, 34: p. 5-7.

Guillamon J.M., Sabate J., Barrio E., Cano J., Querol A., 1998. Rapid identification of wine yeast species based on RFLP analysis of the ribosomal internal transcribed spacer (ITS) region. Archives of Microbiology, 169: p. 387-392.

Hernández A., Martín A Córdoba M.G., Benito M.J., Aranda E., Pérez-Nevado F., 2008. Determination of killer activity in yeasts isolated from the elaboration of seasoned green table olives. International Journal of Foof Microbiology, 121(2): p. 178-188.

Jeyaram K., Singh M.W., Capece A., Romano P., 2008, Molecular identification of yeast species associated with 'Hamei'. A traditional starter used for rice wine production in Manipur, India. International Journal of Food Microbiology, 124: p. 115-125.

Karigar C.S., Rao S.S., 2011. Role of microbial enzymes in the bioremediation of pollutants: a review. Enzyme Research, article ID: 805187.

Kasubuchi M., Hasegawa S., Hiramatsu T., Ichimura A., Kimura I., 2015. Dietary gut microbial metabolites, short-chain fatty acids, and host metabolic regulation. Nutrients, 7: p. 2839-2849.

Lock L.L., Corbellini V.A., Valente P., 2007. Lipases produced by yeasts: powerful biocatalysts for industrail purposes. TECNO-LÓGICA, Santa Cruz do Sul, 11 (n. 1 e 2): p. 18-25.

Magliani W., Conti S., Gerloni M., Bertolotti D., Polonelli L., 1997. Yeast killer systems. Clinical Microbiology Reviews, 10: p. 369-400.

Magliani W., Conti S., Travassos L.R., Polonelli L., 2008. Minireview. From yeast killer toxins to antibiobodies and beyond. FEMS Microbiology Letters, 288: p. 1-8.

Marquina D., Santos A., Peinado J.M., 2002. Biology of killer yeasts. International Microbiology, 5: p. 65-71.

Mirhendi S.H., Kazemi B., Kordbacheh P., Samiei S., Khoramizadeh M.R., Pezeshki M., 2001. A PCR-RFLP method to identification of the important opportunistic fungi: Candida species, Cryptococcus neoformans, Aspergillus famigatus and Fusarium solani. Iranian Journal of Public Health, 30(3,4): p. 103-106.

Mirhendi H., Makimura K., Khoramizadeh M.R., Yamaguchi H., 2006. One enzyme PCR-RFLP assay for identification of six medically important Candida species. Japanese Journal of Medical Mycology, 47: p. 225-229.

Naumova E.S., Sukhotina N.N., Naumov G.I., 2004. Molecular-genetic differentiation of the dairy yeast Kluyveromyces lactis and its closest wild relatives. FEMS Yeast Research, 5, p. 263-269.

Palmer G.E., Kelly M.N., Sturtevant J.E., 2005. The Candida albicans vacuole is required for differentiation and efficient macrophage killing. Eukaryotic Cell, 4 (10): p. 1677-1686.

Park K., Lee S., Maken S., Koh W., Min B., Park J., 2006. Modification of lipase from Candida rugosa with poly(ethylene oxide-co-maleic anhydride) and its separation using aqueous two-phase partition system. Korean Journal of Chemical Engineering, 23(4): p. 601-606.

Pérez-Brito D., Tapia-Tussell R., Quijano-Ramayo A., Larqué-Saavedra A., Lappe P., 2007. Molecular characterization of Kluyveromyces marxianus strains isolated from Agave fourcroydes (Lem.) in Yucatan, Mexico. Molecular Biotechnology, 37: p. 181-186.

Pham T., Wimalasena T., Box W.G., Koivuranta K., Storgårds E., Smart K.A., Gibson B.R., 2011. Evaluation of ITS PCR and RFLP for differentiation and identification of brewing yeast and brewery 'wild' yeast contaminants. Journal of the Institute of Brewing, G-2011-1130-1177: p. 556-568.

Saxena R.K., Ghosh P.K., Gupta R., Sheba Davidson W., Bradoo S., Gulati R., 1999. Microbial lipases: Potential biocatalysts for future industry. Current Science, 77 (1): p. 101-116.

Schaffrath R., Breunig K.D., 2000. Genetics and molecular physiology of the yeast Kluyveromyces lactis. Fungal Genetics and Biology, 30: p. 173-190.

Schmitt M.J., Breinig F., 2006. Yeast viral killer toxins: lethality and self-protection. Nature Reviews Microbiology, 4: p. 212-221.

Shirazi S.H., Rahman S.R., Rahman M.M., 1998. Short communication: Production of extracellular lipases by Saccharomyces cerevisiae. World Journal of Microbiology & Biotechnology, 14: p. 595-597.

Slifkin M., 2000. Tween 80 opacity test responses of various Candida species. Journal of Clinical Microbiology, 38 (12): p. 4626-4628.

Stergiou P.Y., Foukis A., Sklivaniti H., Zacharaki P., Papagianni M., Papamichael E.M., 2012. Experimental investigation and optimization of process variables affecting the production of extracellular lipase by Kluyveromyces marxianus IFO 0288. Applied Biochemistry and Biotechnology, 168 (3): p. 672-680.

Tsuboi R., Komatsuzaki H., Ogawa H., 1996. Induction of an extracellular esterase from Candida albicans and some of its properties. Infection and Immunity, 64: p. 2936-2940.

Vakhlu J., Kour A., 2006. Yeast lipases: enzyme purification, biochemical properties and gene cloning. Electronic Journal of Biotechnology, 9 (1): p. 69-85.

van Asbeck E.C., Clemons K.V., Stevens D.A., 2009. Candida parapsilosis: a review of its epidemiology, pathogenesis, clinical aspects, typing and antimicrobial susceptibility. Critical Reviews in Microbiology, 35 (4): p. 283-309.

Vassu T., Stoica I., Csutak O., Muşat F., 2001. Genetica microorganismelor şi inginerie genetică microbiană. Note de curs şi tehnici de laborator, Ed. Petrion, Bucharest.

Verdugo Valdez A., Segura Garcia L., Kirchmayr M., Ramirez Rodriguez P., González Esquinca A., 2011. Yeast communities associated with artisanal mezcal fermentations from Agave salmiana. Antonie van Leeuwenhoek, 100: p. 497-506.

Yonezawa H., Osaki T., Hanawa T., Kurata S., Zaman C., Hoong Woo T.D., Takahashi M., Matsubara S., Kawakami H., Ochiai K., Kamiya S., 2012. Destructive effects of butyrate on the cell envelope of Helicobacter pylori. Journal of Medical Microbiology, 61: p. 582-589.

Zhang M., Shi J., Jiang L., 2015. Modulation of mitochondrial membrane integrity and ROS formation by high temperature in Saccharomyces cerevisiae. Electronic Journal of Biotechnology, 18: p. 202-209.

3

RESEARCH ON OBTAINING, CHARACTERIZATION AND USE OF EDIBLE FILMS IN FOOD INDUSTRY

Floricel CERCEL[1], Mariana STROIU[1], Daniela IANIȚCHI[2], Petru ALEXE[1]

[1] "Dunarea de Jos" University of Galati, Faculty of Food Science and Engineering,
111 Domneasca Street, 800201, Galati, Romania
[2]University of Agronomic Sciences and Veterinary Medicine of Bucharest,
59 Mărăști Blvd., District 1, 011464, Bucharest, Romania

Corresponding author email: dianitchi@yahoo.com

Abstract

This review study aimed to give information about the use of plant extracts in meat product processing as antimicrobial and antioxidant agent. Microbial spoilage and lipid oxidation are the major causes of the deterioration and reduction of shelf-life in meat products. Lipid oxidation in meat products results in formation of off-flavors and undesirable chemical compounds such as aldehydes, ketones, alcohols and hydrocarbons. Growth of microorganisms in meat products causes not only microbial spoilage but also development off oodborne diseases. To inhibit lipid oxidation and growth of microorganisms, especially pathogenic microorganisms in meat products, several preservation techniques, such as pasteurization, reduction of water activity (salting, drying, freezing etc.), acidification, fermentation, synthetic and natural antimicrobial and antioxidant additives have been used in meat industry. Many synthetic and natural food additives such as butylated hydroxytoluene (BHT), butylated hydroxyanisole (BHA), propyl gallate, α-tocopherol, nisinand organic acidsare commonly used in the meat industry to inhibit or delay the oxidation process and reduce the microbial growth. In recent years, consumer demands for natural food additives have increased because of negative and toxic effects of synthetic food additives on human health. Herbs, spices, fruits and vegetables, and their powders, oils and extracts have been reported to be a good source of various phenolic compounds, such as flavonoids, terpenoids, carotenoids, could therefore be incorporated in meat products as a source of natural antioxidants and antimicrobials to extend shelf-life and safety of meat products.

Key words: edible films, myofibrillar proteins, water vapor permeability, moisture content, color of the films.

INTRODUCTION

The films are products obtained from food biopolymers and additives having food purity. Film-forming biopolymers can be proteins, polysaccharides and lipids.

The preparation of edible films and edible membranes, based on muscle protein, can be accomplished using:

a) *myofibrillar protein concentrates*:
- fish surimi (Cuq et al., 1995, 1997; Monterrey-Q, 1988);
- beef surimi (Souza et al., 1997);
- surimi from mechanically deboned poultry, heart muscle (Ionescu et al., 2008). The following categories of surimi can be used: wet surimi, frozen or thawed surimi, surimi dried by lyophilisation (Monterrey-Q, 1998) or surimi dried in the air (Cuq et al., 1997d);

b) *sarcoplsamatic proteins* (Iwata, 2001; Tanaka et al., 2001). Sarcoplasmatic proteins, unlike the myofibrillar proteins, are globular proteins that require an initial heat denaturation to form a continuous matrix (Iwata et al., 2000). By the heat treatment, the globular protein structure is changed, causing an exposure of the -SH groups, and concequently -S-S- links are produced between the adjacent protein chains and also hydrophobic interactions are occuring (Perez-Gogo and Krochta, 2001; Sobral et al., 2004, 2005; Garcia and Sobral, 2005);

c) *fish muscle* that contains both myofibrillar and sarcoplasmatic proteins (Nile Tilapia, Paschoalich et al., 2003).

MATERIALS AND METHODS

Bighead carp was procured fresh from the local fish store.

The fish was transported to the laboratory in a cool bag and then stored at 4°C until processing.

Determining the approximate chemical composition

The contents of water, protein, fat and ash were determined using standard method of analysis (AOAC, 1990; Ionescu et al., 1992). Also, moisture was determined by fast drying to constant weight using the thermobalance "Precisa XM 60" Total nitrogen was determined by Kjeldahl semimicro method, mineralization being performed in the "Trade Raypa" facility. Total proteins were calculated by multiplying the total nitrogen content by a factor of 6.25. All chemical analyzes were carried out in duplicate.

The pH was measured potentiometrically, using the pH meter type "Hanna" using protein dispersions with a concentration of 10% (G/V)), at a temperature of 22 ± 1^0C.

Samples were ran in duplicate.

The formation of biodegradable/edible films

In order to obtain the protein films, two methods are used:

- *the solvent process* involves the protein dispersion or solubilization in the film-forming solution. This procedure has been extensively studied and applied to produce edible / biodegradable films and membranes from diferent proteins and, in particular myofibrillar proteins (Cuq et al., 1995, 1998; Monterrey-Q, 1988);

- *dry process* is based on the thermoplasmatic properties of the proteins to a low water content (Hernandez-Izquierdo et al., 2008; De Graaf, 2000).

Properties of edible films and coatings based on proteins
Films solubility

The proteins with high molecular weight are generally insoluble in water and thus have a high potential to form water-resistant films (Cuq et al., 1998). Protein films do not lose integrity after 24 hours of immersion in water (Cuq et al., 1998b). Plasticizers used (sorbitol, glycerin or sucrose) in the manufacture of protein-based films increase the content of dry substance soluble in water. In general, hydrophilic plasticizers improve the solubility, which increases when the levels of added plasticizers is increased. Monterey-Q (1998) reported that a significant part of glycerol remains insoluble in water, suggesting the production of protein-glycine interactions.

Protein monomers and low molecular weight peptides, formed during the conditioning of film forming solutions and immobilized in the network, may be water-soluble protein components (Cuq et al., 1995).

Water vapor permeability

The values of water vapor permeability of films based on fish myofibrillar proteins are bigger ($3.8-3.9 \times 10^{-12}$ mol.m/m²sPa) compared to synthetic films: cellulose acetate ($0.28 - 0.90 \times 10^{-12}$ mol.m/m²sPa), high density polyethylene (0.014×10^{-12} mol.m/m²sPa) and low density polyethylene ($0.04-0.054 \times 10^{-12}$ mol.m/m²sPa), but lower than in case of films based on corn zein (6.5×10^{-12} mol.m/m²sPa) or soybean proteins (194×10^{-12} mol.m/m²sPa). Water vapor permeability of protein based films is limited due to the inherent hydrophobicity of the proteins. Hydrophilic plasticizers, such as glycerin, facilitate the transfer of water vapor through protein based films.

RESULTS AND DISCUSSIONS

In our experiment, edible / biodegradable films were created using bighead carp myofibrillar proteins.

Bighead carp myofibrillar proteins were obtained by the conventional procedure of repeated washing with cold water of the minced meat, followed by centrifugation and refining to remove water, water-soluble substances, sarcoplasmatic proteins, lipids, skin residues and bones (surimi procedure).

Table 1. Film-forming solutions composition

Film	pH solution	Protein, g%	Glycerine, %	Gelatine, g%	Cyclodextrin, g%
a	2.7	1.0	50	-	-
b	2.7	1.5	50	-	-
c	2.7	2.0	50	-	-
d	2.7	2.5	50	-	-
e	2.7	2.0	30	-	-
f	2.7	2.0	70	-	-
g	2.7	1.0	30	2.0	-
h	2.7	1.5	40	-	1.0

Several film types were made using different percentages of fish myofibrillar protein, different levels of glycerin in strongly acidic medium. In addition, composite films were obtained by adding gelatin or cyclodextrins, together with the basic constituents (protein, glycerine, water). The compositions of film forming solutions (FFS) are indicated in Table 1. Film types were designated with letters.

The approximate composition of bighead carp myofibrillar proteins

The approximate chemical composition of myofibrillar fish proteins is shown in Table 2. According to Sikorski (1981), the protein isolate obtained by the conventional procedure contains myosin as main protein, which represents 50-60% of the myofibrillar proteins.

Table 2. Approximate composition of bighead myofibrillar proteins

Constituent	Quantity	
	Wet weight, g%	Dry substance, g/100 g s.u.
Water content	83.82829	
Total protein	14.68942	90.83406
Lipids	0.1951	1.206428
Ash	0.1698	1.049982

Films appearance

All the films we obtained were transparent, flexible and uniform. The films had smooth surfaces, without pores or cracks visible to the naked eye. When gelatine and cyclodextrin were used for film formulations, they showed a slightly yellowish color compared to the films based only on fish proteins.

The appearance of the two parts of the film was slightly different for all films made. The lower part of the film that came into contact with the casting plate was brighter, while the top part of the film was dull, possibly due to phase separation that occurs in solution during the drying process.

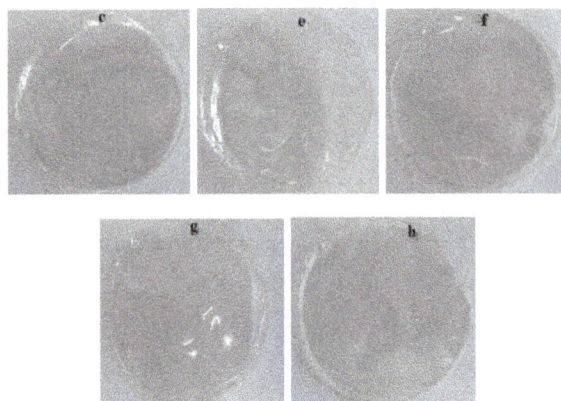

Figure 1. Edible/biodegradable films

All kinds of films we obtained were easily detached manually from the pouring plates, except for the films with a higher addition of glycerol (70 g per 100 g protein), these being slightly sticky.

Film thickness

The film thickness was measured using the micrometer in 10 randomly chosen areas. Film thickness measurement accuracy was ±5%. For each type of film, there were made 8 films, their thickness being measured after drying and conditioning. The average thickness and standard errors of the films made for this paper are shown in Table 3. In case of film-forming solutions at pH 2.7, in Figures 2-5 are shown the average thickness variations based on the type of film, on the region of measurement, on the protein concentration of the film forming solution, on the level of glycerin added and on the addition of gelatin or cyclodextrin. As it can be seen, the films showed some nonuniformity of thickness, depending on the areas where the measurements were made, the thickness differences were not statistically significant ($p < 0.05$) (Figure 2). The average thickness of the films increased from 0.030 ± 0.001 mm to 0.067 ± 0.002 mm with the increase of fish myofibrillar protein level of from 1% to 2.5%, in the presence of 50% glycerol per 100 g protein (Figure 3). The average thickness of the films was strongly positively correlated with the levels of fish myofibrillar protein from the film forming solutions, Pearson correlation coefficient is 0.94.

Table 3. Thickness of the films basen on bighead carp myofibrillar proteins

Films	Film thickness, mm	Films	Film thickness, mm
a	0.030±0.001	i	0.066±0.03
b	0.052±0.001	j	0.098±0.09
c	0.066±0.002	k	0.100±0.04
d	0.067±0.002	l	0.061±0.03
e	0.066±0.002	m	0.088±0.04
f	0.075±.002	n	0.041±0.04
g	0.055±0.001	o	0.082±0.04
h	0.061±0.003	p	0.105±0.05

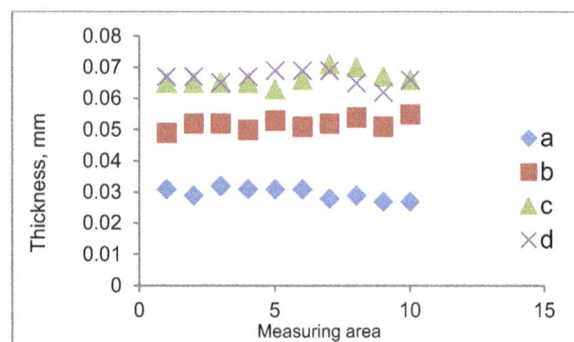

Figure 2. The variation of film thickness based on bighead carp proteins depending on the measuring location

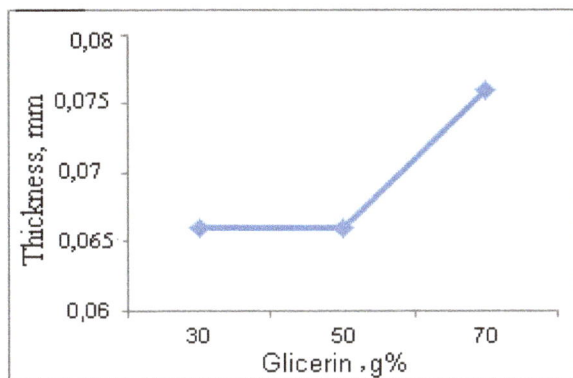

Figure 3. The influence of the protein concentration on the film thickness (pH 2.7)

In case of using various levels of glycerol, the average thickness of the 8 films depending on the measurement areas, varied within the limits of 0.061 to 0.078 mm (Figure 4). The highest average thickness was found in films prepared with an addition of 70 g glycerin / 100 g protein (0.075 ± 0.002) mm. At a level of 2% of protein there ware no differences in thickness depending on the level of glycerol, when it was within the range 30-50% compared with the protein (Figure 5).

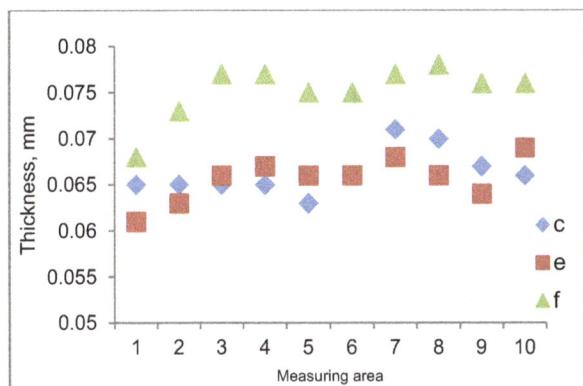

Figure 4. Film thickness depending on the addition of glyceroland on the measuring area (pH 2.7)

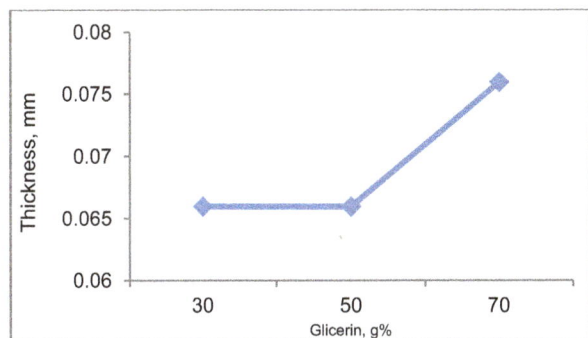

Figure 5. Film thickness variation depending on the level of glycerin (pH 2.7)

The water content, water solubility and water vapor permeability of the films
The water content and water solubility

The contents of water and the water solubility of the films based on fish myofibrillar proteins (FMP) shown in Table 4, varied depending on the type of film, within the limits of 19.94 ± 0.09 and 30.80 ± 0.09%, respectively 12.65 ± 0.08 and 24.58 ± 0.06%. The lowest water content was found on the film with 1% cyclodextrin addition (film h), and the biggest on the film with 2.5% fish myofibrillar protein content (film d). The lowest solubility in water was found in case of film d, and the highest solubility corresponded to the film with 2% protein, pH 2.7 and 30% glycerol / 100 g protein (film e). The water content and films solubility depended on the compositions of films forming solutions (FFS) (Figures 6, 7).

In the case of solutions with pH 2.7, the water content correlated poorly positive with the level of protein (r = 0.199), and negatively with the solubility (r = -0.465). We found a strong negative correlation between water solubility and moisture content of the films (r = -0.959). By increasing the protein content, polymer networks more dense and more resistant to water are formed. It is possible that at strongly acid pH, the fish myofibrillar proteins to undergo some structural changes that influence physical parameters of the films.

Table 4. The water content, water solubility and water vapor permeability of the films

Film type	Water content, g%	Solubility, g%	Permeability X 10^{-10}g m^{-1} s^{-1}Pa^{-1}
a	29.31±0.08	19.35±0.03	0.41±0.007
b	26.11±0.08	18.23±0.1	0.71±0.050
c	25.81±0.07	16.55±0.05	0.81±0.003
d	30.63±0.09	12.65±0.08	0.89±0.020
e	27.72±0.05	17.21±0.06	0.72±0.003
f	28.76±0.06	17.88±0.06	0.96±0.003
g	29.09±0.1	21.32±0.03	0.43±0.013
h	19.94±0.09	22.51±0.05	0.46±0.003

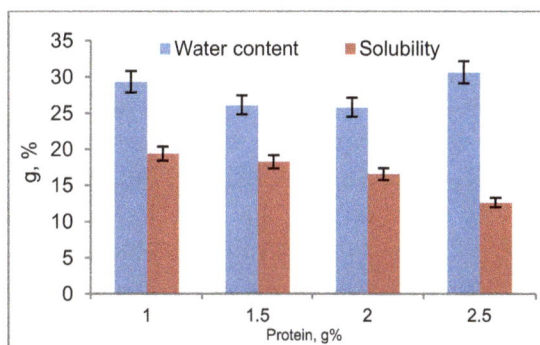

Figure 6. Content of the film forming solution with pH 2.7

The values obtained for moisture content (MC) and for the solubility in water (SW), depending on the level of glycerin and on the pH are shown graphically in Figure 7. For both values of pH, an increase of glycerin content from 30 to 70% (g/100 g protein) resulted in a significant increase in the moisture content and in the amount of solubilized dry substance, in particular for the addition of glycerin > 50%.

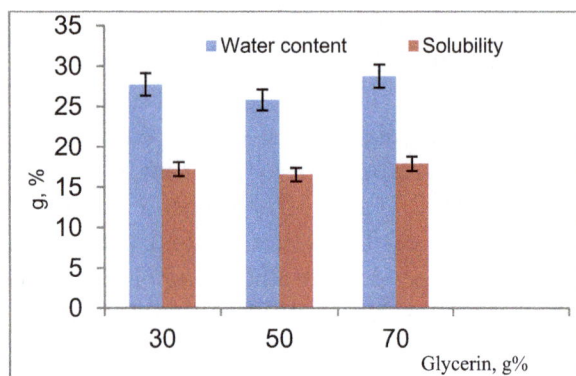

Figure 7. The variation of the water content and solubility depending on the level of glycerin (pH 2.7)

The solubility in water (SW) is regarded as an indicator of edible/biodegradable films resistance in water, which is an important factor in the prepackaging of food products because of the high water activity and the possibility of contamination in the presence of water (Bourtoom and Chinnan, 2005). In general, higher solubility indicates less resistance to water. The solubility in water for the films made by us, was similar to other research reports or lower. Thus, Wengo et al. (2007) reported for films based on Alaska Pollack surimi a solubility in water of 21%; Rostamzad et al. (2015) reported a solubility in water of 19.1% for films made from myofibrillar proteins from silver carp (*Hypophthalmichthys molitrix*); Tao et al. (2015) have found a value for solubility of $31.42 \pm 0.89\%$ for the films consisting of silver carp surimi. We consider, the same as Orliac et al. (2002) and Artharn et al. (2007) that cross linked proteins from the film were insoluble, whereas the most part of glycerol has been released into the water. The films based on fish myofibrillar proteins were stabilized by different links, which mainly include intermolecular disulphitic covalent links causing a lower solubility (Chinabhark et al., 2007).

Lower solubility in water for films made in strongly acidic medium can be justified by the fact that at pH <3.0 the protein degradation is more pronounced and Maillard reactions are favored, which would lead to the formation of strong cross links, stabilized by covalent links. In conclusion, we can appreciate that the bighead carp myofibrillar proteins, obtained by the surimi procedure, led to the formation of stable networks and that only low molecular weight hydrophilic substances were soluble in water.

Water vapor permeability
Water vapor permeability (PVA) is another important and widely studied property of the biodegradable/edible flexible films. This property of the protective films covers their ability to preserve, as appropriate, a dry product (chips, pretzels, candies) or a wet product (cheese, muffins, chewing gum). Without proper protective packaging the products can lose or gain moisture until the relative equilibrium to the environment humidity is achieved, this resulting in products consistency and stability modification.
Water vapor permeability values for films based on fish muscle proteins listed in Table 4. shows the dependence of these values on several factors such as, the composition of film-forming solutions, the concentration of fish myofibrillar proteins, the level of added glycerin. The lowest value for permeability $(0.41 \ \text{gm}^{-1}\text{s}^{-1}\text{Pa}^{-1})$ was found on the film containing 1% proteins, 30 g glycerin/100 g protein, and pH 2.7 (film a).
The variations of permeability for the films we tested over time are shown in Figures 8 and 9. Relatively high values of PVA indicate that films based on fish muscle proteins (bighead carp) are poor barriers to water vapor, due to the hydrophilic/hydrophobic nature of the polymer from the film matrix. As can be seen in Figure 9, film permeability values varied with the concentration of the protein, the highest values of PVA corresponding to the level of 2.5% proteins, both for acid environment (pH 2.7). This finding may be explained by the fact that the fish myofibrillar proteins contain significant levels of amino acids with polar nature, such as aspartic acid, glutamic acid, arginine and lysine (Shahidi,

1994; Paschoalick et al., 2003; Tongnuanchan et al., 2011).

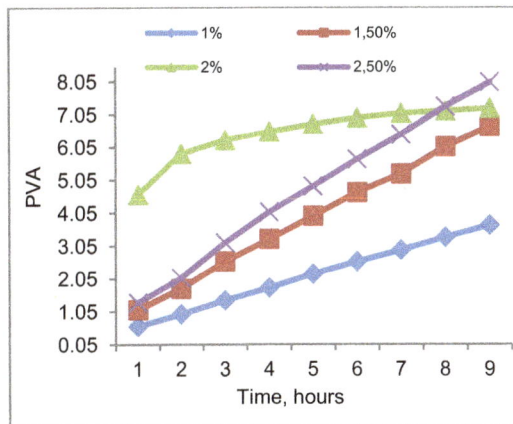

Figure 8. The variation of water vapor permeability of the films depending on the time and on the level of protein (pH 2.7)

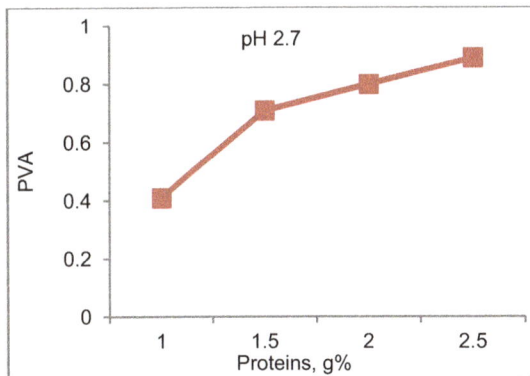

Figure 9. Water vapor permeability of the films depending on the level of myofibrillar proteins from the film forming solutions with pH 2.7

The properties of barrier to water vapor of the films we tested were affected by the level of incorporated plasticizer. The plasticizer agent used by us was glycerol, a hydrophilic polyol. Additions of 30 or 50 g glycerin/100 g protein caused insignificant increases for the PVA values of the films made in strongly acidic medium (Figure 10).

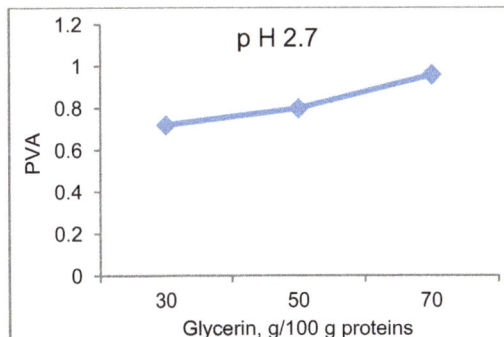

Figure 10. Water vapor permeability of the films depending on the addition of glycerin in the film forming solutions with pH 2.7

Protein films without plasticizer are often brittle and rigid, due to extensive interactions between the polymer molecules (Krochta, 2002). The addition of plasticizer affects not only the flexibility and other properties of the films, but additionally the film strength at water vapor and gases permeability (Sothornvit and Krochta, 2000 and 2001).

The variation in time of the water vapor transfer through comopozite films is shown in Figure 11, where you can observe greater resistance to water vapor for the films with gelatin addition.

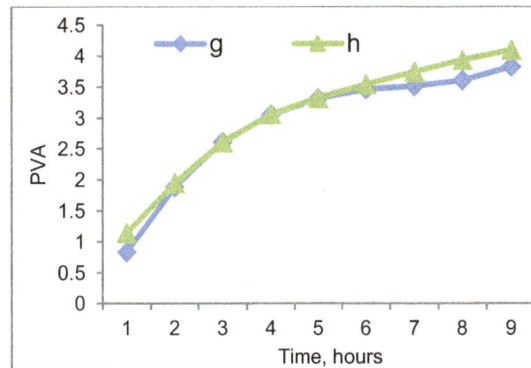

Figure 11. The permeability of the composite films according to the time

Increases in the permeability of films based on the level of glycerol were found as well by Nuthong et al. (2009) for the films based on porcine plasma (3%), by McHugh et al (1994) with films made from whey protein isolate, by Nemet et al. (2010) for the films obtained from myofibrillar proteins derived from chicken breast, who found values for the film permeability much lower than those obtained by us (0.21 to $0.29 \cdot 10^{-6} gm^{-1}s^{-1}Pa^{-1}$; myofibrillar protein concentration of 2%, pH 3.0, glycerin 25 to 65 g/100 g protein). Vanin et al. (2005) found no differences in the amounts of PVA on the gelatin based films where the levels of glycerin added to the film forming solutions varied between 10-30 g/100 g protein.

In our experimental conditions, the thickness of the films was relevant for PVA, in line with the increase of PVA for the films with greater thickness. The results listed in Table 4, indicate that PVA for edible protein films obtained by us are far greater than the values of PVA for plastic films [LDPE (low density polyethylene) $0.002 \cdot 10^{-10}$; high density polyethylene $0.008 \cdot 10^{-10}$; oriented polypropylene $0.038 \cdot 10^{-10}$; polyester $0.198 \cdot 10^{-10}$; polyvinylidene chloride $0.002 \cdot 10^{-10}$ gm^{-1}s^{-1}Pa^{-1}] (Krochta et al., 1997; Fairley et al., 1997; Shiku et al., 2003; Garcia and Sobral, 2005).

Films color

Color is an important feature of the edible/ biodegradable films because it affects consumer acceptability of potential food applications. In our study, films color was assessed by determining the color parameters L* [(radiance, brightness, white/black; 100/0)], a* [red (+60)/ Green (-60)], b* [yellow (+60)/blue (-60) and ΔE] of food films which were prepared under both acidic and alkaline conditions.

Color parameters are registered in Table 5. Negative values of a* indicate, for the films made by us, a slightly green tint, which intensifies with the increase of protein content from the film forming solution. The same behavior has been reported by Sobral (1999) as well. For this specific parameter, slight differences were obseved between created acidic and alkaline conditions. The positive values of b* indicate yellow tones for films, the color intensity increasing along with the protein level increase from 1% to 2.5%, the increase being 39.4% for the film forming solution with pH 2.7.

Acidic conditions are inducing myofibrillar protein degradation which leads to the dismissal of amino groups for browning reactions during heat treatment and drying process (Prodpan and Benjakiti, 2005; Chinabhark et al., 2007). The values of L* parameter varied with the level of protein and with the addition of glycerin. The films have become less bright by increasing the protein concentration of the film forming solutions.

Table 5. Color parameters of edible films

Film	L*	a*	b*	ΔE
a	93.13±0.32	-1.22±0.05	10.19±0.46	93.69
b	92.95±0.39	-1.50±0.10	13.24±1.54	93.90
c	92.34±0.34	-1.73±0.03	15.97±0.33	93.73
d	89.95±0.47	-2.05±0.04	21.85±1.69	92.49
e	94.01±0.24	-0.94±0.05	7.47±0.39	94.30
f	93.57±0.29	-1.08±0.09	8.83±0.91	93.81
g	93.93±0.29	-1.03±0.08	8.59±0.53	94.33
h	93.66±0.21	-1.07±0.06	8.77±0.49	94.34

As shown in Table 5, parameter L* and b* were positively correlated with the level of protein from the film forming solutions, while

a* parameters and color difference were negatively correlated, in all the cases the linear regression coefficients were greater than 0.975. Relatively weak correlations were recorded between L* and a * and between ΔE and a* parameters.

The increase of glycerin concentration causes the reduction of total color difference (ΔE) of the films, possibly due to dilution effect, basically in an independent manner of the fish myofibrillar protein concentration (Sobral et al., 2005); glycerin being a colorless substance.

The films obtained in this work showed, in general, colors comparable to films produced from muscle protein of Nile Tilapia (Paschoalick et al., 2003), but were more colorful than films based on myofibrillar proteins of Nile Tilapia (Sobral and et al., 2000), of egg albumen (Gennadios et al., 1996) and of pig skin gelatin (Sobral et al., 1999). Prodpran et al. (2005) have obtained films with the following color parameters: L = 86.1 ± 3; a* = 0.4 ± 0.1; b* = 39 ± 0.3 and 0.98 ± E = 86.18. Nuthong et al. (2009) have found the folowing values for films based on porcine plasma: L* = 87.12 ± 0.33; a* = - 2.37 ± 0.04; b* = 7.47 ± 0.13.

CONCLUSIONS

There were made 8 types of films based on fish myofibrillar proteins (bighead carp) by the solvent procedure by incorporation into the film forming solutions of different levels of protein, plasticizer and other additives, such as gelatin and cyclodextrin.

The films obrained by us were transparent, flexible and uniform, had smooth surfaces without pores or cracks visible to the naked eye. All kinds of films were easily detached manualy from the molding plates.

The thickness of the films varied depending on the measuring area, on the film forming solutions composition and on the environment characteristics. The increase of protein concentration and of the level of added glycerin caused an increase almost linear of the mean values of film thickness.

The increase of the glycerin level results in reduced color difference of the films due to the dilution effect of glycerin, regardless of the protein concentration. The transparency of the films having higher protein concentration was lower in case of reduced concentration of glycerin, which becomes much more lower for high concentrations of plasticizer.

All the films studied in this work maintained their integrity after 24 hours of immersion in water, which indicates that the bighead carp myofibrillar proteins led to the formation of stable networks and that only low molecular weight hydrophilic substances were solubilized in water.

Water vapor barrier properties of the films were dependent on the levels of protein, of glycerin and of the addition of gelatin or cyclodextrin; PVA values and water content of the films increased with protein levels.

REFERENCES

Artharn A., Benjakul S., Prodpran T. and Tanaka M., 2007. Properties of a protein-based film from round scad (*Decapterus maruadsi*) as affected by muscle types and washing. Food Chemistry 103: p. 867-874.

Bourtoom T. and Chinnan M.S., 2008. Preparation and properties of rice starch - chitosan blend biodegradable film. Lebensm-Wiss. U-Technol., 2008, 41: p. 1633-1641.

Chinabhark K., Benjakul S., Prodpran T., 2007. Effect of pH on the properties of protein-basedfilm from bigeye snapper (*Priacanthus tayenus*) surimi. Bioresour.Technol., 98(1): p. 221-225.

Cuq B., Aymard C., Cuq J.L. and Guilbert S., 1995. Edible packaging films based on fish myofibrillar proteins: formulation and functional properties. Journal of Food Science, 60, p. 1369-1374.

Cuq B., Gontard N., Cuq J.L. and Guilbert S., 1997. Selected functional properties of fish myofibrillar protein-based film as affected by hydrophilic plasticizers. Journal of Agriculture and Food Chemistry 45: p. 622-626.

Fairley P., Krochta J.M. and German J.B., 1997. Interfacial interactions in edible emulsion films from whey protein isolate. Food hydrocolloids, 11: p. 245-252.

García F.T., Sobral P.J.A., 2005. Effect of the thermal treatment of the film-forming solution on the mechanical properties, color and opacity of films based on muscle proteins of two varieties of tilapia. LWT - Food Science and Technology, 38: p. 289-296.

Gennadios A., Weller C.L., Hanna M.A. and Froning G.W., 1996. Mechanical and barrier properties of edible wheat gluten-based films. J. Food Sci., 61: p. 585-589.

Hernandez-Munoz P., Villalobos R. and Chiralt A., 2004. Effect of cross-linking using aldehydes on properties of glutenin-rich films. Food Hydrocolloids 18: p. 403-411.

Ionescu A., Berza M., Banu C., 1992. Metode şi tehnici pentru controlul peştelui şi produselor din peşte. Editura Universităţii din Galaţi, p. 238.

Ionescu A., Aprodu I., Daraba A., Porneala L., 2008. The effects of transglutaminase on the functional properties of the myofibrillar protein concentrate obtained from beef heart. Meat Science, 79(2), p. 278-284.

Iwata K., Ishizaki S., Handa A., & Tanaka M., 2000. Preparation and characterization of edible films from fish-water soluble proteins. Fisheries Science, 66, p. 372-378.

Krochta M. and Johnston C.D., 1997. Edible and biodegradable polymer films: Challenges and opportunities. Food Technol., 51: p. 61-74.

Krochta J.M., Films, edible. 1997. In The Wiley Encyclopedia of Packaging Technology, 2nd Edition. A.L. Brody and K.S. Marsh (Eds.) John Wiley & Sons, Inc. New York.

Krochta J.M., 2002. Proteins as raw materials for films and coatings: Definitions, current status and opportunities. In: "Protein-based films and coatings". A. Gennadios (Ed.), p. 1, CRC Press Inc. Publishing Co., New York, CT, p. 1-39.

McHugh T.H. and Krochta J.M., 1994. Dispersed phase particle size effects on water vapor permeability of whey protein-beeswax edible emulsion films. J. Food Process. Preserv., 1994, 18(3): p. 173-188.

Nemet N.T., Šošo V.M. and Lazić V.L., 2010. Effect of glycerol content and pH value of film-forming solution on the functional properties of protein-based edible films. APTEFF, 41, p. 57-67.

Nuthong P., Benjakul S., Prodpran T., 2009. Effect of some factors and pretreatment on the properties of porcine plasma protein-based films. Food Science and Technology, 42: p. 1545-1552.

Orliac O., Rouilly A., Silvestre F. and Rigal L., 2002. Effects of additives on the mechanical properties, hydrophobicity and water uptake of thermo-moulded films produced from sunflower protein isolate. Polymer 43: p. 5417-5425.

Paschoalick T.M., Garcia F.T., Sobral P.J.A. and Habitante A.M.Q.B., 2003. Characterization of some functional properties of edible films based on muscle proteins of Nile tilapia. Food Hydrocolloids 17: p. 419-427.

Perez-Gago M.B. & Krochta J.M., 2001. Denaturation time and temperature effects on solubility, tensile properties, and oxygen permeability of whey protein edible films. Journal of Food Science, 66(5), p. 705-710.

Prodpran T. and Benjakul S., 2005. Acid and alkaline solubilization on properties of surimi based film. Songklanakarin. Journal of Science and Technology, 27(3): p. 563-574.

Rostamzad H., Paighambari S.Y., Shabanpour B. and Ojagh S.M., 2015. Characteristics of a biodegradable protein based films from Silver carp (Hypophthalmichthys molitrix) and their application in Silver carp fillets. International Food Research Journal 22(6): p. 2318-2326.

Shiku Y., Hamaguchi P.Y. & Tanaka M., 2003. Effect of pH on the preparation of edible films based on fish myofibrillar proteins. Fisheries Science, 69, p. 1026-1032.

Sikorski Z.E. and Noczk M., 1981. Modification of tehnological properties of fish proteins concetrates. Crit. Rev. Food Sci. Nutr., 1981, 14: p. 202-230.

Sothornvit R., Krochta J.M., 2000. Water vapor permeability and solubility of films from hydrolyzed whey protein. Journal of Food Science, Vol. 65, No. 4, p. 700-703.

Sothornvit R. and Krochta J.M., 2001. Plasticizer effect on mechanical properties of b-lactoglobulin films. Journal of Food Engineering, 2001, 50: p. 149-155.

Sobral P.J.A., 1999. Propriedades funcionais de biofilmesde gelatin em função da espessura. Ciência & Engenharia, 1999, 8(1): p. 60-67.

Sobral P.J.A., 2000. Influência da espessura sobre certas propriedade de biofilmes á base de proteínas miofibrilares. Pesquisa Agropecuária Brasileira, 35(6): p. 251-1259.

Sobral P.J.A., Santos J.S.S., García F.T., 2005. Effect of protein and plasticizer concentrations in film forming solutions on physical properties of edible films based on muscle proteins of a Thai Tilapia. Journal of Food Engineering, 70: p. 93-100.

Tanaka M., Shoichiro Ishizaki, Toru Suzuki and Rikuo Takai, 2001. Water Vapor Permeability of Edible Films Prepared from Fish Water Soluble Proteins as Affected by Lipid Type 1. Journal of Tokyo University of Fisheries, 87: p. 31-37.

Tao Z., Weng W.Y., Cao M.J., Liu G.M., Su W.J., Osako K. &Tanaka M., 2015. Effect of blend ratio and pH on the physical properties of edible composite films prepared from silver carp surimi and skin gelatin. J. Food Sci. Technol., 52(3): p. 1618-1625.

Tongnuanchan P., Benjakul S., Prodpran T., Songtipya P., 2011. Characteristics of film based on protein isolate from red tilapia muscle with negligible yellow discoloration. International Journal of Biological Macromolecules 48, p. 758-767.

Vanin F.M., Sobral P.J.A., Menegalli F.C., Carvalho R.A. and Habitante A.M.Q.B., 2005. Effects of plasticizers and their concentrations on thermal and functional properties of gelatin-based films. Food Hydrocolloids, 19: p. 899-907.

***AOAC, 1990. Moisture in Meat. Official Methods of Analysis 950.46, 11, 931.

***AOAC, 1990. Crude protein in Meat. Offcial Methods of Analysis 981.10, 11, 937.

4

RESULTS REGARDING BIOMASS YIELD AT MAIZE UNDER DIFFERENT PLANT DENSITY AND ROW SPACING CONDITIONS

Georgeta DICU[1], Viorel ION[2], Daniela HORHOCEA[1], Daniel STATE[1], Nicoleta ION[3]

[1]SC Procera Agrochemicals Romania SRL, 47 Muncii Street, 915200, Fundulea, Calarasi, Romania
[2]University of Agronomic Sciences and Veterinary Medicine of Bucharest, 59 Marasti Blvd, District 1, 011464, Bucharest, Romania
[3]Apiculture Research and Development Institute of Bucharest, 42 Ficusului Blvd, District 1, 013975, Bucharest, Romania

Corresponding author email: georgeta.dicu@procera.ro

Abstract

Maize (Zea mays L.) is important for supplying biomass to be used as substrate for biogas production either as energy crop or as crop residues. As energy crop for producing biomass, maize is recognized as being one of the most used crops. Maize as energy crop need a specific crop technology, with specific technological features among which there are counting the using of the suitable plant density and row spacing. From this perspective, the aim of the present paper is to present the results we have obtained regarding the biomass yield at several maize hybrids studied under different plant density and row spacing conditions. In this respect, four maize hybrids (H1, H2, Cera 400, and Cera 430) were studied at different plant densities (70,000; 80,000; 90,000; 100,000; 110,000; 120,000 and 125,000 plants.ha^{-1}) and at two row spacing conditions (75 cm and 37.5 cm). Researches were performed in a field experiment in the year 2015, under rainfed conditions. The field experiment was located in South Romania, respectively at Fundulea, Calarasi County (44°28' N latitude and 26°28' E longitude). The biomass determinations were performed in the early dough - dough plant growth stages, with the purpose to calculate the yields of above-ground biomass (expressed in tons.ha^{-1}) in the growth stages of the maize plants when the biomass could be used as raw material for biogas production, but also as fodder (silo) for animals. In our field experiment, increasing in plant density was associated with an increase of above-ground biomass yield up to a threshold beyond which the biomass yield is decreasing, this threshold being different according to row spacing. Smaller plant densities favoured the biomass yields at wide rows (row spacing of 75 cm), while higher plant densities favoured the biomass yields at narrow rows (row spacing of 37.5 cm). The biomass yields registered higher values at row spacing of 37.5 cm than at row spacing of 75 cm.

Key words*: biomass, yield, maize, plant density, row spacing.*

INTRODUCTION

Biomass is the most common form of renewable energy (McKendry, 2002), this being used as source of energy since ancient times. In a time when the human society has to rely more and more on renewable sources of energy, the importance of biomass became even greater. Within the agricultural systems, biomass could be used for energetic purposes either as crop residues or as energy crop.

Among the energy crops, maize (*Zea mays* L.) has a great potential to produce biomass which could be used for different energy purposes (Ion et al., 2016). Thus, maize is considered to be very suitable for producing biomass that can be used as substrate for biogas production (Amon et al., 2006; Balodis et al., 2011; Băşa

et al., 2013). It seems that the most efficient utilization of maize is supplying of green maize biomass directly to biogas plants for heat and power energy production (Dubrovskis et al., 2010).

Energy crops used for biogas production have to be easy to be cultivated and they have to be not too much demanding for inputs. Apart these characteristics, they should provide high dry matter yield and high methane output per area unit (Dubrovskis et al., 2010). From this respect, maize as a C4 plant has a high capacity to produce biomass and it is already the most used crop for biogas production in several countries. For instance, maize is the main crop grown for biogas production (biomass crop) in Germany (Brauer-Siebrecht et al., 2016) and maize is the most important energy crop grown for co-digestion in Flanders (De Vliegher et al., 2012).

Energy crops are demanding appropriate cultivation technologies according to the growing conditions as high biomass yields to be possible to be obtained. Within the cultivation technologies, plant density and row spacing could contribute to the production of biomass in an efficient way (Ion et al., 2015).

Biomass yield is significantly affected by different plant population densities (Abuzar et al., 2011). Plant densities can be increased to provide maximum dry matter production (Yilmaz et al., 2007). In this respect, planting density in maize has kept an upward trend for several decades, increasing at a rate of approximately 1000 plants $ha^{-1}.yr^{-1}$, and may continue to increase in the foreseeable future (Duvick, 2005).

Increasing plant density is one of the ways of increasing the capture of solar radiation within the canopy (Moderras et al., 1998). However, the efficiency of the conversion of intercepted solar radiation into maize yield decreases with a high plant population density because of mutual shading of plants (Sharifi and Zadeh, 2012). The increasing of plant density determines a decreasing of dry biomass of the maize plant (Ion et al., 2014). Despite the decreasing of dry biomass of the maize plant, the increasing of plant density is associated with an increase of the biomass yield ($kg.ha^{-1}$) up to a certain level. When the plant density is too high, the maize plants compete with each other for available resources. That is why an increase in plant population density in excess is not expected to have an effect on biological yield (Van Averbeke and Marais, 1992). Anyway, the relationship between planting density and biomass increase is not linear, particularly at high densities (Dhugga, 2007). Plant densities which are lower than optimum values are leading to low yields and less efficient use of the resources available to plants.

As maize do not have tillering capacity to adjust to variation in plant stand, optimum plant population is important (Azam et al., 2007). Optimum population levels should be maintained to exploit maximum natural resources, such as nutrients, sunlight, soil moisture and to ensure satisfactory yield

(Sharifi and Zadeh, 2012). But optimum plant density is affected by the genetic properties and vegetation time of the given hybrid, by the conditions of the production area, by the crop year and the extent of water and nutrient supply (Murányi, 2015). So, the optimum plant density is different according to environmental and technological conditions.

Generally, maize is cultivated in wide spaced rows (Nik et al., 2011). But, maize produced in narrow rows can increase yields and result in a quicker canopy closure (Satterwhite et al., 2006). Narrow rows provide advantages through earlier row closure (lessening erosion), better plant distribution (utilisation of nutrients) and higher yield (Reckleben, 2011). That is why narrow compared with standard row spacing had positive effects on whole-plant yield of one hybrid (Baron et al., 2006).

Especially for energy maize production, where the energy must not strictly be contained in the cob, other row distances such as 55 cm, 50 cm, 37.5 cm or even 25 cm are conceivable in order to achieve better utilisation of space and fertiliser (Reckleben, 2011). Also, twin-row planting systems in maize have been proposed as an alternative spatial arrangement that should theoretically decrease plant-to-plant competition, alleviate crop crowding stress and improve yields (Robles et al., 2012).

As row spacing is narrowed seeding rate should be towards the higher end of what is recommended for the area; in this respect, researches have shown that higher yields can be achieved with higher seeding rates, but must have adequate water (Bean and Marsalis, 2012).

The aim of the present paper is to present the results we have obtained regarding the biomass yield at several maize hybrids studied under different plant density and row spacing conditions.

MATERIALS AND METHODS

Researches were performed is a field experiment located within the experimental field belonging to Procera Company from Fundulea area in South Romania (44°28' N latitude and 26°28' E longitude). The field experiment was under rainfed conditions, in the year 2015, and had three experimental factors, respectively: maize hybrid, with four variants (H1 and H2, which are hybrids specially designed for high plant densities; Cera 400 and Cera 430, which are hybrids designed for usual plant densities); row spacing (75 cm and 37.5 cm); plant density, with the following variants: 70,000; 80,000; 90,000; 100,000; 110,000 and 120,000 plants.ha^{-1} for row spacing of 75 cm; 80,000; 90,000; 100,000; 110,000; 120,000 and 125,000 plants.ha^{-1} for row spacing of 37.5 cm. The field experiment was designed in split plots with 4 replications.

The preceding crop was maize and the crop fertilization was performed with 83 kg.ha^{-1} of nitrogen (200 kg.ha^{-1} of complex fertilizer of type 10:20:0 + 300 kg.ha^{-1} of ammonium sulphate fertilizer with 21% of nitrogen as active substance) and 40 kg.ha^{-1} of phosphorous (from the 200 kg.ha^{-1} of complex fertilizer of type 10:20:0). The fertilizers were applied before seedbed preparation. The sowing was performed on 15th of April 2015. The weed control was realised by the help of herbicide Frontier Forte (720 g.l^{-1} of dimetenamid-P), which was applied in a rate of 1 l.ha^{-1} in pre-emergence (before emergence).

The soil from Fundulea area is of chernozem type and it has a humus content of 2.8-3.2%, texture of type loam to clay loam, and pH of 6.4-6.8. The climatic conditions registered in the studied area in 2015, for the period April-August, are the following: 19.9°C the average temperature, while 18.6°C is the multiannual average; 260 mm sum of rainfall, while 327.9 mm is the multiannual average rainfall for the analysed period. The year 2015 is characterised as being warmer and drier (even drought) than normal years for the studying area.

The biomass determinations were performed in the early dough - dough plant growth stages, this being the moment when the maize biomass can be used as substrate for biogas production, but also as fodder (silo) for animals. For this purpose, the maize plants from one square meter were cut at soil level. The plant samples were weighed immediately in the field, as the fresh biomass to be calculated. From each sample, one average maize plant was weighed separately, and then it was cut into pieces and taken into the laboratory where it was dried in the oven for 24 hours at 80°C, as the dry biomass content to be determined and the dry biomass yield to be calculated. The fresh and dry biomass yields are presented in the present paper as average values for the four studied maize hybrids, they represent the above-ground biomass, and they are expressed in tons per hectare.

RESULTS AND DISCUSSIONS

It is known and it is expected that the increasing of plant density determines a decreasing of dry biomass of the maize plant (Ion et al., 2014). However, the increasing of plant density in our field experiment was associated with an increase of fresh and dry biomass yield (Figures 1 and 2).

The increase of fresh and dry biomass yield is happening up to a threshold beyond which the biomass yield is decreasing. It has to be highlighted that the increasing of biomass yield with the increasing of plant density takes place for both row spacing, respectively 75 cm and 37.5 cm between rows. But, the threshold plant density is different according to row spacing, being of 100,000 plants.ha^{-1} for row spacing of 75 cm and of 120,000 plants.ha^{-1} for row spacing of 37.5 cm.

Regarding the highest values of the fresh biomass yield, the highest fresh biomass yield for row spacing of 75 cm was of 30.7 tons.ha^{-1} at plant density of 100,000 plants.ha^{-1}, while for row spacing of 37.5 cm the highest fresh biomass yield was of 32.5 tons.ha^{-1} at plant density of 120,000 plants.ha^{-1} (Figure 1).

Regarding the highest values of the dry biomass yield, it was registered the same situation as in the case of fresh biomass, respectively the highest dry biomass yield for row spacing of 75 cm was of 14.7 tons.ha^{-1} at plant density of 100,000 plants.ha^{-1}, while for row spacing of 37.5 cm the highest dry biomass yield was of 15.3 tons.ha^{-1} at plant density of 120,000 plants.ha^{-1} (Figure 2).

The smaller plant densities (70,000; 80,000, and 90,000 plants.ha^{-1}) favoured the biomass yields at row spacing of 75 cm, while higher plant densities (over 100,000 plants.ha^{-1}) favoured the biomass yields at row spacing of 37.5 cm.

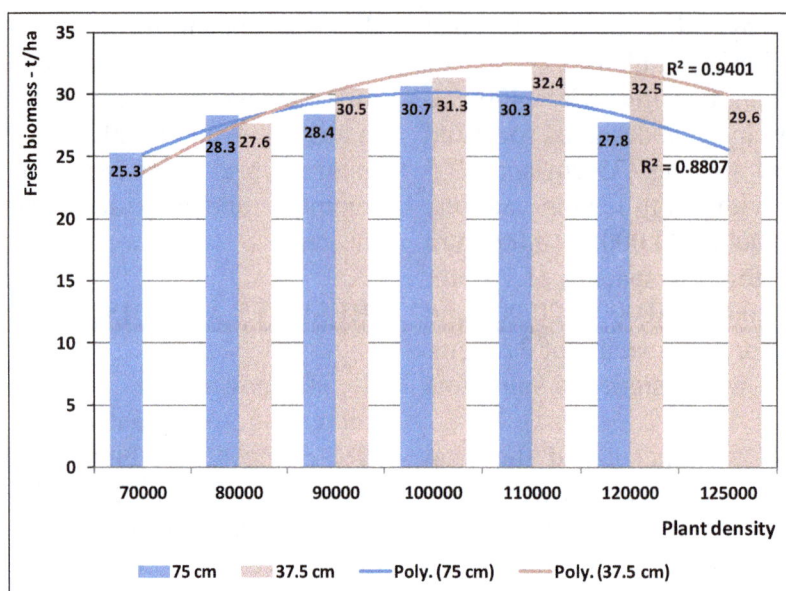

Figure 1. Fresh biomass yields at maize under different row spacing and plant density conditions

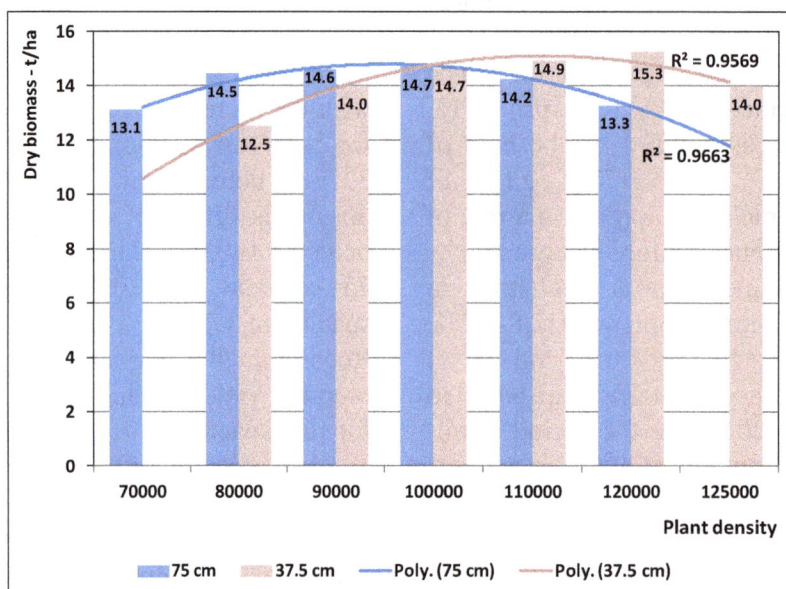

Figure 2. Dry biomass yields at maize under different row spacing and plant density conditions

The fresh and dry biomass yields registered higher values at row spacing of 37.5 cm than those registered at row spacing of 75 cm (Figure 3).

Thus, the average fresh biomass yield registered at row spacing of 37.5 cm was of 30.67 tons.ha^{-1}, while that registered at row spacing of 75 cm was of 28.43 tons.ha^{-1}.

The average dry biomass yield registered at row spacing of 37.5 cm was of 14.23 tons.ha^{-1}, while at row spacing of 75 cm the average dry biomass yield was of 14.07 tons.ha^{-1}.

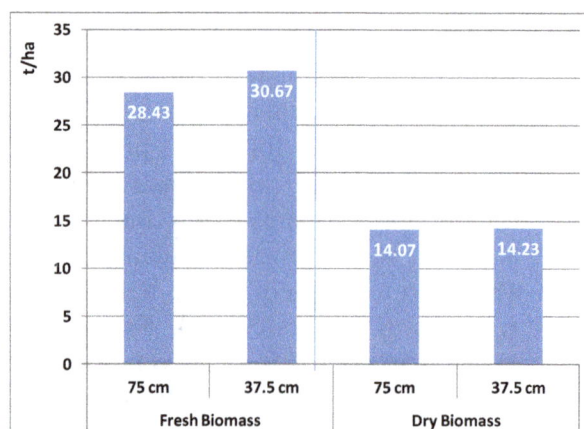

Figure 3. Average fresh and dry biomass yield at maize under different row spacing

CONCLUSIONS

In our field experiment, increasing in plant density was associated with an increase of above-ground biomass yield up to a threshold beyond which the biomass yield is decreasing. This threshold was different according to row spacing. Thus, for the studied conditions, the threshold in plant density was of 100,000 plants.ha^{-1} for row spacing of 75 cm and 120,000 plants.ha^{-1} for row spacing of 37.5 cm. The smaller plant densities (up to 100,000 plants.ha^{-1}) favoured the biomass yields at row spacing of 75 cm, while higher plant densities (over 100,000 plants.ha^{-1}) favoured the biomass yields at row spacing of 37.5 cm. The biomass yields registered higher values at narrow rows (row spacing of 37.5 cm) than at wide rows (row spacing of 75 cm).

ACKNOWLEDGEMENTS

The researches carried out for the elaboration of the present paper were financed by Romanian Program "Partnerships for Priority Domains", project PN-II-PT-PCCA-2011-3.2-1778 "OPTImization of BIOMass and Approach to Water conservation" (OPTIBIOMA-W), Contract no. 45/2012.

REFERENCES

Abuzar M.R., Sadozai G.U., Baloch M.S., Baloch A.A., Shah I.H., Javaid T., Hussain N., 2011. Effect of plant population densities on yield of maize. The J. of Animal & Plant Sciences, 21(4): p. 692-695.

Amon T., Kryvoruchko V., Amon B., Bodiroza V., Zollitsch W., Boxberger J., 2006. Biogas Production from Energy Maize. Landtechnik 61, No. 2, p. 86-87.

Azam S., Ali M., Amin M., Bibi S., Arif M., 2007. Effect of plant population on maize hybrids. Journal of Agricultural and Biological Science, Vol. 2, No. 1, January, p. 13-20.

Balodis O., Bartuševics J., Gaile Z., 2011. Biomass yield of different plants for biogas production. Proceedings of the 8th International Scientific and Practical Conference, Vol. 1, p. 238-245.

Baron V.S., Najda H.G., Stevenson F.C., 2006. Influence of population density, row spacing and hybrid on forage corn yield and nutritive value in a cool-season environment. Canadian Journal of Plant Science, p. 1131-1138.

Băşa A.Gh., Ion V., Dicu G., State D., Epure L.I., Ştefan V., 2013. Above-ground Biomass at Different Hybrids of Maize (Zea mays L.) Cultivated in South

Romania in Drought Conditions. Scientific Papers. Series A, Agronomy, Vol. LVI, p. 177-184.

Bean B., Marsalis M., 2012. Corn and Sorghum Silage Production Considerations. High Plains Dairy Conference, Amarillo, Texas, p. 1-7.

Brauer-Siebrecht W., Jacobs A., Christen O., Götze P., Koch H.J., Rücknagel J., Märländer B., 2016. Silage Maize and Sugar Beet for Biogas Production in Rotations and Continuous Cultivation: Dry Matter and Estimated Methane Yield. Agron., 6(2): p. 2-12.

De Vliegher A., Van Waes C., Baert J., Van Hulle S., Muylle H., 2012. Biomass of annual forage crops for biogas production. Grassland Science in Europe, Vol. 17, Proceedings of the 24th General Meeting of the European Grassland Federation, Lublin, Poland, 3-7 June, p. 463-465.

Dubrovskis V., Plume I., Bartusevics Ja., Kotelenecs V., 2010. Biogas production from fresh maize biomass. Engineering for Rural Development, Jelgava, 27-28.05, p. 220-225.

Dhugga K.S., 2007. Maize Biomass Yield and Composition for Biofuels. Crop Science, Vol. 47, November-December, p. 2211- 2227.

Duvick D.N., 2005. Genetic progress in yield of United States maize (Zea mays L.). Maydica 50: p. 193-202.

Ion V., Băşa A.Gh., Temocico G., Dicu G., Epure L.I., State D., 2014. Maize plant biomass at different hybrids, plant populations, row spacing and soil conditions. Romanian Biotechnological Letters, Vol. 19, No. 4, p. 9551-9560.

Ion V., Basa A.Gh., Dicu G., Dumbrava M., Epure L.I., State D., 2015. Biomass yield at maize under different sowing and growing conditions. Book of Proceedings, Sixth Internat. Scientific Agricultural Symposium „Agrosym 2015", p. 285-290.

Ion V., Băşa A.Gh., Dumbravă M., Ion N., 2016. Results regarding biomass yield at maize under different crop technology conditions and in a situation of drought. 16th International Multidisciplinary Scientific GeoConference (SGEM 2016), Energy and Clean Technologies Conference Proceedings, Book 4 - Renewable Energy Sources and Clean Technologies, Vol. I, 30 June - 6 July, Albena, Bulgaria, p. 603-608.

McKendry P., 2002. Energy Production from Biomass (part 1): Overview of Biomass. Bioresource Technology, 83, p. 37-46.

Moderras A.M., Hamilton R.L., Dijak M., Dwyer L.M., Stewart D.W., Mather D.E., Smith D., 1998. Plant population density effects on maize inbred lines grown in short-season environments. Crop Science 34: p. 104-108.

Murányi E., 2015. Effect of plant density and row spacing on maize (Zea mays L.) grain yield in different crop year. Columella - Journal of Agricultural and Environmental Sciences, Vol. 2, No. 1, p. 57-63.

Nik M.M., Babaeian M., Tavassoli A., Asgharzade A., 2011. Effect of plant density on yield and yield components of corn hybrids (Zea mays). Scientific Research and Essays, 6(22): p. 4821-4825.

Reckleben Y., 2011. Cultivation of maize - which sowing row distance is needed? Landtechnik 66, No. 5, p. 370-372.

Robles M., Ciampitti I.A., Vyn T.J., 2012. Responses of Maize Hybrids to Twin-Row Spatial Arrangement at Multiple Plant Densities. Agronomy Journal, Vol. 104, Issue 6, p. 1747-1756.

Satterwhite J.L., Balkcom K.S., Price A.J., Arriaga F.J., van Santen E., 2006. Hybrid, row pattern, and plant population comparisons for conservation tillage corn production. Southern Conservation Systems Conference, Amarillo TX, June 26-28, p. 104-112.

Sharifi R.S., Zadeh N.N., 2012. Effects of plant density and row spacing on biomass production and some of physiological indices of corn (*Zea mays* L.) in second cropping. Journal of Food, Agriculture & Environment Vol. 10 (3&4): p. 795-801.

Van Averbeke W., Marais J.N., 1992. Maize response to plant population and soil water supply: I. Yield of grain and total above-ground biomass. South African Journal of Plant and Soil, 9(4): p. 186-192.

Yilmaz S., Gozubenli H., Konuskan O., Atis I., 2007. Genotype and plant density effects on corn (*Zea mays* L.) forage yield. Asian Journal of Plant Sciences, 6 (3): p. 538-541

5

COMPARISON OF FOUR GENOMIC DNA ISOLATION METHODS FROM SINGLE DRY SEED OF WHEAT, BARLEY AND RYE

Daniel CRISTINA[1,2], Matilda CIUCĂ[2], Călina-Petruța CORNEA[1]

[1]University of Agronomic Sciences and Veterinary Medicine of Bucharest,
59 Mărăști Blvd., District 1, Bucharest, Romania
[2]National Agricultural Research and Development Institute Fundulea,
1 Nicolae Titulescu Street, 915200, Fundulea, Călărași, Romania

Corresponding author email: danielcristina89@gmail.com

Abstract

Modern breeding programs are based on the differences that distinguish one plant from another one, differences encoded in the plant's genetic material, the DNA. Genotypic selection, particularly at the DNA level, can be exploited in Marker Assisted Selection (MAS) to identify desirable recombinants among segregating populations. Successful DNA amplification is vital for the detection of specific DNA targets, and this depends on the ability of DNA isolation methods to produce good quality DNA.
DNA isolation from plant tissues remains difficult because of the presence of a rigid cell wall surrounding the plant cells. DNA isolation methods are affected by several factors like the amount of tissue needed and its availability, the number of steps involved and the chemicals used.
In this study four different DNA isolation methods, based on CTAB and SDS (three methods), applied on single dry seed of wheat, barley and rye, were tested and compared. The quality of DNA was assessed by spectrophotometric measurements, gel electrophoresis and PCR reactions. The results of the experiments showed that DNA isolated by all four isolation methods is not excessively fragmented, A260/A280 ratio was between 1.6-1.9 and the concentration ranged between 20-194 ng/μl. The quality of the DNA was good and allowed the amplification of specific fragments by PCR. CTAB method had better A260/A280 ratio followed by SDS1, SDS2 and SDS3. Electrophoretic pattern showed better results with SDS2 method, followed by SDS3, SDS1 and CTAB. Furthermore, the CTAB and SDS1 methods need more time than SDS2 and SDS3, even though the number of steps of each method are almost equal (±2 steps). An estimative cost per sample showed that the cheapest method is the SDS3, the other three having similar costs.

Key words: DNA isolation, single dry seed, wheat, barley, rye.

INTRODUCTION

Modern plant breeding, due to advancements in genetics, molecular biology and tissue culture, is being carried out by using molecular genetics tools. Genotypic selection, particularly at the DNA level, can be exploited in Marker Assisted Selection (MAS) to identify desirable recombinants among segregating populations.
PCR allows the selective amplification of specific segments of DNA in a mixture of other DNA sequences. Isolation of DNA would be the first step in such analytical methods. Successful DNA amplification is vital for the detection of specific DNA targets, and in return this depends on the ability of DNA isolation methods to isolate DNA of reasonable quantity, purity, integrity and quality and is often the most time consuming step of a DNA-based detection method (Singh, 2009).

Typical plant DNA isolation methods must go through some basic steps like breaking the cell wall, usually done by grinding the tissue, disruption of the cell membrane, using a detergent like SDS (sodium dodecyl sulfate) or CTAB (cetyltrimethyl ammonium bromide), protection of DNA from the endogenous nucleases with EDTA (chelating agent that binds magnesium ions, generally considered a necessary cofactor for most nucleases), removal of proteins from the buffer/tissue mixture using chloroform or phenol to denature and separate the proteins from DNA, precipitation of DNA with either ethanol or isopropanol (Rogers and Bendich, 1989).
The presence of a rigid cell wall surrounding the plant cells, polysaccharides, proteins, and DNA polymerase inhibitors (tannins, alkaloids,

and polyphenols) makes the DNA isolation from cereals a difficult task. Polysaccharides, the most common contaminants found in the plant DNA isolation, make DNA pellets slimy and difficult to handle. The anionic contaminants inhibit restriction enzymes and affect enzymatic analysis of the DNA. The presence of these compounds reduces the quality and quantity of DNA which often makes the sample non-amplifiable. Furthermore, DNA isolation methods are affected by several factors like the amount of tissue needed and its availability, the number of steps involved, and the chemicals used (Chaves et al., 1995).

DNA isolation performed on seed instead of leaf tissue allows MAS to be carried out independently of the growth season, and the time and glasshouse space needed for growing the plants are saved. Most importantly, the seed can be analyzed during the non-field season, selected and prepared for the next breeding cycle. Furthermore it is possible to send seed samples internationally for comparative studies, this being difficult for leaf samples which have to be kept on ice or lyophilized (Von Post, 2003).

The isolation of high-quality DNA from plant tissue is time consuming, laborious, and quite expensive due to multiple steps and the cost of reagents used. High quality DNA is characterized by predominantly high molecular weight fragments with an A260/280 ratio between 1.8 and 2.0 and the lack of contaminating substances (Abdel-Latif and Osman, 2017).

A fast, simple, and reliable DNA isolation method, which does not require long incubations, multiple steps or expensive commercial kits, that could meet in the PCR, sequencing and next-generation library preparation requirements, will be invaluable to plant research. Therefore, the aim of this study was to compare quality and quantity of DNA isolated using four different isolation methods from a single dry seed of three cereal species, wheat, barley and rye. Furthermore, we compared the isolation methods regarding the number of steps, time and price/sample.

MATERIALS AND METHODS

Plant material was obtained from NARDI Fundulea, Romania, and consisted of seeds from three cereal species, wheat (Izvor – cultivar with the *Lr34* resistance allele), barley (Scânteia cultivar) and rye (Harkovskaya cultivar). 16 individual seeds (four seeds / isolation method, DNA was isolated from each seed - four repetitions) were used from each cultivar. The seeds were dry crushed using a mortar and pestle. The amount of crushed sample obtained from each seed ranged between 30-60 mg.

DNA isolation was performed using four different modified methods (CTAB; SDS 1; SDS 2 and SDS 3).

Table 1. The isolation buffers for each method are shown in

CTAB	SDS 1	SDS 2	SDS 3
100 mM Tris	100 mM Tris	100 mM Tris	100 mM Tris
700 mM NaCl	-	500 mM NaCl	-
50 mM EDTA	50 mM EDTA	50 mM EDTA	50 mM EDTA
2% CTAB	1.5% SDS	1% SDS	1.25% SDS
-	-	2.5% D-Sorbitol	-
-	-	2% N-Lauroylsarcosine sodium salt	-
140 mM β-mercaptoethanol	-	-	-

- CTAB - based on Murray & Thompson, 1980;
- SDS 1 - based on Mohammadi, http://shigen.nig.ac.jp/ewis/article/html/118/article.html;
- SDS 2 - Cristina et al., 2015;
- SDS 3 - based on Chao & Somers, 2012.

The protocols steps for DNA isolation using the above mentioned methods are shown in Table 2. Removal of proteins from the buffer/tissue mixture, in case of CTAB and SDS 1 methods, was done with dichloromethane:isoamyl alcohol (24:1) instead of chloroform:isoamyl alcohol.

Dichloromethane offers a cheaper, less toxic alternative to chloroform in protocols for DNA isolation (Chaves, 1995).

Table 2. Protocols for DNA isolation

Step	CTAB	SDS 1	SDS 2	SDS 3
Isolation buffer	600 µl (freshly made)	500 µl	500 µl (freshly made)	500 µl
Incubation 65°C	60'	60'	-	30'
Cooling samples	2-3'	2-3'	-	-
Incubation 4-7°C	-	-	-	15'
Vortex	-	-	1'	-
DNA purification	Mixed with 1:1 vol. Dichloromethane:Isoamyl alcohol (24:1, V:V)	Mixed with 270 µl Potassium acetate 3M (final conc. 1.62M)	200 µl Potassium acetate 3M (final conc. 1.2M)	Mixed with 250 µl Ammonium acetate 6M (final conc. 3M)
Vortex	-	-	2'	-
Incubation 4-7°C	-	-	-	15'
Centrifugation	12' 9500 RCF	15' 7690 RCF	15' 16055 RCF	15' 7690 RCF
DNA purification	>500 µl supernatant mixed with 1:1 vol. Dichloromethane:Isoamyl alcohol (24:1, V:V)	>400 µl supernatant mixed with 1:1 vol. Dichloromethane:Isoamyl alcohol (24:1, V:V)	-	-
Centrifugation	12' 9500 RCF	10' 7690 RCF	-	-
RN-ase treatment	45'			
DNA precipitation	5µl NaCl (5M) per 100µl sample + 2 vol. EtOH (kept at -20°C) *optional, for better DNA precipitation, incubate samples on ice for 2-3'			
Centrifugation	6' 16055 RCF			
DNA pellet wash	200-300 µl Wash buffer (76% EtOH, 10 mM NH$_4$OAc)			200 µl EtOH 70%
Centrifugation	5' 18620 RCF			
Pellet drying	10-30' at room temperature			
DNA dissolving	100 µl TE (TE should be adjusted according to pellet dimension) > samples are kept overnight for dissolving at 4-7°C			
Optional step	If the pellet is not dissolved in the next day the upper faze could be transferred into new tubes before quality and quantity checking.			

Gel electrophoresis

Both the genomic DNA and PCR products were analysed by agarose gel electrophoresis using 0.8% agarose gel for the genomic DNA and 1.2-1.5% agarose gels for PCR products.

Spectrophotometric measurements.

DNA purity (A260/A280 ratio) and quantity analysis were performed with a Beckman Coulter Life Sciences DU 730 spectrophotometer.

Evaluation of DNA amplification

In order to evaluate if the DNA is amplifiable, all of the samples analysed were subjected to PCR amplification using the following markers: cssfr5 – functional marker for Lr34 gene selection in wheat (Lagudah, 2009), HvBM5A-exon2 – barley VRN-H1 genotype assays (Zlotina, 2013) and SCM9 – rye SSR (Saal and Wricke, 1999). Also, we used ISSR 17898B, UBC818 and UBC 876 for all samples.

One sample of wheat, barley and rye from each method has been chosen and diluted to a working concentration 25 ng/µL. PCR amplifications were performed in ABI ProFlex™ 3 x 32-well PCR System.

PCR amplification with functional marker cssfr5 was performed using KAPA2G Fast Multiplex PCR Kit (KAPA Biosystems) in a 10 µL final reaction volume containing 1X Multiplex Mix, 0.2 mM each primer, 2 µL DNA sample (40-50 ng). PCR programme was: initial denaturation at 95°C for 3 min, followed by 30 cycles of (95°C – 15 s, 62°C – 30 s, 72°C - 30 s) and a final extension at 72°C for 7 min. PCR product was analyzed on 1.2% agarose gel.

PCR reactions performed with primers HvBM5A-exon2 (barley), SCM9 (rye), 17898B and UBC 818 were carried out using MyTaq™ Red DNA Polymerase (Bioline).

PCR conditions were as follows:
- HvBM5A-exon 2 -15 µL final reaction volume containing 1X reaction buffer, 0.5 mM primers, 0.6U DNA polymerase and 3 µL DNA sample (60-80 ng). PCR programme: initial denaturation at 95°C for 1 min, followed by 35 cycles of (95°C – 15 s, 60°C – 15 s, 72°C - 10 s)

and a final extension at 72°C for 5 min. PCR product was analyzed on 1.2% agarose gel.

- SCM9 -10 µL final reaction volume containing 1X reaction buffer, 0.5 mM primers, 0.3U DNA polymerase and 2 µL DNA sample (40-50 ng). PCR programme: initial denaturation at 95°C for 1 min, followed by 40 cycles of (95°C – 15 s, 60°C – 15 s, 72°C - 10 s) and a final extension at 72°C for 5 min. PCR product was analyzed on 1.5% agarose gel.

- 17898B -25 µL final reaction volume containing 1x buffer, 0.28 mM primer, 2U DNA polymerase, 1 µL wheat and rye DNA sample (20-30 ng), 1.5 µL barley DNA sample (30-40 ng). PCR programme: initial denaturation at 95°C for 3 min, followed by 35 cycles of (95°C – 15 s, 44°C – 15 s, 72°C - 30 s) and a final extension at 72°C for 5 min. PCR product was analyzed on 1.2% agarose gel.

- UBC 818 -25 µL final reaction volume containing 1x buffer, 0.28 mM primer, 1U DNA polymerase, 0.5 µL DNA sample (10-15 ng). PCR programme: initial denaturation at 95°C for 3 min, followed by 40 cycles of (95°C – 15 s, 50°C – 15 s, 72°C - 30 s) and a final extension at 72°C for 5 min. PCR product was analyzed on 1.2% agarose gel.

RESULTS AND DISCUSSIONS

The quality and quantity of the template DNA are critical factors for the successful PCR analysis. The efficiency of the DNA extraction steps can be critical for successful amplification since there are many compounds that inhibit DNA amplification that can be co-purified with the DNA, such as polysaccharides, lipids and polyphenols or extraction chemicals.

The comparative analysis of electrophoretic patterns of genomic DNAs (Figure 1) revealed visible DNA bands for all isolation methods applied. DNA isolation methods SDS 2 and SDS 3 overall had better electrophoretic profile, followed by SDS 1 and CTAB.

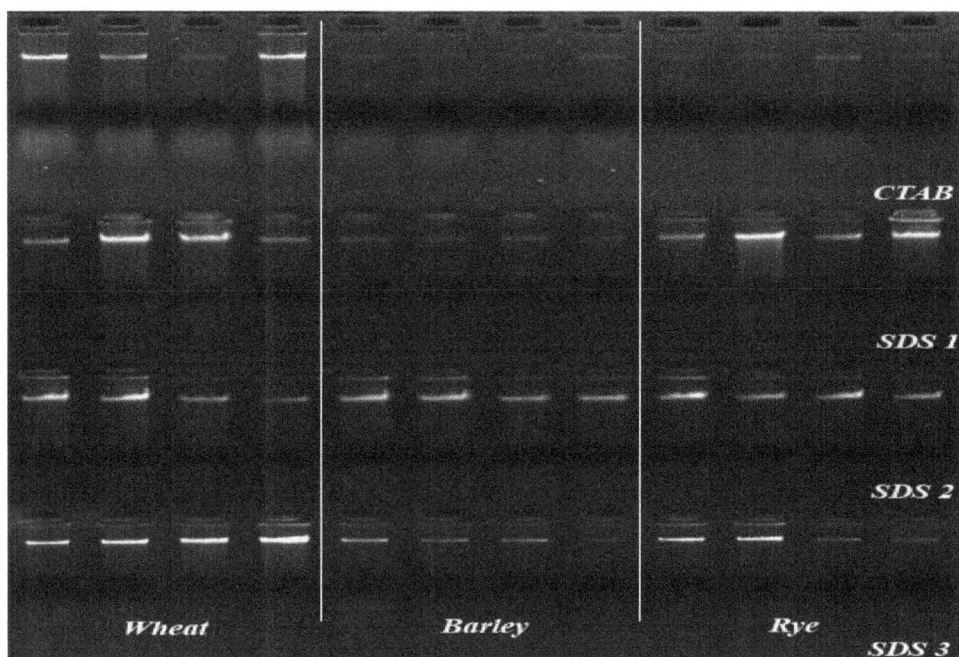

Figure 1. Agarose gel electrophoresis of genomic DNA

Spectrophotometric assessment of DNA quality revealed good results for all isolation methods applied (Table 3).

Table 3. Spectrophotometric results

Sample	CTAB		SDS 1		SDS 2		SDS 3	
	Ratio	ng/µl	Ratio	ng/µl	Ratio	ng/µl	Ratio	ng/µl
Wheat 1	1.767	89	1.693	46	1.617	155	1.736	49
Wheat 2	1.934	76	1.758	113	1.623	195	1.746	34
Wheat 3	1.883	85	1.814	77	1.784	35	1.613	54
Wheat 4	1.859	70	1.701	67	1.612	37	1.701	30
Min.	1.767	70	1.693	46	1.612	35	1.613	30
Max.	1.934	89	1.814	113	1.784	195	1.746	54
Barley 1	1.878	82	1.742	59	1.836	115	1.702	85
Barley 2	1.928	112	1.744	63	1.671	140	1.635	50
Barley 3	1.850	60	1.718	51	1.843	168	1.735	49
Barley 4	1.858	80	1.731	43	1.792	120	1.663	20
Min.	1.850	60	1.718	43	1.671	115	1.635	20
Max.	1.928	112	1.744	63	1.843	168	1.735	85
Rye 1	1.890	59	1.850	63	1.733	136	1.669	47
Rye 2	2.146	57	1.860	48	1.702	40	1.615	36
Rye 3	1.846	64	1.766	38	1.778	50	1.569	21
Rye 4	1.859	39	1.849	57	1.763	66	1.638	27
Min.	1.846	39	1.766	38	1.702	40	1.569	21
Max.	2.146	64	1.860	63	1.778	136	1.669	47

According to the results, CTAB method had overall better A260/A280 ratio (1.767-2.146), followed by SDS 1 (1.693-1.860), SDS 2 (1.612-1.843) and SDS 3 (1.569-1.746). DNA concentrations ranged between 39-112 ng/µl with CTAB method, 38-113 ng/µl - SDS 1, 35-195 ng/µl - SDS 2 and 20-85 ng/µl - SDS 3.

Electrophoretic and spectrophotometric results showed some differences regarding the DNA purity and concentration: in some samples the DNA concentration determined spectrophotometrically was registered as high, but the electrophoretic analysis did not confirm the results, suggesting the presence of contaminants that affect the measurements and/or the electrophoresis results.

PCR amplification

Cultivar Izvor carries *Lr34* resistance allele to leaf rust (Ciuca et al., 2015) meaning that PCR with cssfr5 primers results in a PCR product amplification of 751bp (Figure 2).

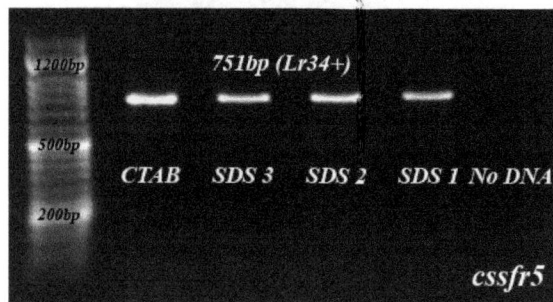

Figure 2. Agarose gel electrophoresis of PCR product obtained with cssfr5 functional marker

All DNA isolation methods applied for wheat cultivar gave good amplification results and no significant differences were observed.

PCR for barley DNA samples (cultivar Scânteia), amplified a 616bp PCR product with HvBM5A-exon 2 primers (Figure 3).

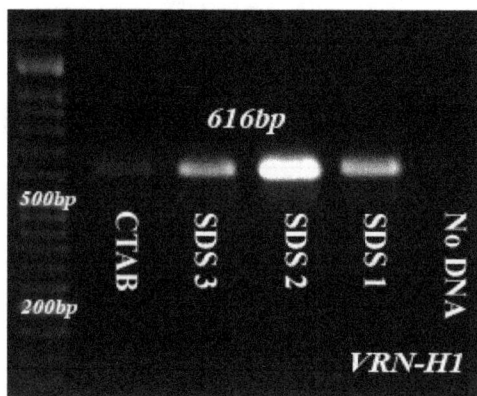

Figure 3. Agarose gel electrophoresis of PCR product obtained with HvBM5A-exon 2 primers

Barley DNA obtained with CTAB isolation method had weaker amplification signal compared to the other methods. DNA obtained with SDS 1 and SDS 3 methods gave good and similar amplification products. The best results seem to be obtained with DNA isolated by SDS 2 method, but the concentration of DNA template was higher (Figure 3).

PCR with SCM9 primers amplified in rye DNA samples (Harkovskaya cultivar) a 220bp product and no product in wheat cultivar Izvor (no-rye reference sample) (Figure 4).

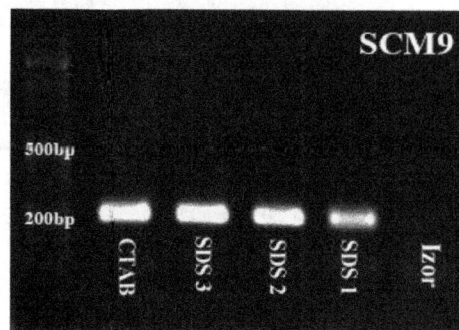

Figure 4. Agarose gel electrophoresis of PCR product obtained with SCM9 primer

All rye DNA samples were amplified but DNA sample obtained with SDS 1 method had a weaker amplified product than all other isolation methods. Nevertheless, DNA obtained with SDS 1 method gave good amplification if we compare it with DNA ladder intensity. As expected, wheat reference sample had no PCR product.

ISSR PCR with 17898B primer gave better results for barley DNA samples, but the amplification for wheat and rye DNA was weaker (Figure 5).

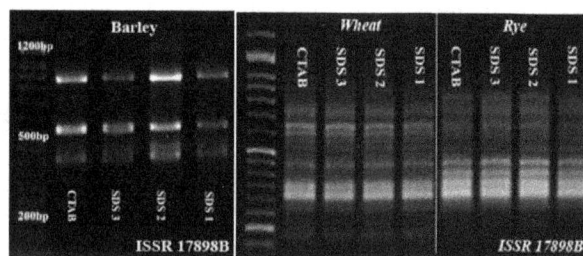

Figure 5. Agarose gel electrophoresis of PCR product obtained with 17898B primer

The next PCR using UBC 818 ISSR primer has shown good amplification for wheat DNA samples whatever the method used. The poorest results were observed in case of barley DNA obtained by CTAB method: no clear amplicons were detected. The best results were recorded with barley DNA samples obtained by SDS 2 and SDS 3 methods.

For rye DNA samples, weaker amplification was detected in the case of samples isolated with SDS 2 method (Figure 6).

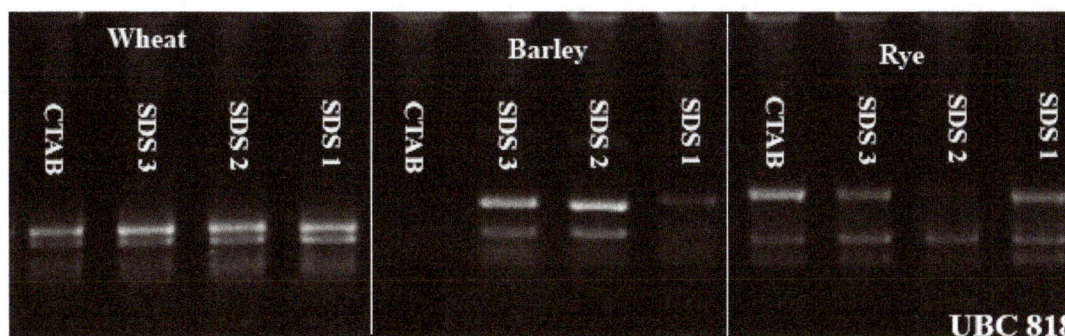

Figure 6. Agarose gel electrophoresis of PCR product obtained with UBC 818 primer

Regarding the cost/sample for each method it was found that SDS 3 method is the cheapest variant (Table 4). The cost/sample was calculated using the online available prices for reagents from Sigma-Aldrich (http://www.sigmaaldrich.com). Furthermore, we took in consideration the time needed to process 10-16 DNA samples: SDS 2 and SDS 3 are the shortest methods. For SDS 3 method it can be added that the extraction buffer does not need to be freshly made (Table 4), making this method suitable for large number of samples.

Table 4. Time, number of steps and the cost/sample

	CTAB	SDS 1	SDS 2	SDS 3
Time (average)	3-4h	3-4h	**2-3h**	**2-3h**
Number of steps	14	14	**12**	13
Cost/sample (€)	0.113	0.113	0.114	**0.107**

CONCLUSIONS

The methods used for the extraction of genomic DNA have a great influence on both quality and quantity of the recovered DNA. Furthermore, time, cost of reagents, amount of biological material needed and its availability are key factors that must be taken into consideration before choosing the best DNA isolation method. Differences among the methods tested related to the plant species were observed.

Barley DNA obtained with CTAB method, presented in this study, had weaker amplification in some cases making this method not suitable for this type of cereal.

The quality of DNA obtained with the methods applied allowed good amplification results when specie-specific primers were used. Contrary, depending on the isolation method used, differences among the DNA samples were detected with ISSR primers.

The best results regarding the DNA quality and quantity, the amplification results, the cost and the time it takes to DNA extraction protocol were obtained with SDS 3 method, making this method suitable for large number of samples.

ACKNOWLEDGEMENTS

The present work was funded through the Ministry of Agriculture and Rural Development – Romania, Research Project ADER116 (2015-2018).

REFERENCES

Abdel-Latif A. & Osman G., 2017. Comparison of three genomic DNA extraction methods to obtain high DNA quality from maize. Plant Methods, 13(1), 1.

Chao S.M. & Somers D., 2012. Wheat and barley DNA isolation in 96-well plates. MAS wheat. http://maswheat. ucdavis. edu/protocols/general_protocols/DNA_isolation_003. htm.

Chaves A.L., Vergara C.E., & Mayer J.E., 1995. Dichloromethane as an economic alternative to chloroform in the extraction of DNA from plant tissues. Plant Molecular Biology Reporter, 13(1), p. 18-25.

Ciuca M., Cristina D., Turcu A.G., Contescu E.L., Ionescu V. & Saulescu N.N., 2015. Molecular detection of the adult plant leaf rust resistance gene Lr34 in Romanian winter wheat germplasm. Cereal Research Communications, 43(2), p. 249-259.

Cristina D., Turcu A.G. & Ciuca M., 2015. Molecular Detection of Resistance Genes to Leaf Rust Lr34 And Lr37 in Wheat Germplasm. Agriculture and Agricultural Science Procedia, 6, p. 533-537.

Lagudah E.S., Krattinger S.G., Herrera-Foessel S., Singh R.P., Huerta-Espino J., Spielmeyer W., Brown-Guedira G., Selter L.L., Keller B., 2009. Gene-specific markers for the wheat gene Lr34/Yr18/ Pm38 which confers resistance to multiple fungal pathogens. Theor Appl Genet 119: p. 889-898. DOI:10.1007/s00122-009-1097-z.

Mohsen Mohammadi, Davoud Torkamaneh, Majid Hashemi, Rahim Mehrabi, Amin Ebrahimi, 2012. Fast and inexpensive DNA isolation from wheat (Triticum aestivum) and other small grains. Wheat Inf. Serv. 114, 2012. Pag.17-20. Available online http://shigen.nig.ac.jp/ewis/article/html/118/article.html.

Murray M.G. & Thompson W.F., 1980. Rapid isolation of high molecular weight plant DNA. Nucleic acids research, 8(19), p. 4321-4326.

Rogers S.O. & Bendich A.J., 1989. Extraction of DNA from plant tissues. In Plant molecular biology manual Springer Netherlands, p. 73-83.

Saal B., Wricke G., 1999. Development of simple sequence repeat markers in rye (Secale cereale L.). Genome 42: p. 964-972.

Singh K., Kaur J., Radu S., Mohamad Ghazali F. & Cheah Y.K., 2009. Real-time PCR evaluation of seven DNA extraction methods for the purpose of GMO analysis. International Food Research Journal, 16(3), p. 329-341.

Von Post R., Von Post L., Dayteg C., Nilsson M., Forster B.P. & Tuvesson S., 2003. A high-throughput DNA extraction method for barley seed. Euphytica, 130(2), p. 255-260.

Zlotina M.M., Kovaleva O.N., Loskutov I.G. & Potokina E.K., 2013. The use of allele-specific markers of the Ppd and Vrn genes for predicting growing-season duration in barley cultivars. Russian Journal of Genetics: Applied Research, 3(4), p. 254-264.

6

ECOLOGY ISSUES OF SOYBEAN CROP PLANTS

Nicolae IONESCU[1], Aurelian PENESCU[2], Pompiliu CHIRILĂ[3]

[1]Agricultural Research and Development Station Pitesti, Pitesti-Slatina Road no. 5, 117030, Pitesti, Romania, Email: nicolaeionescu50@gmail.com
[2]University of Agronomic Sciences and Veterinary Medicine of Bucharest, 59 Mărăşti Blvd, 011464, Bucharest, Romania, Email: a_penescu@yahoo.com
[3]National Agricultural Research and Development Institute Fundulea, N. Titulescu Road no. 1, 915200 Fundulea, Romania, Email: chirilapompiliu@yahoo.com

Corresponding author email: nicolaeionescu50@gmail.com

Abstract

Being a plant with considerable agronomic qualities, soybean returns to farmers attention (Dencescu et al.,1982b; Ionescu, 1985). Indeed the new varieties have improved by characters: high production potential, quality and constantly over time (Gallais & Bennerot, 1992). In exchange for use of these qualities are needed informations about how to adapt plants in different ecological areas (Giosan et al., 1986). Given white luvicsoil in South the plant finds a good regime of sufficient rainfall and temperature (Ionescu et al., 1986; 1994). Against this background proved to be important the specific crop system. 2.0-3.0 t.ha^{-1} yields were obtained while during the filling of the grains fell 150-200 mm water. During the growing season of 100-140 days have accumulated 1200-1400^{0}C temperatures above 10^{0}C. Among during the growing season and grain production achieved a positive correlation (r=0.591***) so that varieties with higher period during the growing season formed the best production. Plant morphology: size, number of pods per plant and grains were positively correlated with grain yield. Negative correlations were obtained between the thousand grains weight-TGW with production of grains, TGW with grain number per plant and between fat content with the content of crude protein of the grain. Production levels achieved over the years has been between 1.06 and 3.49 t.ha^{-1} like limits. Among the crop items highlighted technological density peaks at 60-80 seeds.sq.m^{-1}.

Key words: ecology, morphology, quality, soybean plants, albic luvisol.

INTRODUCTION

Due to its importance (Bîlteanu & Bîrnaure, 1989), soybean [*Glycine max* (L) Merrill, *G. hispida* (Moench) Max a.s.o*] is required increasingly more so in the world (Cregan, 2008), and us (David, 2002). To obtain high yields of grains is recommended primarily cultivation of new varieties adapted (Haş, 2006). These new varieties are accepted under the rules of priority established specific eco-system (Ionescu & Ionescu, 2012). In one such eco-soy system will not miss sustainability, environmental protection and the cultivation technological links as such. Thus, each ecological zone of soybean eco-system adapts its own rules after. The aim is that the new variety to produce as much grain, both quantitatively and qualitatively.

Given soybean genetics (Dencescu, 1980; 1982a; Cregan, 2008), obtaining maximum production of soy beans in a variety is based on

a real complex characters. Some of these characters is based on the additive gene action, such as: the number of seeds per plant, grain size and height (size) of the plant. Other characters besides the additive action, is also based on dominance and epistasis of genes, for example: the number of nodes per plant, number of pods per plant and the number of beans in the pod (David, 2002). The quality of soybeans is expressed mainly through the protein and fat content. Along with the production of grain, the two characters have a wider complexity of both the structure and the respective genetic factors (Wilcox & Shibles, 2001; Yin & Vyn, 2005). Meanwhile, grain yield and quality varieties and new lines are greatly influenced by environmental factors.

Expressing characteristics of soybean adaptability to eco-culture medium can be done through the study of correlations (Yin & Vyn,

2005). Some ie between content of proteins, the fat and morphological characters of soybean were positive. However, correlations between proteins and fats found in most grains varieties were negative (Wilcox & Shibles, 2001; Ifrim & Haş, 2008).

In this paper we present the adaptation aspects of soybean plants through the correlations in several directions. The first direction refers to grain yield response to climatic factors: rainfall and temperatures assets. Another line shows the connection between morphological elements: size, number of nodes on the stem, number of pods per plant and number of grains per plant, on grain yield formation. A third line shows the correlations between absolute grain weight (TGW) with grain yield, number of grains per plant, and between the fat content with protein. Grain yield obtained over the years has high-lighted the influence of year and density, as two elements which best expresses the ecology of some new soybean varieties and lines.

MATERIALS AND METHODS

Soybean grown to normal after the technology developed by the resort. We used a relatively long period of time, lines and new varieties recommended for areas that belong white luvic-soils of southern territory. The data represent the average periods of research. Measurements and determinations were made both in the field (Photo 1) and laboratory follow several parameters.

During the vegetation period (VP) of soy, precipitations were noted in two periods and namely the entire period between sunrise and maturity, and during the submission of the reserve substances period (July and August). Temperatures active: Σtn^0 C$>10^0$C were accumulated throughout the growing season, as well as the number of vegetation days.

Soybean plants were measured: size from the crop, the number of nodes on the main stem, number of pods on the whole plant, grain number per plant, and grain from production area. Experiences with lines and new varieties were made after the block method in five repetitions and variants had each 25 m^2. In experiments with the density method was all the blocks in 5 replications, with 20, 40, 60, 80, and 100 grains/m^2, 25 m^2 each variants.

Soybean were determined: thousand grains weight (TGW), the average number per plant, the fat content and protein content.

Between the different measurements and determinations settled most important correlations.

Photo 1. Daciana variety (0)(David, 2006)

RESULTS AND DISCUSSIONS

On the climate. Soy is generally high demands for water and warm. From the beginning, soybeans need water for germination of 150% of their dry weight. The specific consumption is expressed by the transpiration coefficient between 300 and 700. Sweating period for water occurs during the formation of reproductive organs, flowering and grain filling.

The correlations obtained between the amount of rain fallen, so the entire growing season and in the months of July- August shows positive upper results (Figure 1). If the entire growing season, the correlation coefficient was pasitive, but not significant (r=0.142), one from July to August was significant (r=0.695*). The chart shows that the rains have fallen throughout the soybean growing season were between 180 and 480 mm, while the July-August were between 50 and 220 mm.

Compared to soy light behaves as a short-day plant, so that integration as early seeding ensures claims for medium varieties photoperiod district here. Warm factor recorded in vegetation soybean varieties was between 1200-1400^{0}C (Σtn^0 C$>10^0$C).

Correlations between vegetation period (in days) with Σtn^0C$>10^0$C and between vegetation period with grain yield were positive and highly secured statistic (Figure 2). Thus, during the growing season of 95 to 140 days were cumulative 1200-1400 Σtn^0 C$>10^0$C and average yields were between 22 and 28 q/ha (according to the regression line).

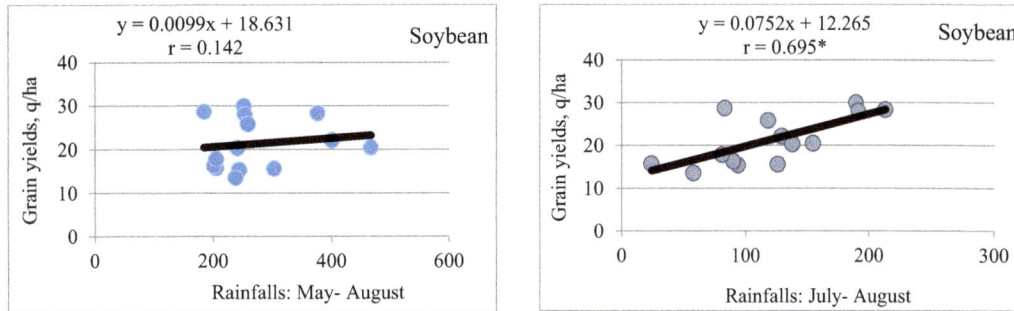

Figure 1. Correlation between rainfalls sum (two periods) and soybean grain yields

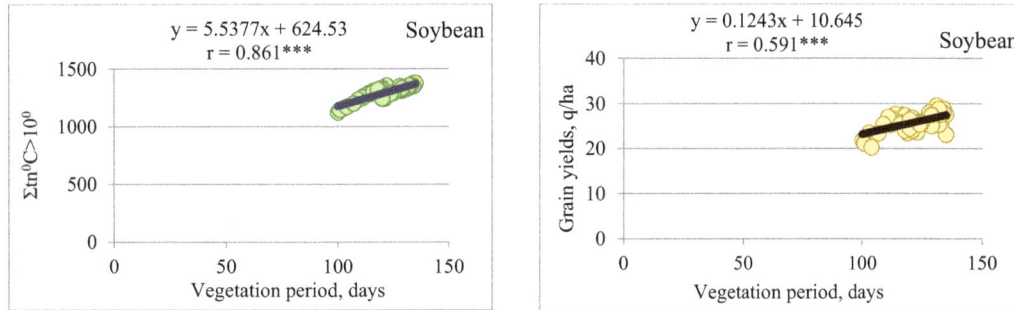

Figure 2. Correlations between vegetation period (VP) with active temperatures sum
($\Sigma tn>10^0C$) (left) and VP with soybean grain yields

Characteristics of soybean plant. The correlation between plant height obtained with the number of nodes on the main stem was negative (r=-0.377). This means that varieties and lines formed more pods and beans at fewer nodes. Plants 60-90 cm high were formed between 13 and 10 knots. The correlation between plant tall with grain production is positive, slightly increasing and no significant (r=0.205) (Figure 3). The data show that the plants were between 60-80 cm tall and produced between 24 and 28 q/ha grain.

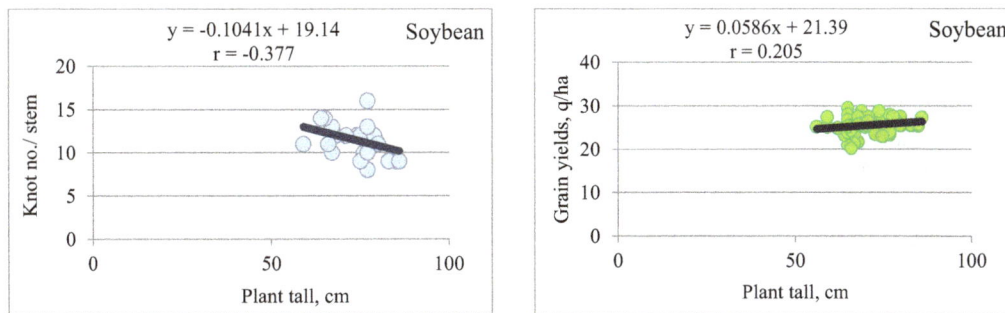

Figure 3. Correlations between plant size (tall) and knot number per stem (left)
and plant size (tall) with soybean grain yields

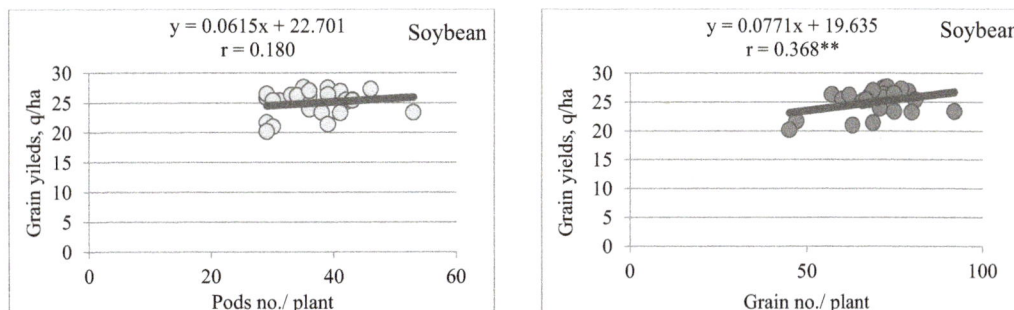

Figure 4. Correlations between pods no. per plant with grains yields (left)
and grains no. per plant with soybean grain yields

Other characteristics: the number of soybean pods per plant and the number of grains per plant were positively correlated with the grain productions (Figure 4). The data showed that 25-46 pods per plant resulted in the formation of about 25 q/ha grain, with r=0.180. The number of grains of between 40-90 on a soybean plant ensured production of 23-27 q/ha, with r=0.368** (insured statistic). By comparing the two characters showed that the number of grains formed on a soybean plant was decisive in the formation of higher production.

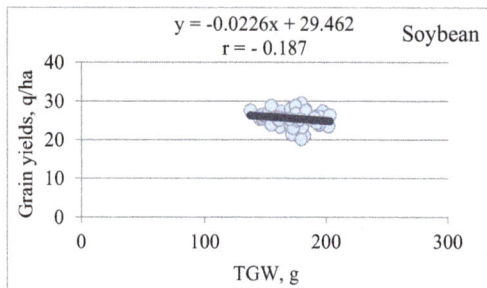

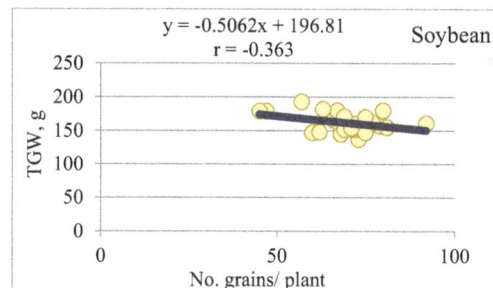

Characteristics of soybean grain. The correlations obtained between the thousand grains weight (TGW) with production of grain and between the number of grains per soy plant and TGW were negative (r=-0.187, and r=-0.363 respectively, Figure 5).

Between beans fat content and protein content was obtained a negative correlation, with r=-0.833***.

Data show that crude fat varied between 16% and 19%, and crude protein between 45% and 36% (Figure 6 and Photo 2).

Figure 5. Correlations between TGW with grain yields (left) and between grain number/ plant and TGW

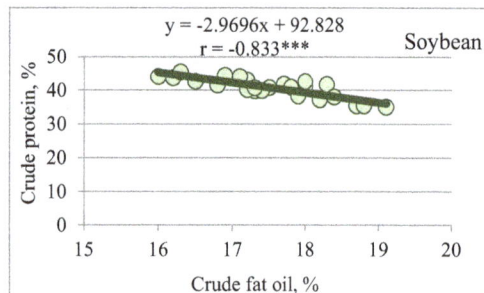

Figure 6. Correlation between fat oil and protein contents from soybean grain

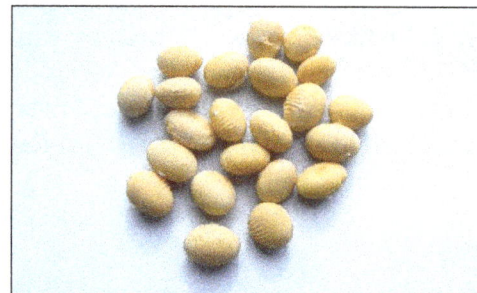

Photo 2. Daciana grains

Soybean grain yield. Production levels achieved over the years were located at different levels. Greatest influence was the year of culture (Table 1). In new soybean varieties (Daciana also) and lines were markedly different. Absolute values of soybean yields were 30.1-34.9 q/ha like maximum, and 10.6-11.8 q/ha like minimum. Among the technological elements of soybean crop, density plays an important role. The high yields obtained over time were obtained from 60 to 80 grains/m^2 (Table 2).

CONCLUSIONS

Rainfall that fell in soybean vegetation were between 200 and 500 mm, and in July- August period of 40 to 220 mm. However, production levels were 15-30 q/ha, as it has contributed to the formation of cross-system practiced. The amount of active degrees of temperature (>10^{0}C) was between 1200-1400^{0}C, which allows the cultivation of early varieties (000, 00) and middle (0). The growing season was 95- 140 days. Higher production was achieved with more grains per plant, followed by higher tall and a greater number of pods per plant. In contrast, the tall (high) plants are knots less. Negative correlations were obtained between the TGW and production, the number of grains on a plant and TGW, and between grains fats and proteins.

Average grain yields were obviously influenced by climatic conditions. The levels of more than 30 q/ha were obtained for 5 years, 11-18 q/ha were in 3 years, and the following years were intermediate. Between technological measures, 80 grains/m^2 density ensured 2/3 of years in the achievement of great productions.

Table 1. The grain yileds of differing soybean varieties and lines, q.ha^{-1}

Var./years	1	2	3	4	5	6	7	8	9	10	11	12	13	14
A$_1$	15.7	**30.1**	29.7	17.5	28.6	23.0	28.9	**30.6**	13.8	13.0	13.0	16.7	16.5	19.1
A$_2$	18.7	23.3	25.0	18.3	27.1	20.8	24.8	29.9	15.3	16.0	13.8	19.5	16.8	17.5
A$_3$	15.3	26.5	22.5	22.9	**30.2**	16.4	20.6	28.0	15.3	14.6	15.0	19.9	15.8	14.9
A$_4$	14.4	27.9	19.4	22.4	27.3	22.2	27.9	28.3	17.7	14.9	13.5	15.5	15.8	16.3
A$_5$	16.1	**34.0**	28.9	29.9	27.9	21.4	28.6	**30.7**	16.0	18.0	14.3	19.9	17.0	14.2
A$_6$	17.0	25.9	29.1	22.8	**30.7**	22.1	**30.1**	**30.5**	15.2	14.8	15.3	18.7	20.0	18.6
A$_7$	13.8	**32.3**	29.8	24.4	29.4	20.9	25.6	29.7	14.0	14.9	14.8	20.7	18.3	17.5
A$_8$	16.8	29.2	23.6	21.2	**30.9**	23.3	28.8	26.9	16.5	17.8	15.3	16.4	15.8	18.8
A$_9$	18.1	29.1	26.7	21.8	28.4	22.5	26.1	**30.7**	12.1	16.6	16.5	20.6	14.3	18.1
A$_{10}$	**10.6**	27.5	**33.8**	25.3	29.4	21.2	25.8	29.6	**10.9**	14.2	13.5	18.5	17.8	17.6
A$_{11}$	11.4	27.0	**30.2**	18.0	29.1	21.9	25.8	**31.1**	15.9	14.1	13.3	19.1	14.4	19.9
A$_{12}$	20.5	**32.9**	**31.3**	24.7	26.6	21.1	**31.5**	29.4	17.4	13.7	16.0	19.3	14.3	17.3
A$_{13}$	13.5	**30.5**	**32.8**	21.5	29.6	21.9	26.5	28.0	17.6	15.5	14.8	19.2	16.3	14.5
A$_{14}$	16.9	**30.5**	**31.5**	25.4	29.4	21.5	27.7	28.1	14.6	14.6	13.0	20.3	17.0	19.2
A$_{15}$	16.9	26.0	28.4	22.7	29.4	21.4	26.2	**31.9**	17.0	16.1	14.5	20.2	15.0	16.6
A$_{16}$	15.8	**32.4**	25.7	21.8	27.1	22.9	26.6	**31.6**	17.6	13.6	11.8	18.8	15.8	20.1
A$_{17}$	10.7	**34.9**	23.6	19.4	28.6	22.7	25.0	28.5	16.1	18.4	12.8	17.8	15.3	16.2
A$_{18}$	15.0	**33.1**	**30.1**	22.7	29.6	21.7	26.3	28.2	16.1	16.6	11.8	21.9	14.3	14.7
A$_{19}$	17.0	**31.6**	29.2	19.2	27.9	19.8	23.9	26.5	15.3	15.9	11.3	19.6	15.3	18.2
A$_{20}$	19.9	**31.9**	27.0	25.7	29.6	21.0	23.6	27.1	18.4	14.1	11.5	17.1	17.5	18.6
A$_{21}$	13.2	**31.2**	29.9	17.6	25.8	15.6	28.0	26.5	16.1	12.3	13.0	22.7	18.0	19.9
A$_{22}$	18.6	27.0	29.2	23.7	24.8	15.3	20.6	28.2	14.9	14.8	13.8	20.7	14.8	18.8
A$_{23}$	**10.8**	**31.6**	29.9	21.4	25.8	16.4	20.1	26.6	17.0	17.9	12.8	19.1	17.8	20.2
A$_{24}$	16.3	**32.0**	25.5	21.2	29.1	16.4	21.8	22.4	17.0	15.3	11.8	18.7	16.3	19.6
A$_{25}$	18.1	**32.0**	29.8	22.6	26.8	15.7	25.4	28.7	17.2	17.7	12.8	18.0	17.3	20.0
DL 5%	*2.41*	*2.97*	*5.75*	*1.71*	*1.55*	*4.12*	*3.17*	*2.22*	*3.33*	*3.46*	*2.96*	*4.44*	*2.37*	*3.35*
DL 1%	*3.22*	*4.03*	*7.81*	*2.32*	*2.09*	*5.47*	*4.19*	*2.95*	*4.44*	*4.59*	*3.93*	*5.89*	*3.22*	*4.47*
DL 0.1%	*4.20*	*5.40*	*10.46*	*3.11*	*2.80*	*7.05*	*5.39*	*3.80*	*5.79*	*5.91*	*5.07*	*7.61*	*4.31*	*5.84*
MEDIA	**15.6**	**30.0**	**28.1**	**22.2**	**28.4**	**20.4**	**25.9**	**28.7**	**15.8**	**15.4**	**13.6**	**20.5**	**16.3**	**17.9**

Table 2. The density influence on soybean grain yields

Var./years	1	2	3	4	5	6	7	8	9	10	11	12
20	10.9	22.2	15.1	10.4	16.0	10.3	14.2	14.1	28.0	11.3	11.2	23.8
40	17.7	25.0	**17.9**	13.6	18.0	11.5	16.3	14.2	30.0	12.7	13.2	**26.8**
60	18.3	26.9	17.3	14.0	**18.8**	13.8	**19.1**	17.0	30.5	13.2	14.7	25.2
80	**20.8**	**27.8**	15.5	14.4	18.7	**15.3**	17.1	**18.9**	**31.0**	**14.8**	**15.5**	24.4
100	18.6	25.1	15.3	14.1	17.2	14.9	16.0	16.6	29.3	11.6	13.8	24.0
DL 5 %	*4.01*	*5.82*	*2.75*	*1.87*	*4.97*	*3.77*	*3.93*	*2.64*	*5.60*	*3.23*	*2.30*	*3.93*
DL 1 %	*5.50*	*7.98*	*3.77*	*2.56*	*7.53*	*5.43*	*5.66*	*3.79*	*8.06*	*4.65*	*3.15*	*5.65*
DL 0.1%	*7.49*	*10.86*	*5.14*	*3.49*	*12.10*	*7.98*	*8.32*	*5.57*	*11.85*	*6.84*	*4.27*	*8.01*
MEDIA	**17.3**	**25.4**	**16.2**	**13.3**	**17.7**	**13.2**	**16.5**	**16.2**	**29.8**	**12.7**	**13.7**	**24.8**

REFERENCES

Bîlteanu G., Bîrnaure V., 1989. Fitotehnie. Ed. Ceres, Bucureşti, 1: p. 350-392.

Cregan P.B., 2008. Soybean molecular genetic diversity. Genetics and Genomics of Soybean, 2(1): p. 17- 34.

David I., 2002. Comportarea unor genotipuri de soia sub aspectul stabilităţii producţiei în perioada 1997-1999. Probleme de genetică teoretică şi aplicată, XXXIV, 1-2: p. 17-36.

David I., 2006. Soiul semitimpuriu de soia Daciana. INFO-AMSEM 2: p. 39-41.

Dencescu S., 1980, Cercetări privind ereditatea conţinutului de substanţe proteice, a conţinutului de substanţe grase şi a elementelor de producţie la soia. Teză Doctorat, ASAS Bucureşti.

Dencescu S., 1982. Corelaţii între principalele caractere agronomice la soia. Probleme de genetică teoretică şi aplicată, XIV, 5: p. 563-589.

Dencescu S., Miclea E., Butică A., 1982. Cultura soiei. Ed. Ceres, Bucureşti, p. 1-227.

Gallais A. & Bannerot H., 1992. Amélioration des espèces végétables cultivées. INRA, Paris, p. 1-147.

Giosan I., Nicolae I., Sin G., 1986. Soia. Ed. Academiei Române, Bucureşti.

Haş I., 2006. Producerea seminţelor la plantele agricole. Ed. Academic Press, Cluj Napoca, p. 63- 72.

Ifrim S.E., Haş I., 2008. The relationship between some morphophysiological plant characters and the quality of the soybean Glycine max (L.) production. Bulletin UASVM, Agriculture, 65(1): p. 135-140.

Ionescu N., 1985. Soia-posibilități de cultură în zona solurilor argilo- iluviale din sudul României. Producția vegetală-cereale și plante tehnice, București, 2: p. 9-14.

Ionescu N., Dencescu S., Stoica V., Preoteasa C., 1986. Comportarea soiurilor și liniilor de soia în condițiile solurilor podzolice din sudul țării. Analele ICCPT Fundulea, cat. B, 53: p. 73-82.

Ionescu N., Popescu A., Popa A., 1994. Rezultate privind influența temperaturii și a densității asupra soiei bacterizate în condițiile solurilor podzolice. Analele ICCPT Fundulea, cat. B, 61: p. 141-148.

Ionescu N., Ionescu S.G., 2012. Cercetări privind reducerea gradului de îmburuienare din cultura soiei prin metode chimice și nechimice. Analele INCDA Fundulea, cat. B, 80: p. 161-172.

Wilcox J.R. & Shibles R.M., 2001. Interrelationships among seed quality attributes in soybean. Crop Science, 41: p. 11-14.

Yin X.A. & Vyn T.J., 2005. Relathionships of isoflavone, oil and protein in seed with yield of soybean. Agronomy Journal, 97: p. 1314- 1321.

7

PREDICTION OF DROUGHT RESISTANT LINES OF WINTER WHEAT USING CANOPY TEMPERATURE DEPRESSION AND CHLOROPHYLL CONTENT ANALIZIS

Doru-Gabriel EPURE, Marius BECHERITU, Cristian-Florinel CIOINEAG

Probstdorfer Saatzucht Romania SRL, 20 Siriului Street, District 1, Bucharest, Romania

Corresponding author email: doru.epure@probstdorfer.ro

Abstract

The main objective of the study was to validate the use of canopy temperature depression (CTD) as a rapid early generation screening tool for drought tolerance in wheat breeding. CTD was measured at 3 drought sites in Romania (Valu lui Traian, Modelu, Drăgăneşti-Vlaşca) on F6. Measurements of chlorophyll content (CHL) on F6 individual plants showed significant correlations with yield. Since a reliable yield estimate requires a plot approximately three times bigger than that needed for an estimate of CTD and CHL, the use of these methods instead of yield estimates may be considerably more efficient. Using alone either CTD or CHL could not provide enough data for drought resistant selection on winter wheat lines. Alternatively, both yield and CTD and CHL could be combined in a selection index as a more powerful indicator of drought tolerance. During the analysed period of three years, the climatic conditions were very different from one year to another. Forty-five lines of winter wheat in two repetition have been used for experiments, conducted in 3 locations in south of Romania. The study shows that genotype with high CTD values are correlated with high CHL values, but has lower drought tolerance. The lines with lower CTD values and high CHL values represent genotypes with high yield on drought conditions.

Key words: Canopy Temperature Depression, Chlorophyll content, drought resistance, winter wheat.

INTRODUCTION

High CTD has been used as a selection criterion to improve tolerance to drought and heat (Amani et al., 1996; Ayeneh et al., 2002; Blum, 1996; Blum et al., 1989; Pinter et al., 1990; Rashid et al., 1999; Reynolds et al., 1994, 2001; Fischer et al., 1998) and has been associated with yield increase among wheat (*Triticum aestivum* L.) cultivars at CIMMYT (Fischer et al., 1998). The suitability of CTD as an indicator of yield and stress tolerance have been reported also by literature (Reynolds et al., 2001). CTD frequently shows a better association with yield and grain number than with total biomass (Reynolds et al., 1997, 1998).

CTD effected by biological and environmental factors like wind, evapotranspiration, cloudiness, conduction systems, plant metabolism, air temperature, relative humidity, and continuous radiation (Reynolds et al., 2001), has preferably been measured in high air temperature and low relative humidity because

of high vapour pressure deficit conditions (Amani et al., 1996).

CTD has been used as a selection criterion for tolerance to drought and high temperature stress in wheat breeding and the used breeding method is generally coming by mass selection in early generations like F3 (Reynolds et al., 2001).

Canopy temperature depression is highly suitable for selecting physiologically superior lines in warm, low relative humidity environments where high evaporative demand leads to leaf cooling of up to 10 °C below ambient temperatures. This permits differences among genotypes to be detected relatively easily using infrared thermometry. However, such differences cannot be detected in high relative humidity environments because the effect of evaporative cooling of leaves is negligible (Reynolds et al., 2001).

Chlorophylls are a dominant factor controlling leaf properties of healthy green vegetation and are thus an essential part of the photosynthetic process. They harness light energy from the sun to store it as chemical energy (Richardson et al., 2002). For optical methods for measuring leaf Chlorophyll content, index values (e.g. SPAD-Value) are commonly used to specify the relative leaf Chlorophyll content (Richardson et al., 2002; Suess et al., 2015).

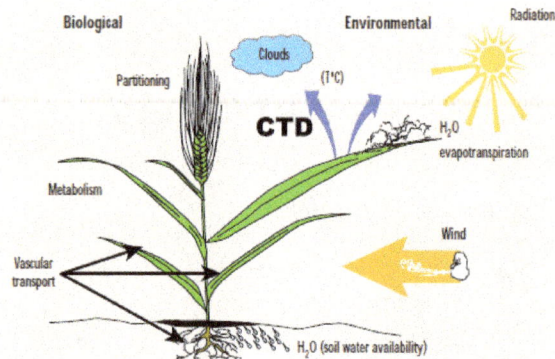

Figure 1. Factors affecting canopy temperature depression (CTD) in plants (Reynolds et al., 2001)

Chlorophyll content is one of the indices of photosynthetic activity (Larcher, 1995). There is usually 4-5 mg of chlorophyll per unit of leaf surface in wheat flag-leaves and it is an indicator of photosynthetic activity (Bojovic et al., 2005)

Leaf chlorophyll content is often highly correlated with leaf N status, photosynthetic capacity (Evans, 1983; Seemann et al., 1987).

It has been reported in several studies that there widely exists a difference of chlorophyll content among different wheat genotypes under the identical climatic, soil and farming conditions (Paknejad et al., 2007; Tas et al., 2007; Guóth et al., 2009; Keyvan, 2010; Kiliç and Yagbasanlar, 2010).

Leaf chlorophyll content was positively correlated with photosynthetic capacity (Araus et al., 1997), high chlorophyll content in leaves was considered as a favourable trait in wheat crop production (Teng et al., 2004).

Chlorophyll content parameter is a good indicator for predicting yield and drought resistance for wheat (Ping Li et al., 2012).

Drought resistance genotypes of wheat had much higher Chlorophyll content evaluated at anthesis wheat development stage (Ping Li et al., 2012).

Chlorophyll content is directly related to nitrogen status, considering that most of the leaf nitrogen is integrated in chlorophyll. (Richardson et al., 2002).

The greatest chlorophyll content in plants occurs at the outset of the flowering phase, and chlorophyll is believed to take part in the process of organogenesis (Simova et al., 2001). Nitrogen concentration in wheat leaves is related to chlorophyll content, and therefore indirectly to one of the basic plant physiological processes: photosynthesis (Haboudane, 2002; Amaliotis et al., 2004; Lelyveld et al., 2004; Cabrera, 2004).

The content of chlorophyll content and levels of other leaf biochemical constituents can be used as indicators of crop stress under conditions of nutritional deficiencies (Tejada - Zarco, 2004).

MATERIALS AND METHODS

Trial was conducted in 2014-2016 years at Valu lui Traian, Modelu and Drăgăneşti-Vlaşca Research Station situated in south of Romanian Plane. Soil texture and type as well as climatic condition varies from one trial station from another, but were generally drought conditions. The lower annual average temperature during wheat vegetation period was registered on 2014 at Drăgăneşti Vlaşca – 7.56^0C, and maximum average temperature was registered on 2015 at Valu lui Traian – 10.58^0C. The minimum rainfall during wheat vegetation season was 255.7 mm in 2013 at Valu lui Traian and maximum 472.7 mm in 2016 at Drăgăneşti-Vlaşca. In this study, 30 wheat lines in two replicates each for one station have been analysed using a randomized block. Plots were planted at a seeding rate of 300 seed per m^2. The dimension of one plots was 10 m^2. In order to characterize the chlorophyll content of wheat lines have been used a Chlorophyll Content Meter CCM 200 Plus produced by OPTISCIENCE Hudson, USA, which analyse the optical transmittance of wave of light at 653 nm and 931 nm with precision of repeatability $\pm$ 5%, which could operate at a range of temperature from 0 to 50^0C. Every data is resulted from auto average from 10 measurement points. The measurement have been made in three different phenophases of wheat development: EC 43, EC 52 and EC 75. All measurement for all location have been made at 10:00 to 10:30 hours in the morning.

CTD measurements were made by infrared thermometer (Model Optris Laser Sight- Optris GmbH Berlin, Germany) equipped with an laser class II, using IR radiation with wave lengh of 630-650 nm, with temperature resolution at 0.1^0C and an accuracy IR of 0.75^0C, repeatability of ±0.05 K, range temperature from -35 to 900^0C capable to make measurement at RH from 10 to 95% and at environmental temperature range from -30 to 65^0C. Measurement have been made at late morning cloudless periods (10:30 to 11:00 hours).

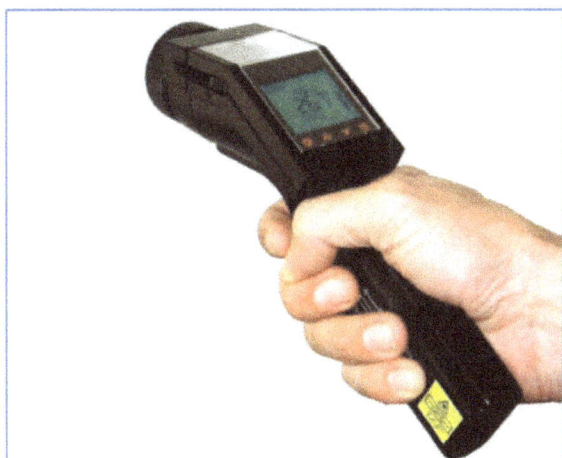

Figure 2. Infrared thermometer Optris Laser Sight used for CTD determination (Optris GmbH, Germany)

As similar to method of Fischer et al. (1998), the data for each plot were the mean of ten readings, taken from the same side of each plot at an angle of approximately 45° to the horizontal in a range of directions such that they covered different regions of the plot and integrated just the flag leaves. It is important to measure the trait when it is best expressed that is, on warm, relatively still, cloudless days. Some environmental flux during the measurement period is inevitable, but correcting data against reference plots, spatial designs, use of replication, and repetition of data collection during the collecting time can compensate for this. Measurements were made at different three periods on different phenophases of wheat development: EC 43, EC 52 and EC 75. Variance analysis of all agronomical traits and CTD measurements on each growth stage were carried out and the significance of cultivar mean square

determined by testing against the error mean square.

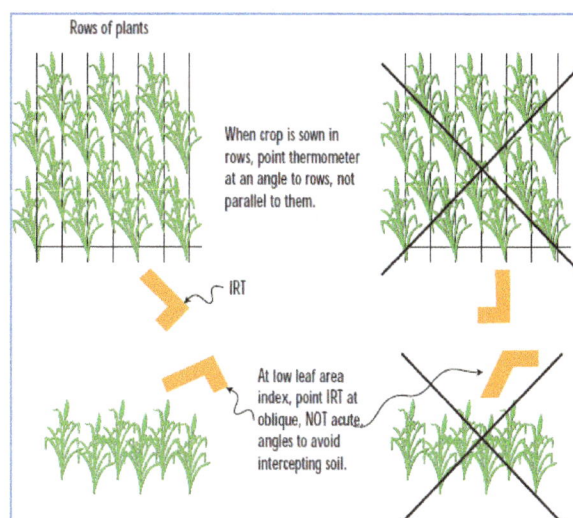

Figure 3. Measuring canopy temperature depression with an infrared thermometer on wheat flag-leaf (Reynolds et al., 2001)

Plots were harvested using a Wintersteiger Classic small plot combine (1998 Wintersteiger Elite). Grain yield was determined from the grain weight with a moisture content of 12% of each plot for each genotype.

RESULTS AND DISCUSSIONS

Measurement of Canopy Temperature Depression have been measured at three intervales of time, respectively CTD 1 at EC 43 stage, CTD 2 at EC 52 and CTD 3 at EC 75 for all three locations (Modelu, Valu lui Traian and Drăgăneşti-Vlaşca). In same time the chlorophyll content have been measured CHL 1 at EC 43, CHL 2 at EC 52 and CHL 3 at EC 75. The measured values have been varying from one measured stage from another and from one location to another for same variety of wheat. Anyhow, the yield has been varying from one location to another and from one year to another for same variety of wheat.

For drought condition from south part of Romanian Plane, the heist yield have been obtained for all research station, and for all experimental years from varieties which have as characteristics lower CTD values and higher values for Chlorophyll content. This is in correlation with literature reports (Araus et al.,

1997; Ping Li et al., 2012; Reynolds et al., 1997, 1998, 2001).

For drought condition the influence of CTD is negatively correlated with yield. The influence is linear. For Modelu' results of the R^2 obtained values have been 0.99 for CTD 1, 0.977 for CTD 2 and 0.979 for CTD 3 (Figure 4).

From data obtained on Modelu has been noticed that all drought resistant varieties have a water consume low per unit. That is indicated by a reduce cooling at leaves surfaces, measured as a low value for Canopy Temperature Depression.

Varieties considered drought resistant for climatic conditions from Modelu presents values of CTD lower than 1.5 K (Figure 4).

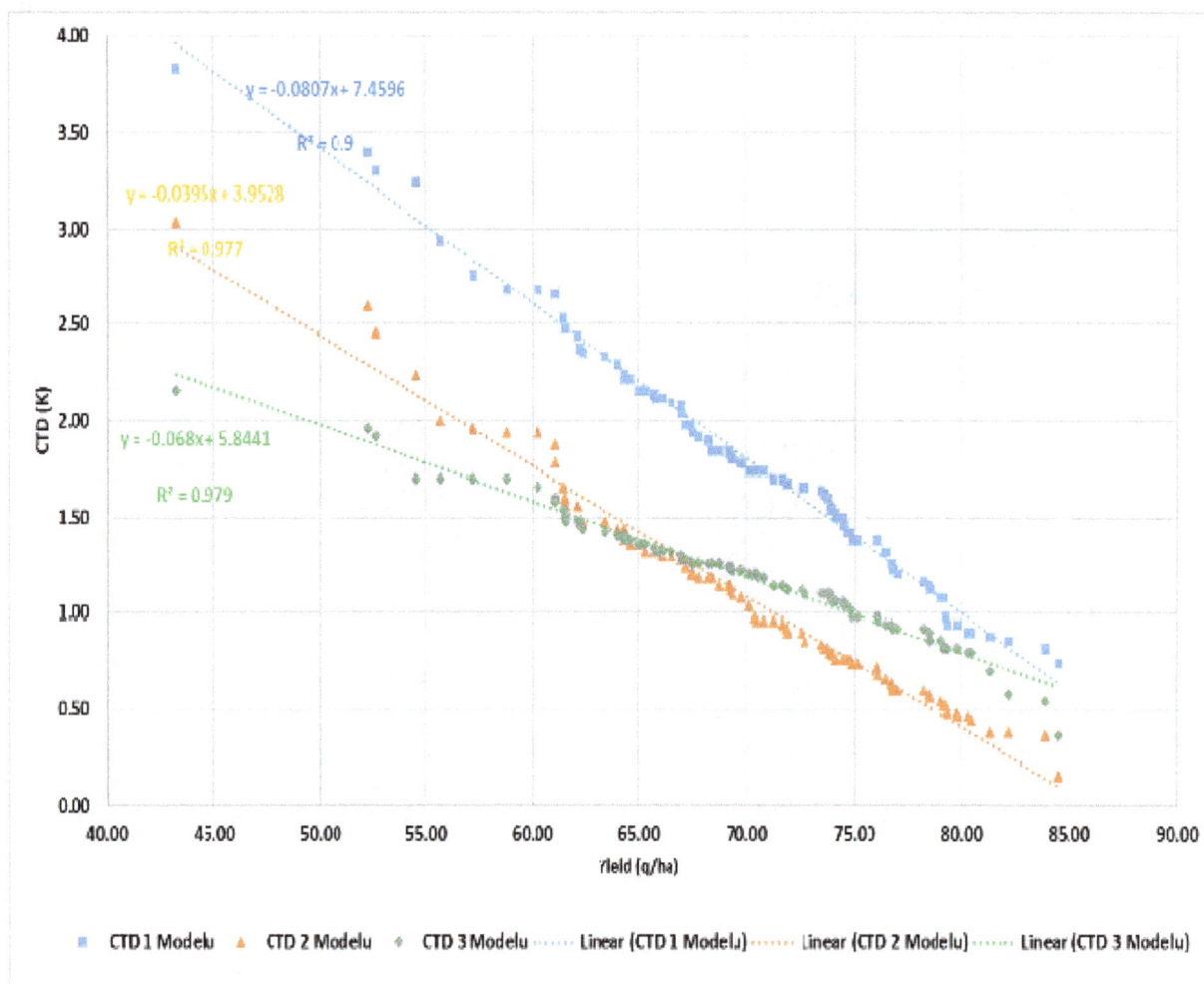

Figure 4. Influence of Canopy Temperature Depression on Yield at Modelu Research Station

From data obtained has noticed that Chlorophyll content of wheat' leaves is exponentially correlated with yield, that means that increasing of Chlorophyll content has as results exponentially increasing of yield (Figure 5). The values of Chlorophyll content vary from 26.36% to 52.94% in phenophases EC 43; from 25.08% to 52.78% at EC 52 and from 27.56% to 46.76% at EC 75. Have been noticed that during rain filling the values of chlorophyll content were decreasing.

Anyhow content of chlorophyll is directly correlated with Nitrogen fertilizations (Evans, 1983; Seemann et al., 1987). From data analysis, the R^2 present for all CHL 1, CHL 2 and CHL 3 data computing value of 0.971 (Figure 5).

The highest values of Chlorophyll content indicate the highest yield for all time of measurement EC 43, EC 52 and EC 75 (Figure 5). Similar values have been reported in literature (Bojovic et al., 2005).

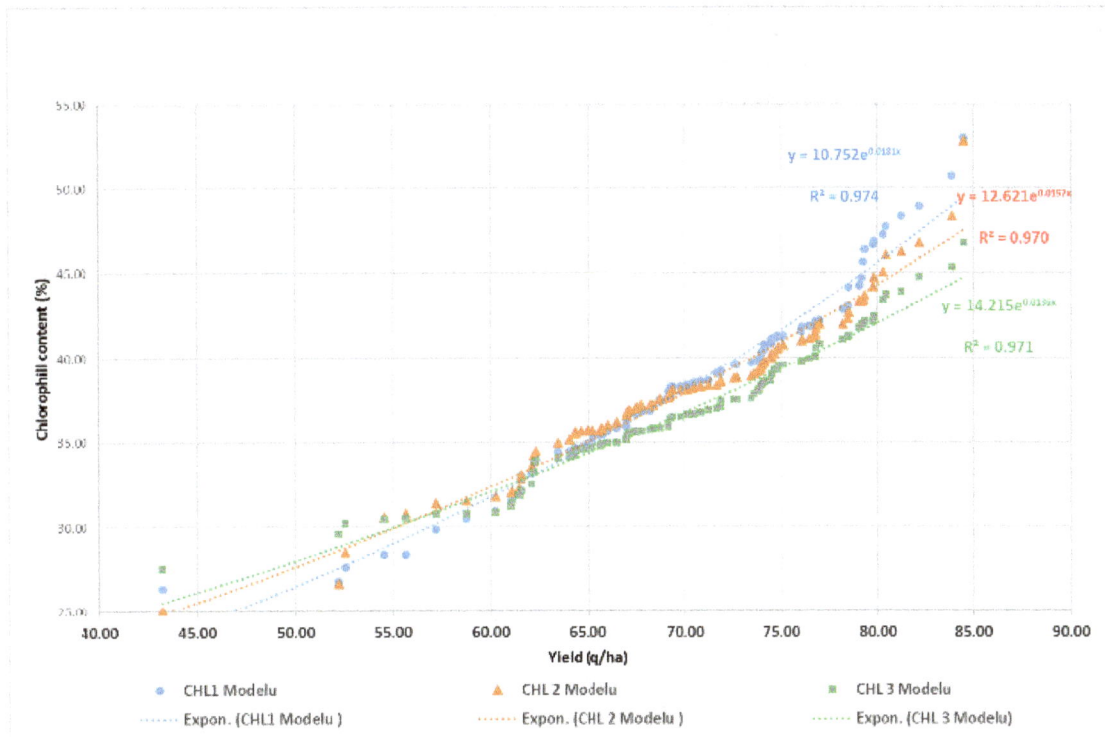

Figure 5. Influence of chlorophyll content on yield at Modelu Research Station

The F6 tested breeding line of wheat, analysed in Modelu for drought resistance present a correlation between Canopy Temperature Depression and Chlorophyll content for all collected data. The lowest CHL 1 is correlated with higher CTD 1, the lowest CHL 2 is correlated with higher CTD 2, the lowest CHL 3 is correlated with higher CTD 3 and has presented by lines with lowest yield obtained in climatic conditions of Modelu. *Ad contrarium*, the highest Chlorophyll content is correlated with lower Canopy Temperature Depression and is presented by wheat line with highest yield obtained in Modelu climatic conditions.

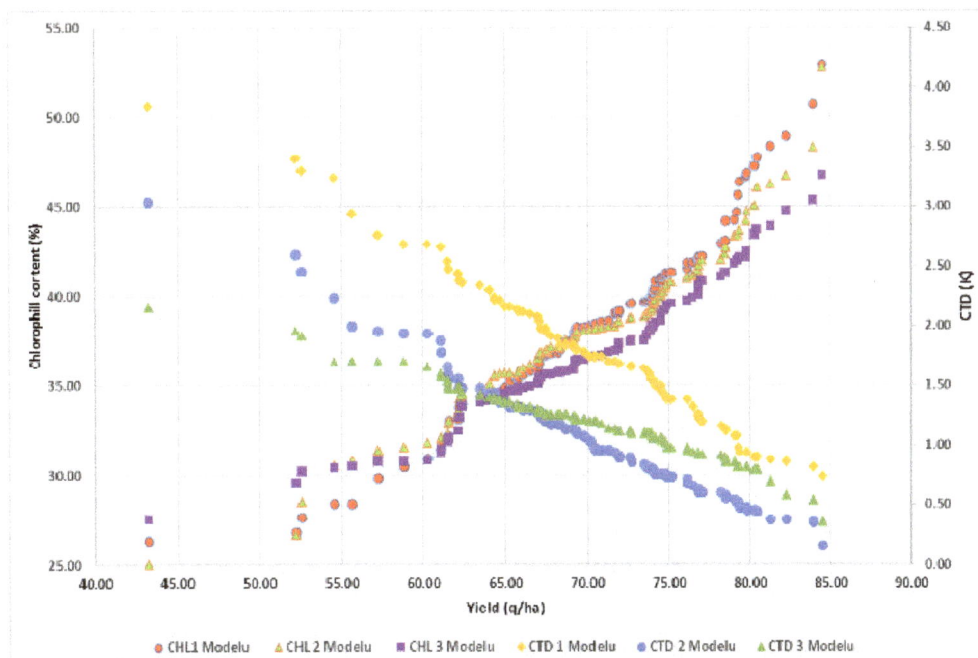

Figure 6. Correlations between Chlorophyll content and Canopy Temperature Depression plotted against yield for Modelu data analysis

At Valu lui Traian Research Station the values obtained for Chlorophyll content and Canopy Temperature Depression presented different values than those obtained at Modelu, also the yields have different values from yields obtained in Modelu, but the highest yield has been obtained by the same wheat line as in Modelu. The highest yield is correlated with highest values of Chlorophyll content and

lowest value of Canopy Temperature Depression (Figure 6).

The Chlorophyll content has exponentially correlated with yield and Canopy Temperature Depression is inverse direct correlated with yield. The R^2 correlation values are 0.94 for CHL 1, 0.94 for CHL2 and 0.83 for CHL 3; 0.91 for CTD 1, 0.97 for CTD 2 and 0.85 for CTD 3 (Figure 7).

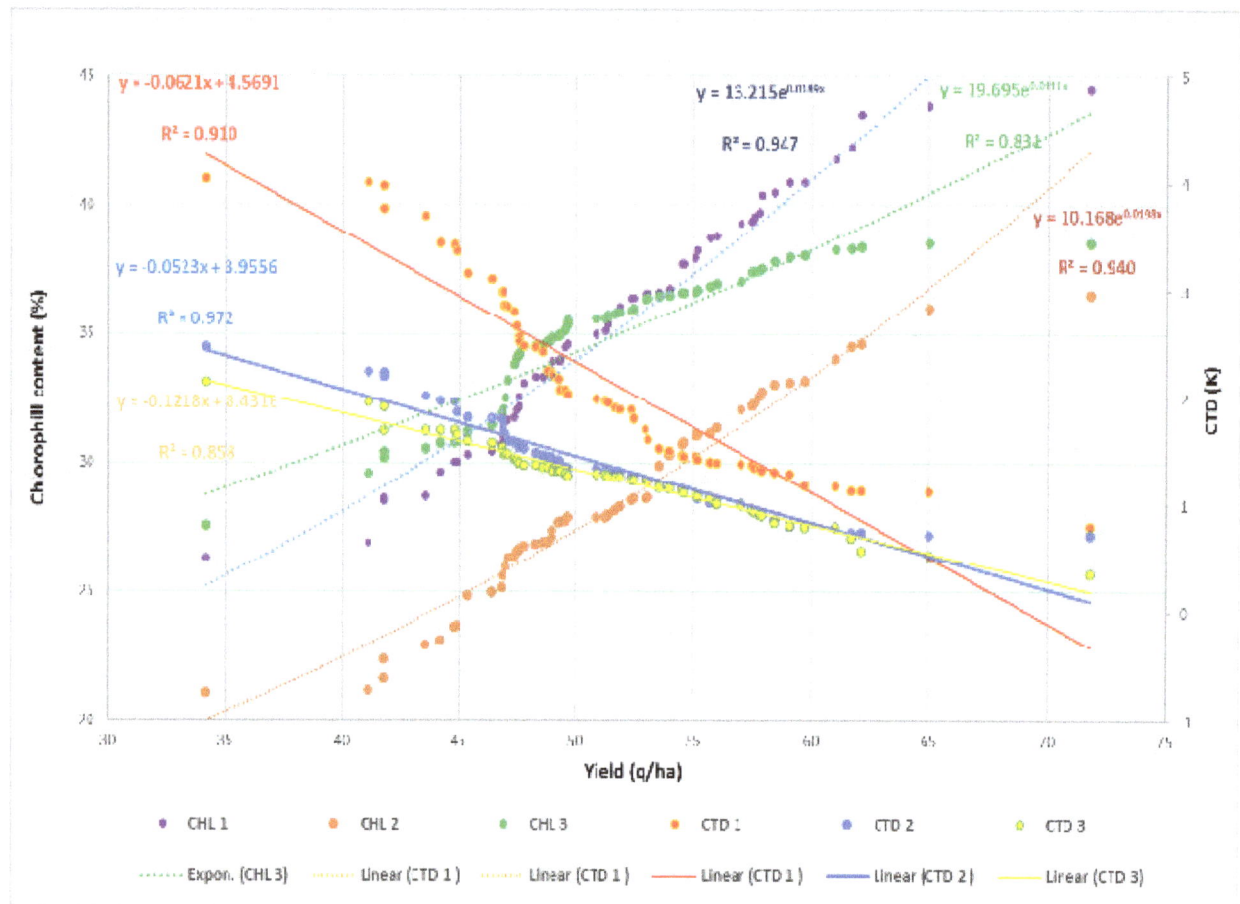

Figure 7. Influence of Chlorophyll content and CTD on yield on Valu lui Traian Research Station

Data obtained on Drăgăneşti-Vlaşca Research Station confirm results obtained in previous locations (Figure 8).

During EC 75 a high heat stress affected experimental location which is recorded on CHL 3 and CTD 3 values.

The analysed varieties with lower CTD and high CHL gave the high yield. Due to a late generation of breeding selection F6, the

analysed wheat lines act as a training population. Even that the yields have been different and also the absolute values of CHL and CTD, the behaviour of lines have been similar to all environment analysed.

However, the behaviour of genetic lines has been correlated with CTD and CHL analysed data and shown repeatability and heritability.

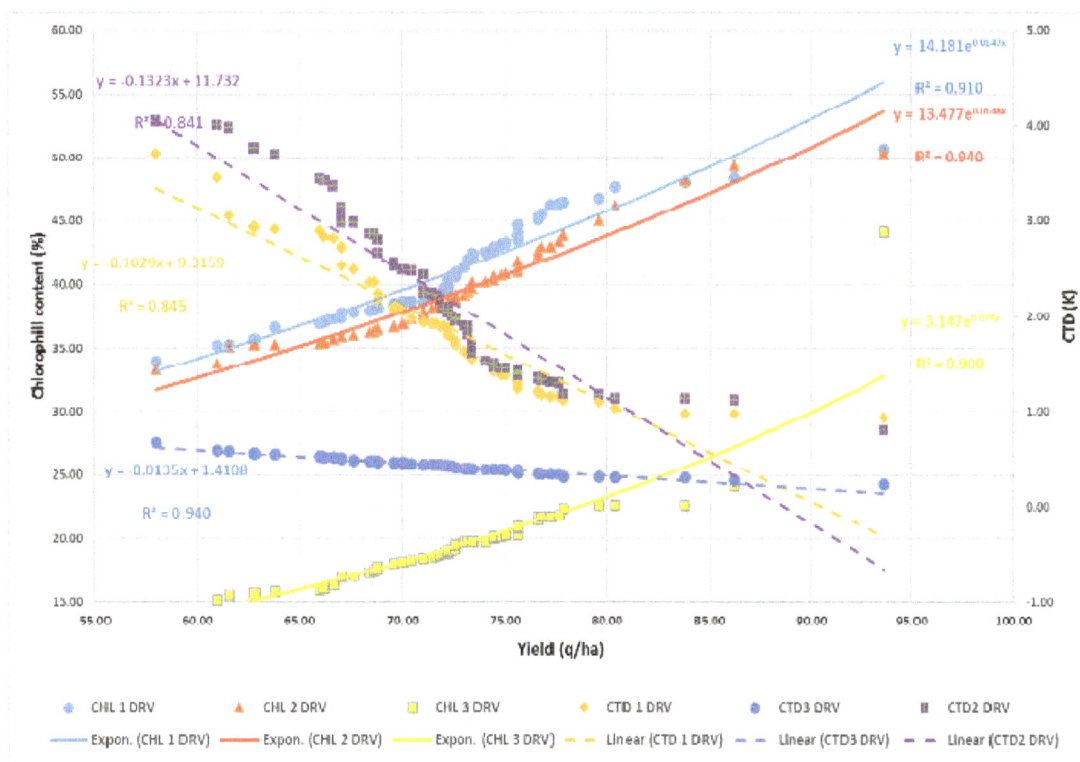

Figure 8. Influence of Chlorophyll content and CTD on yield on Drăgăneşti-Vlaşca Research Station

CONCLUSIONS

Prediction of drought resistant lines of winter wheat in breeding selection population could be made using Canopy Temperature Depression and Chlorophyll content analyses.

Using alone either CTD or CHL could not provide enough data for drought resistant selection on winter wheat lines due to different climatic conditions from one year to another. Alternatively, both yield and CTD and CHL could be combined in a selection index as a more powerful indicator of drought tolerance.

The values of CTD and CHL, for one lines varies from one trial location to another and from one year to another for same location, due to different climatic or soil conditions, but the hierarchy of lower CTD values or higher CHL values are maintained. Screening on F6 breeding generation validate selection on early breeding generation and indicate the appropriate candidates suitable for testing on drought condition on F7 breeding generation. The study shows that genotype with lower CTD values and high CHL values represent genotypes with high yield on drought conditions.

ACKNOWLEDGEMENTS

This work was supported by grant of the Romanian National Authority for Scientific Research and Innovation CCCDI-UEFISCDI, project number 44E/2015. This research was funded by the EU Eurostars projects "E! 8959 Genomic selection for nitrogen use efficiency in wheat".

REFERENCES

Amaliotis D., Therios I., Karatissiou M., 2004. Effect of nitrogen fertilization on growth, leaf nutrient concentration and photosynthesis in three peach cultivars. II International Symposium on Irrigation of Horticultural Crops. ISHS Acta Horticulturae 449, p. 36-42.

Amani I., Fischer RA., Reynolds M.P., 1996. Evaluation of canopy temperature as a screening tool for heat tolerance in spring wheat. Journal of Agronomy and Crop Science 176, p. 119-129.

Araus J.L., Amaro T., Zuhair Y., Nachit M.M., 1997b. Effect of leaf structure and water status on carbon isotope discrimination in field-grown durum wheat. Plant Cell and Environment 20, p. 1484-1494.

Araus J.L., Bort J., Ceccarelli S., Grando S., 1997. Relationship between leaf structure and carbon isotope discrimination in field grown barley. Plant Physiol. Biochem. 35, p. 553-541.

Ayeneh A., Van Ginkel M., Reynolds M.P., Ammar K., 2002. Comparison of leaf, spike, peduncle, and canopy temperature depression in wheat under heat stress. Field Crops Res., 79: p. 173–184.

Blum A., 1986. The effect of heat stresss on wheat leaf and ear photosynthesis. J. Experimental Botany 37, p. 111-118.

Blum A., 1996. Crop responses to drought and the interpretation of adaptation. Plant Growth Regul, 20, p. 135-148.

Blum A., Shipiler L., Golan G., Mayer J., 1989. Yield stability and canopy temperature of wheat genotypes under drought stress. Field Crops Res., 22: p. 289-296.

Bojobic B., Stojanovic J., 2005. Chlorophyll and carotenoid content in wheat cultivars as a function of mineral nutrition. Archives of Biological Sciences Belgrade 57 (4), p. 283-290.

Cabrera R.I., 2004. Evaluating yield and quality of roses with respect to nitrogen fertilization and leaf nitrogen status. XXV International Horticulturae Congress. Acta Horticulturae 511.

Evans J.T., 1983. Nitrogen and photosynthesis in the flag leaf of wheat. Plant Physiol. 72, p. 297-302

Fischer R.A., Rees D., Sayre K.D., Lu Z.-M., Condon A.G., Larque Saavedra A., 1998. Wheat yield progress associated with higher stomatal conductance and photosynthetic rate, and cooler canopies. Crop Sci. 38 (6), p. 1467-1475.

Guóth A., Tari I., Gallé Á., Csiszár J., Horváth F., Pécsváradi A., Cseuz L., Erdei L., 2009. Chlorophyll a fluorescence induction parameters of flag leaves characterize genotypes and not the drought tolerance of wheat during grain filling under water deficit. Acta Biol Szeged. 53, p. 1-7.

Haboudane D., Miller J.R., Tremblay N., Zarco-Tejada P.J., Dextraze L., 2002. Intergrated narrow-band vegetation indices for prediction of crop chlorophyll content for application to precision agriculture. Remote Sensing of Environment 81, p. 416-425.

Keyvan S., 2010. The effects of drought stress on yield, relative water content, proline, soluble carbohydrates and chlorophyll of bread wheat cultivars. J Anim Plant Sci. 8, p. 1051-1060.

Kiliç H., Yagbasanlar T., 2010. The effect of drought stress on grain yield, yield components and some quality traits of durum wheat (Triticum turgidum ssp. durum) Cultivars. Not Bot Hort Agrobot Cluj. 38: p. 164-170

Larcher W., 1995. Physiological Plant Ecology. Third edition. Springer, Berlin.

Lelyveld L.J., Smith B.L., Frazer C., 2004. Nitrogen fertilization of tea: Effect on chlorophyll and quality parameters of processed black tea. International Symposium on the Culture of Subtropical and Tropical Fruits and Crops, ISHS Acta Horticulturae 275, p. 168-180.

Paknejad F., Nasri M., Moghadam H.R.T., Zahedi H., Alahrnadi M.J., 2007. Effects of drought stress on chlorophyll fluorescence parameters, chlorophyll content and grain yield of wheat cultivars. J. Biol Sci. 7, p. 841-847.

Ping L.I., Pute W.U., Jianli C., 2012. Evaluation of flag leaf chlorophyll content index in 30 spring wheat genotypes under three irrigation regimes. Australian Journal of Crop Science 6, p. 1123-1130.

Pinter P.J., Zipoli G., Reginato R.J., Jackson R.D., Idso S.B., Hohman J.P., 1990. Canopy temperature as an indicator of differential water use and yield performance among wheat cultivars. Agric. Water Manag. 18, p. 35-48.

Rashid A., Stark J.C., Tanveer A., Mustafa T., 1999. Use of canopy temperature measurements as a screening tool for drought tolerance in spring wheat. J. Agron Crop Sci. 182, p. 231-237.

Reynolds M.P., Balota M., Delgado M.I.B., Amani I., Fischer R.A., 1994. Physiological and morphological traits associated with spring wheat yield under hot, irrigated conditions. Aust. J. Plant Physiol. 21, p. 717-30.

Reynolds M.P., Singh R.P., Ibrahim A., Ageeb O.A.A., Larqué-Saavedra A., Quick J.S., 1998. Evaluating physiological traits to complement empirical selection for wheat in warm environments. Euphytica 100: p. 84-95.

Reynolds M.P., Nagarajan S., Razzaque M.A., Ageeb O.A.A., 1997. Using canopy temperature depression to select for yield potential of wheat in heat-stressed environments. Wheat Special Report No. 42. Mexico, D.F.: CIMMYT.

Reynolds M.P., Ortiz-Monasterio J.I., McNab A., 2001. Application of Physiology in Wheat Breeding. Mexico, D.F.: CIMMYT.

Richardson A.D., Duigan S.P., Berlyn G.P., 2002. An evaluation of noninvasive methods to estimate foliar chlorophyll content. In: New Phytologist, Vol. 153(1), p. 185-194. http://doi.org/10.1046/j.0028-646x.2001.00289.x

Seeman J.R., Sharkey T.D., Wang J., Osmond C.B., 1987. Environmental effects on photosynthesis, nitrogen-use efficiency, and metabolic pools in leaves of sun and shade plants. Plant Physiol. 84, p. 796-802.

Simova-Stoilova Lj, Stoyanova Z., Demirevska-Kepova, K., 2001. Ontogenic changes in leaf pigments, total soluble protein and Rubisco in two barley varieties in relation to yield. Bulg. J. Plant Physiology 27(1-2), p. 15-24.

Suess S., Van der Linden S., Okujeni A., Schwieder M., Leitão P.J., Hostert P., 2015. Using class-probabilities to map gradual transitions in shrub vegetation maps from simulated EnMAP data. Remote Sens. 7, p. 10668-10688.

Tas S., Tas B., 2007. Some physiological responses of drought stress in wheat genotypes with different ploidity in Turkiye. World J Agric Sci. 3, p. 178-183.

Tejada-Zarco P.J., Miller J.R., Morales A., Berjon A., Aguera J., 2004. Hyperspectral indices and simulation models for chlorophyll estimation in open-canopy tree crops. Remote Sensing of Environment 90, p. 463-476.

Teng S., Qian Q., Zeng D., Kunihiro Y., Fujimoto K., Huang D., Zhu L., 2004. QTL analysis of leaf photosynthetic rate and related physiological traits in rice (Oryza sativa L.). Euphytica 135, p. 1-7.

8

MACHINES FOR GATHERING AND UTILIZATION OF RESIDUAL BIOMASS FROM TOBACCO PRODUCTION

Georgi KOMITOV, Dimitar KEHAJOV

Agricultural University - Plovdiv, 12 Mendeleev Avenue, Plovdiv, Bulgaria

Corresponding author email: gkomitov@abv.bg

Abstract

As biomass from the tobacco production can be regarded the stems of tobacco plants. They have huge market potential as a fuel, because of the available materials, that are not used (they are incinerated or buried into soil) and they have a high energy density.
Developments in technique make it possible to use new technologies and perfection systems, that allow waste biomass from tobacco to become attractive and environmentally friendly energy source with high quality and minimal costs for utilization.
In this paper is presented the technological capabilities for gathering, processing and utilization of residual biomass from the tobacco production. The machines for utilization of residual biomass from tobacco production are applicable in conditions of small and large farmers, who meet certain needs heating.

Key words: biomass, tobacco, machines, heating systems.

INTRODUCTION

Tobacco industry is an important part of agriculture in Bulgaria. The culture is from type "technical" and represented 13.8% of total agricultural exports. In our country are grown as large leaf and small-leaved tobacco. Tobacco is grown in weak soils (mountainous and hilly areas) and without irrigation. Tobacco has large energy potential, but it is used only 50% extremely leaf mass. The other part in form of stems is not used or is used as a mineral fertilizer of the soil, on which it is grown (Komitov, 2014).

Rational use of resources, necessary for the utilization of residual biomass from tobacco production requires the use of highly efficient processing technologies, the cost of which is minimal.

Technologies of use (even small scale) of the energy sources by agricultural and forest origin (biomass) developed quickly and making use of such resources competitive (Komitov, 2015; Failoni, 2006).

Outlook for energy saving and the development of new energy sources are related to economic models, aimed at maintenance and local development. This leads to the following favourable benefits: the development of new regions for agricultural and forestry production and markets, reducing the energy costs for domestic enterprises, economic growth regions, low cost of the energy conversion, stimulate large-scale construction of small installations, favourable impact on the environment (Failoni, 2006).

RESULTS AND DISCUSSIONS

For use of one or another machine for harvesting and utilization of biomass from tobacco industry is necessary advance to create route technology for movement of energy raw materials.

In the main idea of using residues from tobacco industry stand an economically accessible energy source with a large stock (Komitov, 2014).

For its transform in energy is not necessary complicated and expensive processes.

First operation in this transformation of the bioenergy from tobacco stems is collecting from the field. This can be done in two ways:

- By using an self-propelled forage harvester, (Figure 1). The harvester is having large price and usually this machine is used by large manufacturers of tobacco. The possession of this machine is warranted, because it has big efficiency. The principle at harvest is crawled the area from tobacco stems at least possible moves.

Figure 1. Self-propelled forage harvester

- By using the forage harvester, attached to tracktor (Figure 2). The method is suitable for small farmers, because isn't necessary to use expensive equipment.

Figure 2. Forage harvester attached to tracktor

Drawback of this method is double crawl on the tobacco field, primarily with working unit and secondly with transport unit for biomass. This can be avoided by using of trailers, attached to a forage harvester.

Another functin of forage harvester (except function for collecting tobacco stems) is crushing the stems at harvest. The result is heterogeneous mixture (with various sizes).

For transportation of biomass from agricultural field into enterprise for further processing using suitable trailers (Figure 3).

When choosing a trailer should be borne in mind the small size of already crushed tobacco stems (10-20 mm in length and thickness 10-5 mm).

Because of this specificity is necessary the select trailer, as not to lost the biomass when transporting. This can be done in enclosed trailers for grain, type "gondola" with cover.

After transport in factory processing it is necessary to dry the biomass with humidity, suitable for further processing. For this

technological operation used grain dryers, which may be rotary (Figure 4) or band (Figure 5).

Figure 3. Transport trailer for agricultural production

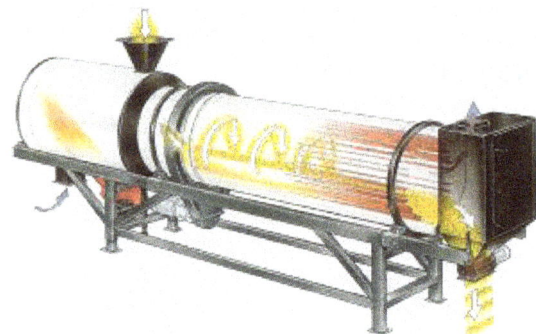

Figure 4. Rotary dryer for biomass

In rotary dryer the raw biomass is fill into inclined rotating bunker. There entrance for blowing hot air. The biomass humidity decreased after contact with hot air.

Band dryer is a conveyor belt, which is placed in a suitable furnace. The hot air blows into biomass to a suitable humidity. The tape must be selected, so as to provide blown the biomass from all sides.

The choice of dryer is determined by the volume of biomass which will be dried. Here can be used dryers with continuous or intermittent operation in drying process. The output humidity can be controlled with a suitable controller.

Suitable option for small farmers are mobile dryers. The drying process is done on the field. Another good option for those producers is natural drying. Disadvantages of natural drying are a long drying time, required large indoor areas and impossible for briquetting and peletizing the same heating season. Advantage of this type of drying is zero price of costs.

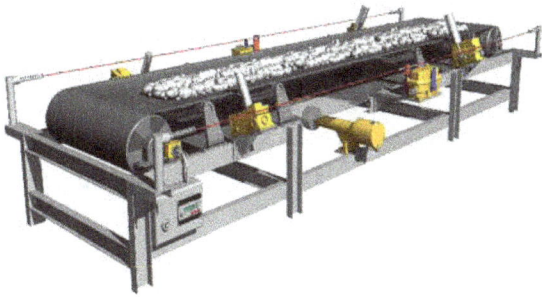

Figure 5. Band dryer for biomass

For ease further processing, drying continues untill receive humidity from 8 to 12%. The temperature of the heating champer should be in range from 650^0 to 700^0 C.

Biomass after drying must be submitted in machines for briquetting or pelletizing with suitable sizes (4 mm length and 1.5 mm thickness). The dimensions are recommended for better functioning of subsequent process. Fragmentation of biomass is carried out in a mill machine (Figure 6). As suitable for simple use and exploitation proved hammer mill. The process helps to align the particles of the biomass. The peripheral speed of the hammers is in the range 55-80 m/s, and the desired particle size can be adjusted within certain limits, by replacing the output grid.

Figure 6. Hammer mill for biomass

By dispenser biomass (after drying and fragmenting) is delivered in the press or granulator for forming granules. The extrusion process may be carried out at high or low pressure and temperature. In biomass are contains lignin and tar. They are natural adhesive at low temperatures and soften at temperature of about 80^0 C. This allow the

material to acquire another form. Thus the elements behave pellets or briquettes in compressed form, without adhesives (potato starch). A favourable effect on the process small amount of lubricant can be added to the product.

Briquetting and pelletizing are technology which compacts biomass. Finished products are solid biogenic fuels. They are ready for combustion in heating installations. The purpose of sealing the biomass compaction is improving indicators (quality and cost) of the original biomass as fuel. Such indicators are calorific value, density, keeping and s.o. (Komitov, 2015).

For pelletization may be applied variations of pelleting machines. They are generally separated into two types depending on the method for introduction biomass into the matrix. The machines for pelletizing are separated through this methods on machines with a flat matrix or a cylindrical matrix (Figure 7).

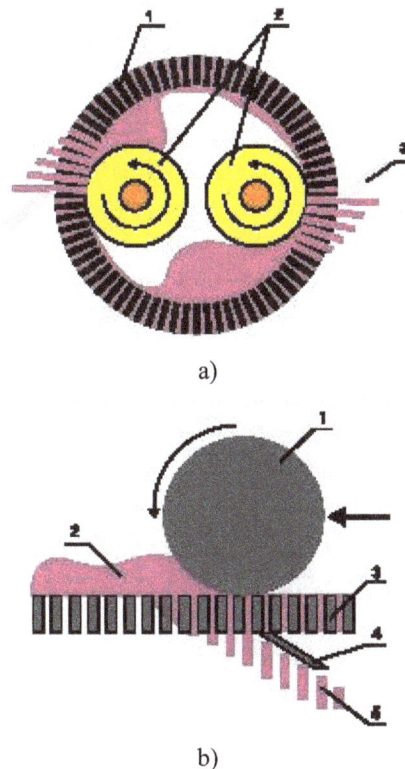

a)

b)

Figure 7. Schemes of pelletizing machines
a) with cylindrical matrix:
1-matrix, 2-rollers, 3-knife;
b) with flat matrix:
1-roller, 2-biomass, 3-matrix, 4-knife, 5-pellet

The choice of pelletizing machine depends on the desired performance. The productivity range is from 15 to 20 kg for pelletizing machines with cylindrical matrix and from 450 to 500 kg for pelletizing machines with flat matrix (per hour).

Principle of pelletization (in both cases) is the sealing of the biomass in the holes of the matrix by pressure with the roller. The pellets, after the customization, leave the matrix from the other side. Suitable length of pellets is formed with knife for cutting. The dimensions of the product are 6-8 mm thick and 10-15 mm in length.

For production of briquettes can be used press - mechanized or manually (Figure 8).

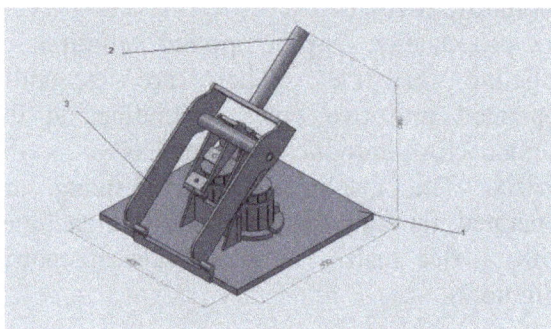

Figure 8. Manual press for briquettes
1-plate, 2-lever drive, 3-lever support

Piston press has small performance and is usually used for supporting the household for heating. Briquettes are produced with a diameter 80 mm and a length of 40 mm.

For the production of the briquettes can be used presses with piston or with extruder (extruder press is shown on Figure 9).

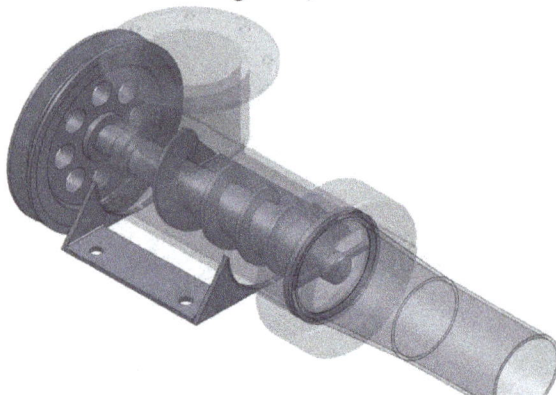

Figure 9. Extruder press

For the combined production of briquettes and pellets is the appropriate use of this extruder press. The briquettes or pellets is carried out with the replacement of the adaptor that is used. The adaptor can be found at the end of the device.

Solid fuels produced from biomass is used in heating installations. On the market there are many types of heating systems. Before choosing a press machine, it is necessary to choose a heating system. When using briquettes do not need adaptation of the heating system. Automating the combustion process is not necessary. They are normally applied to small or family heating installations.

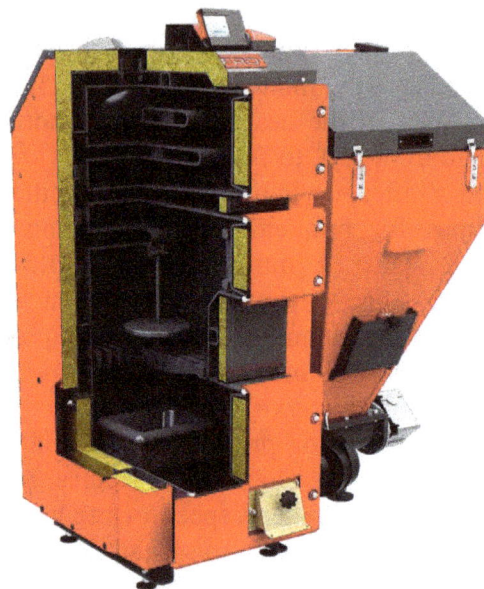

Figure 10. Burning boiler for waste from tobacco

Application of pellets as a source of energy is carried out by applying the automated pellet burners for biomass combustion. Through the use of the burner is avoiding the need for regular charging and adjusting the heat in the heating system. The power of the heating systems of this type varies widely - from 10-15 kW to 1000 kW. For the use of waste from the tobacco industry could use a boiler, shown on figure 10. The fuel is full in appropriate bunker. By screw fuel is fed into the combustion chamber, in the appropriate mode of choice. In boilers with an appropriate selection of the screw it is possible to burn briquettes.

CONCLUSIONS

Technology-route has been created for movement of energy raw materials and utilization of waste biomass in the tobacco industry.

Below are the different variants of machinery for the implementation of various technological operations in utilization of biomass in tobacco.

REFERENCES

Komitov G., Kehajov D., 2014. Distribution of areas occupied with tobacco plantations in our country. Conference with international participation RU&US 14. Scientific proceedings of RU "A. Kanchev", Vol. 53, book 1.1, Rousse, p. 96-99.

Komitov G., Kehajov D., 2015. Structural and technological features of machines for briquetting of residues of tobacco. Agricultural Sciences, Plovdiv.

Failoni, 2006. Renewable energy sources. Trakia University, Stara Zagora.

9

ETHANOL FERMENTATION FROM MICROWAVE-ASSISTED ACID PRETREATED RAW MATERIALS BY *Scheffersomyces stipitis*

Mustafa GERMEC[1,2], Ali OZCAN[1], Cansu YILMAZER[1], Nurullah TAS[1], Zeynep ONUK[1], Fadime DEMIREL[1], Irfan TURHAN[1]

[1]Akdeniz University, Department of Food Engineering, 07058, Antalya, Turkey
[2]Cankiri Karatekin University, Department of Food Engineering, 18100, Cankiri, Turkey

Corresponding author email: nafrinahrut@gmail.com

Abstract

The production of value-added products from renewable resources have been studied by researchers for years and are still being studied due to the fact that there is a point of interest. In this study, ethanol production from microwave-assisted acid pretreated renewable resources including barley husk, wheat bran, and rye bran was performed. The hydrolysis of renewable resources were performed at 700 W of microwave power, 6.92 min of irradiation time, 1:18.26 w/v of solid-to-liquid ratio, and 3.67% v/v of acid ratio for barley husk, 600 W, 6.92 min, 1:16.69 w/v, and 1.85% for wheat bran, and 460 W, 6.15 min, 1:17.14 w/v, and 2.72% for rye bran. The hydrolysates were detoxified with 2% (w/v) activated charcoal at 30°C and 150 rpm for 30 min. Then, the enriched hydrolysates were utilized to produce ethanol in shake flask fermentation by Scheffersomyces stipitis (ATCC 58784) that is a xylose fermenting yeast at 150 rpm and 30°C with 5% (v/v) inoculum size. Results indicated that the highest ethanol production (6.15 g/L) was performed in shake flask fermentation with wheat bran medium. However, the lowest ethanol concentration (1.37 g/L) was obtained from barley husk medium. Also, 4.89 g/L of ethanol was produced in rye bran media. Nonetheless, while the highest ethanol yield was calculated to be 44.11% in wheat bran media, the lowest ethanol yield was 12.78% in barley husk medium. On the other hand, the sugar utilization yields in barley husk, wheat bran, and rye bran mediums were also computed to be 85.33, 94.09, and 94.74%, respectively. Consequently, raw materials used in this study can be utilized as good carbon sources for production of ethanol by fermentation.

Key words: microwave-assisted, pretreatment, detoxification, Scheffersomyces stipitis, shake flask fermentation, ethanol.

INTRODUCTION

Renewable resources such as agricultural residues, agro-industrial residues, food processing wastes, human and animal wastes etc., are the most abundant available and inexpensive materials on the earth. Therefore, these wastes can be evaluated for the production of value-added products/chemicals by biotechnological processes due to their high carbohydrate contents since they consisted of cellulose (40-50%), hemicellulose (25-30%), lignin (15-20%), and the other extractives components (Germec et al., 2016c; Menon & Rao, 2012).

Energy consumption has increased day by day coupled with the world population (Zhu et al., 2006). Fossil fuels have been utilized to meet the energy and organic chemical demand of the world for years. But fossil fuel reserved are reduced with each passing day and thus are becoming expensive. To overcome such problems, the attempts are increased for the production of biofuels from renewable resources. Therefore, alternative resources and approaches required to be investigated to meet the energy requirement of the world (Fatehi, 2013).

An important step for production of biofuels such as bioethanol by fermentation is pretreatment (Menon & Rao, 2012). The pretreatment types are mechanical, physical, chemical, physico-chemical, and biological, which are applied to renewable resources for production of value-added products/chemicals (Fatehi, 2013; Menon & Rao, 2012). Once is microwave-assisted dilute acid pretreatment, which is one of the physico-chemical pretreatment processes (Zhang et al., 2017). This pretreatment process is a promising technology for the production of fermentable sugars from the renewable resources and following by the production of value-added products such as ethanol by biotechnological processes (Zhao et al., 2010). To our knowledge, there are no reports on ethanol production from microwave-assisted dilute acid pretreated barley husk, rye bran, and wheat bran. The objective of this study was to evaluate the suitability of ethanol production from microwave-assisted dilute acid pretreated barley husk, rye bran, and wheat bran by using *S. stipitis*.

MATERIALS AND METHODS

Raw material

Barley husk and rye branwere provided from Health Agricultural Products and Food Ind. Trade. Co. Ltd in Konya, Turkey. Wheat bran was obtained from a local feed factory in Osmancik (a district of Corum), Turkey. Among these, barley husk was milled to increase the hydrolysis efficiency by using a grinder (Bosch MKM6000, Ljubljana, Slovenia). Raw materials were stored at +4°C until used.

Microwave-assisted dilute acid pretreatment of raw materials

Microwave-assisted dilute acid hydrolysis of the raw materials (10 g in the liquid phase) was performed in a microwave oven (Beko MD 1610, voltage 230-240 V, ~50Hz, frequency 2450 MHz, and maximum power 1200 W, Foshan, Guangdong, China). Optimum hydrolysis conditions were determined using Box-Behnken Response Surface Methodology by Germec et al. (2017). The optimum conditions were 700 W, 6.92 min, 1:18.26 w/v, and 3.67% for barley husk, 600 W, 6.92 min, 1:16.69 w/v, and 1.85% for wheat bran, and 460 W, 6.15 min, 1:17.14 w/v, and 2.72% for rye bran, respectively. After hydrolysis, the reaction mixtures were cooled to room temperature and then filtered. The hydrolysates were stored at +4°C until used for fermentation (Germec et al., 2017).

Detoxification with activated charcoal

In order to decrease the concentration of inhibitors liberated during the pretreatment of raw materials, the hydrolysate (100 ml) was detoxified with activated charcoal detoxification method. Briefly, it was performed using a shaking incubator

(CERTOMAT® IS, Gottingen, Germany) at 30°C and 150 rpm with 2% (w/v) activated charcoal for 30 min. Following the detoxification, the activated charcoal was separated from the hydrolysates by using a centrifuge (4000 rpm, 20°C, and 30 min) (VWR Mega Star 3.0R, Osterode am Harz, Germany) and then the supernatants were removed for the fermentation to ethanol (Germec et al., 2016a; Mateo et al., 2013).

Microorganism and medium

Scheffersomyces stipitis ATCC 58784 was obtained from American Type Culture Collection (Manassas, VA, USA). *S. stipitis* ATCC 58784 was grown at 30°C for 48 h in a yeast extract-malt (YM) medium containing 10 g of glucose, 3 g of yeast extract, 3 g of malt extract, and 5 g of peptone per liter of deionized water. The medium pH was adjusted to 6.2 with 4 N NaOH and HCl. The culture was stored at 4°C and sub-cultured bi-monthly in order to maintain viability. For a long-term storage, stock cultures were maintained in 20% glycerol at -80°C. *S. stipites* was grown in 250 mL flasks containing 100 mL of YM at 30°C and 150 rpm for 24 h for inoculation (Germec et al., 2016b; Zhu et al., 2011).

Ethanol fermentation medium

The base-line medium was composed of 10 g of glucose, 3 g of yeast extract, 3 g of malt extract, and 5 g of peptone per liter of deionized water. For fermentations, the detoxified hydrolysates were used as carbon source instead of glucose, but all other ingredients were added in the fermentation medium (Germec et al., 2016b; Germec et al., 2016d).

Shake flask fermentation

Shake flask fermentations were carried out in a shaking incubator (CERTOMAT® IS, Gottingen, Germany) with 250 ml flasks containing 100 ml of the prepared mediums from the detoxified raw material hydrolysates. All fermentation runs were performed in duplicate. The initial pH of mediums was adjusted to 6.2 by adding 8 N NaOH and 4 N HCl. Then, the flasks were autoclaved at 121.1°C for 15 min. After autoclaving and cooling down to room temperature, 5% (v/v) of prepared inoculum at 30°C for 24 h was used to inoculate into the flasks and ethanol fermentations were performed for a period of 120 h. During ethanol fermentations, temperature was maintained at 30°C and agitation speed was kept at 150 rpm. Samples (1 ml) were collected every 4 or 8 h for the first 12 h and every 12 or 24 h for the remainder of the fermentation and analyzed for residual sugar, ethanol production (P) as well as optical cell density for biomass concentration (X) in fermentation broth (Germec et al., 2016a).

Analysis

Ethanol

The ethanol was determined by using a HPLC (Thermo Scientific UltiMate 3000, Dreieich, Germany) equipped with a RefractoMax 520 refractive index detector, autosampler, column oven, and computer controller. Separations were performed on a Transgenomic ICSep ORH-801 column (Apple Valley, MN) at 70°C using 0.01 N H_2SO_4 as the mobile phase with a 20 µL injection volume. The flow rates of 0.5 ml/min was used for analysis of ethanol.

Residual sugar concentration

The residual sugar concentration in the fermentation broth was analyzed by 3,5-dinitrosalicylic acid method (Miller, 1959). Briefly, a measurement of absorbance at 575 nm was recorded. A calibration curve for the spectrophotometric measurements (Thermo Scientific Evolution 201 UV-Vis, Shanghai, China) was created using a glucose solution. Deionized water was used as a blank. Absorbance values were converted to residual sugar concentration by using the obtained standard curve, which was $y=60.401 \times Abs_{575}+0.5751$. Here y is glucose concentration, g/L (Germec et al., 2016a; Germec et al., 2015).

Biomass

The optical cell density was measured using a spectrophotometer (Thermo Scientific 201 UV-Visible Evolution, Shanghai, China) at 600 nm. Uninoculated media was used as a blank. Absorbance values were converted to biomass concentrations by using a standard curve, which was $y=0.3047 \times Abs_{600}-0.2656$, where y

is biomass concentration, g/L (Germec et al., 2016b).

Kinetic parameters

Following kinetics were calculated as follows:
- Sugar consumption (S, g/L) = S_f - S_i
- Ethanol production (P, g/L) = P_f - P_i
- Ethanol yield ($Y_{P/S}$, %) = $(P/S) \times 100$
- Biomass yield ($Y_{X/S}$, %) = $(X/S) \times 100$
- Product yield per biomass ($Y_{P/X}$, g/g) = P/X
- Maximum consumption rate (Q_S, g/L/h) = $(-ds/dt)_{max}$
- Maximum production rate (Q_P, g/L/h) = $(dp/dt)_{max}$
- Sugar utilization yield (SUY, %) = $(S/S_m) \times 100$
- Theoretical ethanol yield (TY, %) = $(Y_{P/S}/51.1) \times 100$

where, S is the sugar consumption (g/L); S_i and S_f are residual sugar concentrations at the beginning and at the end of the fermentation (g/L), respectively; P is the ethanol production (g/L); P_i and P_f are ethanol concentrations at the beginning and at the end of the fermentation (g/L), respectively (g/L); X is biomass production (g/L); Q_S is the slope of the steepest part of sugar consumption profiles (g/L/h); Q_P is the slope of the steepest part of ethanol production profile (g/L/h), the slopes were calculated using at least three points; and S_m is maximum sugar concentration (g/L).

RESULTS AND DISCUSSIONS

In this study, ethanol production from microwave-assisted dilute acid pretreated and detoxified raw material hydrolysates was performed and the results were evaluated in terms of kinetic parameters.

Ethanol production in shake flask fermentation with barley husk medium

Figure 1 depicts the sugar consumption, cell growth, ethanol production plots belong to ethanol fermentation in shake flask fermentation with barley husk medium. Figure 1 shown that the sugar consumption stopped at 96 h of fermentation. At this point, the yeast growth also entered into the stationary phase. However, it indicated that ethanol production is still continuing up to 120 h cultivation. Because, the ethanol concentration produced was both quite low and not economical (Figure 1).

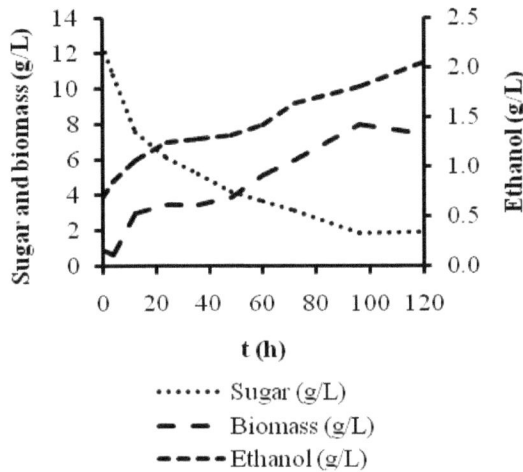

Figure 1. Diagram belong to ethanol production from the detoxified barley husk hydrolysate in shake flask fermentation

Table 1 shows the kinetic parameters for shake flask ethanol fermentation with barley husk medium. The sugar concentration consumed through fermentation was 10.72 g/L. However, 1.84 g/L of sugar was not consumed by the yeast *S. stipitis* (data not shown). On the other hand, although 85.33% of sugar was consumed by the yeast, the ethanol concentration produced was 1.37 g/L, fairly low. Therefore, while the ethanol yield was stayed at 12.78%, the theoretical ethanol efficiency was also 25.01%, which was about 4-times lower than the theoretical efficiency. Additionally, the biomass yield was rather high, which yielded as 68.39%, but the ethanol yield per biomass was 0.19 g/g. In addition, maximum consumption and maximum production rates were also calculated to be 0.41 and 0.03 g/L/h, respectively.

Table 1. The kinetic parameters for detoxified barley husk fermentation

Kinetics	Value	Unit
S	10.72	g/L
P	1.37	g/L
$Y_{P/S}$	12.78	%
$Y_{X/S}$	68.39	%
$Y_{P/X}$	0.19	g/g
Q_S	0.41	g/L/h
Q_P	0.03	g/L/h
SUY	85.33	%
TY	25.01	%

In the literature, no study was carried out the ethanol production from the microwave-assisted dilute acid pretreated and detoxified barley husk hydrolysate by using *S. stipitis*. However, Palmarola-Adrados et al. (2005) was performed the ethanol production from barley husk hydrolysate by using baker's yeast. Our results were not assisted their results. Indeed, they reported that the theoretical ethanol yield was achieved to be 92% and the hydrolysate samples were fermented within the first 5 h. Kim et al. (2008) investigated the bioethanol production from the SAA (soaking in aqueous ammonia)-pretreated barley hulls by using recombinant *Escherichia coli*. Results indicated that the ethanol produced was 24.1 g/L, which corresponded to 89.4% of the maximal theoretical yield depending on the glucan and xylan. Consequently, further researches should be taken place with regard to the ethanol production from microwave-assisted acid pretreated barley husk hydrolysate.

Ethanol production in shake flask fermentation with rye bran medium

Ethanol fermentation performed in shake flask fermentation with rye bran medium was demonstrated in Figure 2.

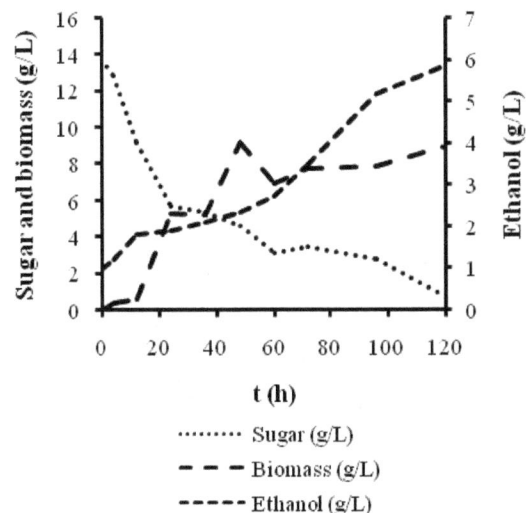

Figure 2. Diagram belong to ethanol production from the detoxified rye bran hydrolysate in shake flask fermentation

The sugar in the media was swiftly consumed in the first 24 h of fermentation and the ethanol was relatively increased. Within the same period, the cell growth also increased rapidly.

After the 24 h of fermentation, the sugar's consumption continued until the end of the fermentation, slowly. At the end of the fermentation, 0.73 g/L of sugar was not consumed by the yeasts. The maximum biomass concentration (9.16 g/L) was at 48 h of fermentation (data not shown). After this point, cell growth entered to the stationary phase. But, interestingly, the ethanol production was swiftly increased when the yeast enter to stationary phase (Figure 2).

According to data from in Table 2, the ethanol yield was 37.36% when the sugar consumption and ethanol production were 13.08 and 4.89 g/L, respectively. On the other hand, the biomass yield was 65.93% since the growing biomass concentration was 8.62 g/L. Therefore, the product yield per biomass was 0.57 g/g.

Table 2. The kinetic parameters for detoxified rye bran fermentation

Kinetics	Value	Unit
S	13.08	g/L
P	4.89	g/L
$Y_{P/S}$	37.36	%
$Y_{X/S}$	65.93	%
$Y_{P/X}$	0.57	g/g
Q_S	0.35	g/L/h
Q_P	0.07	g/L/h
SUY	94.74	%
TY	73.10	%

As the 13.08 g/L of the sugars in the media was consumed by the yeast and the initial sugar concentration was 13.80 g/L, the sugar utilization yield was calculated to be 94.74%, nearly whole sugar. Besides, theoretically, the maximal ethanol yield from the glucose is 51.1%. Since the ethanol yield was 37.36% in this part of the study, the theoretical ethanol yield was 73.10%, which was relatively close to maximal theoretical ethanol yield. In addition, maximum consumption rate and maximum production rate were also calculated, which were yielded to be 0.35 and 0.07 g/L/h, respectively.

In the literature, while there is no study related to ethanol production from microwave-assisted dilute acid pretreated and detoxified rye bran hydrolysate, however, the ethanol was produced from the rye bran. For instance, Wang et al. (1998) was taken place the fermentation of sugars to ethanol obtained from enzymatic hydrolysis of fall rye. The results

indicated that the ethanol production was 409 L/tones, which was equivalent to 90.1% of theoretical ethanol yield. Vidmantiene et al. (2006) was performed the fermentation to ethanol using the baker's yeast of sugars releasing by enzymatic hydrolysis of rye bran and reported that the maximum ethanol concentration and ethanol yield were 44 g/L and 45.7% (89.4% of theoretical yield), respectively. The yields (90.1 and 89.4% of theoretical maximal ethanol yield) were relatively higher than our theoretical yield value.

Ethanol production in shake flask fermentation with wheat bran medium

Figure 3 displays the sugar consumption, cell growth, and ethanol production plots belong to the shake flask fermentation of the microwave-assisted dilute acid pretreated and detoxified wheat bran hydrolysate. According to Figure 3, the sugars in wheat bran hydrolysate was rapidly consumed until the 48 h of fermentation by the yeast S. stipitis.

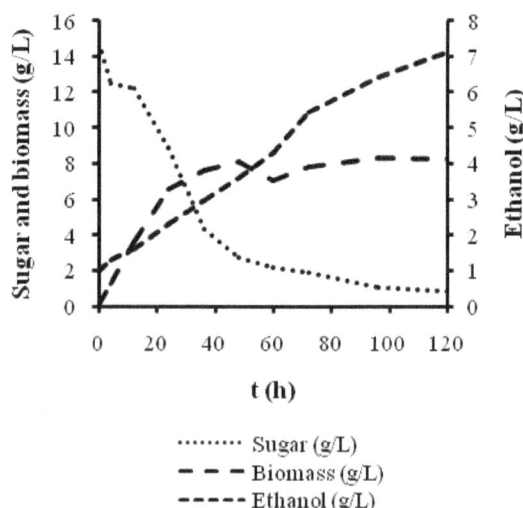

.... Sugar (g/L)
– – – Biomass (g/L)
----- Ethanol (g/L)

Figure 3. Diagram belong to ethanol production from the detoxified wheat bran hydrolysate in shake flask fermentation

Nonetheless, both ethanol production and cell growth was increased with decreasing of the sugars in the media. After the 48 h of fermentation, along with the sugar consumption was continued at a slow rate until the end of the fermentation, while the ethanol concentration was normally increased, the cell growth stopped. The highest biomass concentration was observed at 96 h of fermentation while maximum ethanol production and the lowest sugar concentration were determined at the end of the fermentation (Figure 3).

The kinetic results calculated using the experimental data from the fermentation of wheat bran hydrolysate were given in Table 3. The ethanol concentration produced and the sugar concentration consumed were 6.15 and 13.95 g/L, respectively. Therefore, the results indicated that the ethanol yield was quite close to theoretical ethanol yield, which achieved to be 44.11% (86.32% of the theoretical value).

Table 3. The kinetic parameters for detoxified wheat bran fermentation

Kinetics	Value	Unit
S	13.95	g/L
P	6.15	g/L
$Y_{P/S}$	44.11	%
$Y_{X/S}$	59.09	%
$Y_{P/X}$	0.75	g/g
Q_S	0.33	g/L/h
Q_P	0.10	g/L/h
SUY	94.09	%
TY	86.32	%

However, the growing biomass concentration was also 8.25 g/L, thus, the biomass yield was calculated to be 59.09%. After calculating the ethanol yield and biomass yield, the product yield per biomass was also computed to be 0.75 g/g. Almost whole of the sugar was consumed by the yeast, therefore, the sugar utilization yield was 94.09%. On the other hand, maximum sugar consumption rate and maximum ethanol production rate were 0.33 and 0.10 g/L/h, respectively.

In the literature, many studies were performed with respect to the microwave-assisted extraction of wheat bran, but no study was taken place with regard to the fermentation of microwave-assisted dilute acid pretreated and detoxified wheat bran hydrolysate to ethanol by the yeasts. Okamoto et al. (2011) performed the ethanol production from wheat bran by using the white rot fungus Trametes hirsuta and reported that the highest ethanol concentration was 4.3 g/L, corresponding to 78.8% of the theoretical yield. Investigated the ethanol production from the unfiltered wheat bran hydrolysates by using S. cerevisiae and reported that the ethanol yield was 49%,

corresponding to approximate 96% of theoretical ethanol yield.

CONCLUSIONS

In this study, the ethanol production from the microwave-assisted dilute acid pretreated and activated charcoal detoxified raw material hydrolysates was performed and the results were evaluated in terms of kinetics. Results indicated that, the low kinetic results were obtained from the fermentation of barley husk hydrolysate. Therefore, it should be investigated further why this is so. Good results were achieved from the fermentation of rye bran and wheat bran, especially the latter. Thus, these results can be further improved under the controlled fermentation conditions. In conclusion, the raw materials used in this study can be good, inexpensive, and alternative carbon sources for the production of value-added products by biotechnological processes.

REFERENCES

Fatehi P., 2013. Recent advancements in various steps of ethanol, butanol, and isobutanol productions from woody materials. Biotechnology progress, 29(2): p. 297-310.

Germec M., Kartal F.K., Bilgic M., Ilgin M., Ilhan E., Güldali H., Isci A., Turhan I., 2016a. Ethanol production from rice hull using Pichia stipitis and optimization of acid pretreatment and detoxification processes. Biotechn. progress, 32(4): p. 872-882.

Germec M., Kartal F.K., Guldali H., Bilgic M., Isci A., Turhan I., 2016b. Obtaining growth curves for Scheffersomyces stipitis strains and their modeling. Scientific Bulletin. Series F. Biotechnologies, 20: p. 263-268.

Germec M., Tarhan K., Yatmaz E., Tetik N., Karhan M., Demirci A., Turhan I., 2016c. Ultrasound-assisted dilute acid hydrolysis of tea processing waste for production of fermentable sugar. Biotechnology progress, 32 (2): p. 393-403.

Germec M., Tas N., Ozcan A., Yilmazer C., Onuk Z., Demirel F., Turhan I., 2017. Microwave-assisted acid pretreatment of renewable resources: A process optimization (Submitted). in: The International Conference of the University of Agronomic Sciences and Veterinary Medicine of Bucharest "Agriculture for Life, Life for Agriculture". Romania.

Germec M., Turhan I., Karhan M., Demirci A., 2015. Ethanol production via repeated-batch fermentation from carob pod extract by using Saccharomyces cerevisiae in biofilm reactor. Fuel, 161: p. 304-311.

Germec M., Turhan I., Yatmaz E., Tetik N., Karhan M., 2016d. Fermentation of acid-pretreated tea processing waste for ethanol production using

Saccharomyces cerevisiae. Scientific Bulletin. Series F. Biotechnologies, 20: p. 269-274.

Kim T.H., Taylor F., Hicks K.B., 2008. Bioethanol production from barley hull using SAA (soaking in aqueous ammonia) pretreatment. Bioresource Technology, 99(13): p. 5694-5702.

Mateo S., Roberto I.C., Sánchez S., Moya A.J., 2013. Detoxification of hemicellulosic hydrolyzate from olive tree pruning residue. Industrial Crops and Products, 49: p. 196-203.

Menon V., Rao M., 2012. Trends in bioconversion of lignocellulose: biofuels, platform chemicals & biorefinery concept. Progress in Energy and Combustion Science, 38(4): p. 522-550.

Miller G.L., 1959. Use of dinitrosalicylic acid reagent for determination of reducing sugar. Analytical chemistry, 31(3): p. 426-428.

Okamoto K., Nitta Y., Maekawa N., Yanase H., 2011. Direct ethanol production from starch, wheat bran and rice straw by the white rot fungus *Trametes hirsuta*. Enzyme and microbial technology, 48(3): p. 273-277.

Palmarola-Adrados B., Galbe M., Zacchi G., 2005. Pretreatment of barley husk for bioethanol production. Journal of Chemical Technology and Biotechnology, 80(1): p. 85-91.

Vidmantiene D., Juodeikiene G., Basinskiene L., 2006. Technical ethanol production from waste of cereals and its products using a complex enzyme preparation. Journal of the Science of Food and Agriculture, 86(11): p. 1732-1736.

Wang S., Thomas K., Ingledew W., Sosulski K., Sosulski F., 1998. Production of fuel ethanol from rye and triticale by very-high-gravity (VHG) fermentation. Applied biochemistry and biotechnology, 69(3): p. 157-175.

Zhang Y., Chen P., Liu S., Peng P., Min M., Cheng Y., Anderson E., Zhou N., Fan L., Liu C., 2017. Effects of feedstock characteristics on microwave assisted pyrolysis-a review. Bioresource Technology, 230: p. 143-151.

Zhao X., Song Z., Liu H., Li Z., Li L., Ma C., 2010. Microwave pyrolysis of corn stalk bale: A promising method for direct utilization of large-sized biomass and syngas production. Journal of Analytical and Applied Pyrolysis, 89(1): p. 87-94.

Zhu J., Yong Q., Xu Y., Yu S., 2011. Detoxification of corn stover prehydrolyzate by trialkylamine extraction to improve the ethanol production with *Pichia stipitis* CBS 5776. Bioresource technology, 102(2): p. 1663-1668.

Zhu S., Wu Y., Yu Z., Zhang X., Wang C., Yu F., Jin S., 2006. Production of ethanol from microwave-assisted alkali pretreated wheat straw. Process biochemistry, 41(4): p. 869-873.

10

NATURAL ANTIOXIDANTS USED IN FRYING OILS TO MINIMIZE THE ACCUMULATION OF TOXIC COMPOUNDS

Mihaela GHIDURUŞ, Mioara VARGA

University of Agronomic Sciences and Veterinary Medicine of Bucharest,
59 Mărăşti Blvd., 011464, Bucharest, Romania

Corresponding author email: mihaela_ghidurus@yahoo.com

Abstract

The objective of this experiment was to determine if the quality of sunflower oil enriched with mixtures of two antioxidants of rosemary extracts was improved during frying. The enriched oils have been subjected to frying process at a temperature of 180 ° C ± 1°C and held for about seven hours per day, for a period of 10 days. Samples of frozen cooked potatoes were fried in these oils seven times a day, every hour. Quality evaluation of the oils took place every day of the experiment, as far as refractive index, acidity, p-anisidine value, K232 and K270 and polar and oxidation compounds were concerned.

Key words: rosemary antioxidants, frying process, oil quality, toxic compounds, food safety, sunflower oil, rapid methods.

INTRODUCTION

Fat or oil frying is one of the most common and the oldest methods developed and used by man for the preparation of food. Recent consumer interest in "healthy eating" has raised awareness of the need to limit the consumption of fat and fatty foods (Ghiduruş et al., 2013).

The fast food industry is adopting various methods designed to maintain the quality and increase the useful life of frying oils. Among those they include the use of antioxidants (Paul and Mittal, 1997). As it is known, the phenolic compounds have capacity to act as antioxidants (Zayova et al., 2016). The antioxidant properties of herbal products are mainly attributed to phenolic compounds such as flavonoids and polyphenolic derivates (Nikita et al., 2016). Among the antioxidant compounds, polyphenols have gained importance due to their large array of biological action that include free radical scavenging, metal chelation and enzyme modulation activities (Popa et al., 2016). When added to foods, antioxidants control rancidity development, retard the formation of toxic oxidation products, maintain nutritional quality, and extend the shelf-life of products (Shahidi and Ambigaipalan, 2015).

According to Chammem et al. (2015) the addition of the rosemary extract in the mixture of soybean and sunflower oil reduced the peroxide value by 38% after 30 h of heating. This oil resists to oxidation and conserves the higher amount of unsaturated fatty acids even after 30 h of heating.

The frying temperature recommended for specific foods in different studies varies from 160 to 200^0C, with the optimal frying temperature depending on the type of food, its size, the fat turnover, the size of the frying vat, the number of the frying vats used (Mehta and Swinburn, 2001). Natural antioxidants are usually used in low-temperature conditions. In foods exposed to high temperatures (i.e., potato chips) little information is available on the effectiveness of natural antioxidants. Potato chips oxidize easily, thus losing commercial value and health properties (Lalas and Dourtoglou, 2003). Higher temperatures, especially over 200^0C, accelerate oxidative and thermal alterations and increase the rate of formation of decomposition products (Soriano et al., 2002).

MATERIALS AND METHODS

The experiment was conducted in such a manner to determine if the quality of sunflower oil enriched with mixtures of two antioxidants of rosemary extracts was improved during frying. In this respect the vegetable oil samples were placed in four deep fryers, as follows: 1ˢᵗ deep fryer contained a control sample (M) - normal sunflower oil (reference sample); 2ⁿᵈ contained sample 1 (P1) - sunflower oil with added antioxidant, INOLENS 4 manufactured by Vitiv, 500 ppm rosemary extract; 3ʳᵈ contained sample 2 (P2) - sunflower oil enriched with INOLENS 4 containing 1000 ppm antioxidant as rosemary extract; 4ᵗʰ fryer contained sample 3 (P3) - sunflower oil 1000 ppm antioxidant SyneROX HT as rosemary extract manufactured by Vitiv, which contains citrate as well. The enriched oils have been subjected to frying process at a temperature of 180°C ± 1°C and held for about seven hours per day, for a period of 10 days, and the total number of frying hours being 68. The oils were heated 10 minutes to reach the first predetermined temperature, without frying potatoes. Samples of frozen cooked potatoes weighing 50 g were fried in these oils seven times a day, every hour, each sample, except the first day when frying took place only 5 hours. The methods used for quality evaluation of the oilsare the AOAC standard methods for refractive index, acidity, p-anisidine value, K232 and K270 and a sensor for polar compounds (FOM-Food oil monitor). Fritest and Oxifrit, rapid methods, were used to asses qualitatively the oils as far as the total oxidation compounds are concerned.

The extracts of rosemary used had a polyphenol content of 42 ± 1 mg/g, expressed as carnosic acid and carnosol.

RESULTS AND DISCUSSIONS

As can be seen in Figure 1 both the refractive index and the percentage of solids in the oil samples increased from 1.47 in the fresh oil to about 1.48, however the differences between the three samples of the oil enriched with antioxidant extract and the control oil sample was not significant.

The results are in agreement with those published by other authors in the literature: Yoon et al., (1987), Al-Harbi (1993), Al-Kahtani (1991), who argue that IR values of oils were used to fry, are higher than those of fresh oil. IR values change in relation to the three stages of autooxidation.

During induction the peroxide formation is low, the refractive index remains constant.

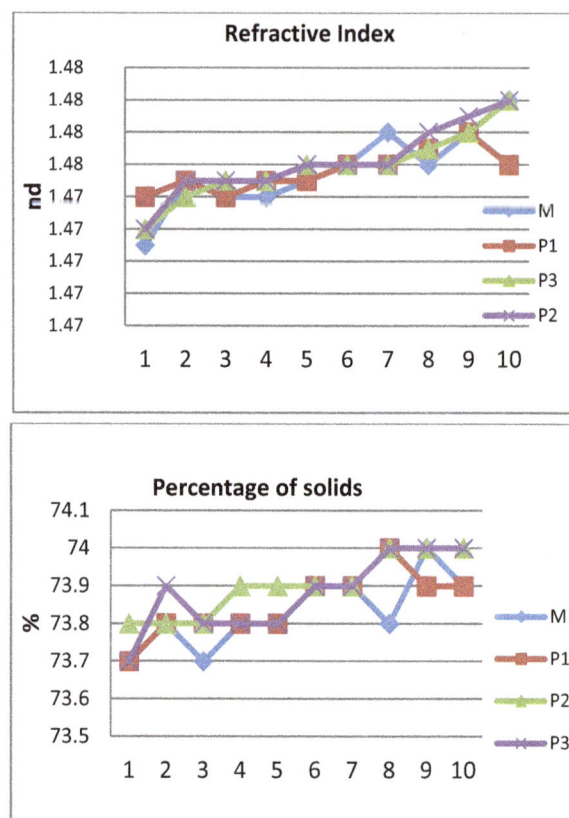

Figure 1. Refractive index and the percentage of solids values of oil samples enriched with antioxidants from rosemary oil, used in 10 days frying process

During the 2ⁿᵈ stage when forming relatively more peroxides, IR increases rapidly until the maximum amount of peroxide is continuing to grow in the third stage peroxides decompose, but not as much as in stage second.

Regarding the acid indexes and the percentage of oleic acid, the highest values were recorded for both parameters in P3, test sample which was containing sunflower oil with 1000 ppm antioxidant SyneROX HT, who reached a value of index acid of 1.253 after 10 days of experiment, the initial value being 0.269 so we observed an increase of 4.66 fold as compared to day one.

The minimum value was recorded in P1 sample, which contained sunflower oil with added antioxidant INOLENS 4, as rosemary extract, 500 ppm, where the acid value was 0.981 on the tenth day (Figure 2).

Free acidity

% Oleic acid

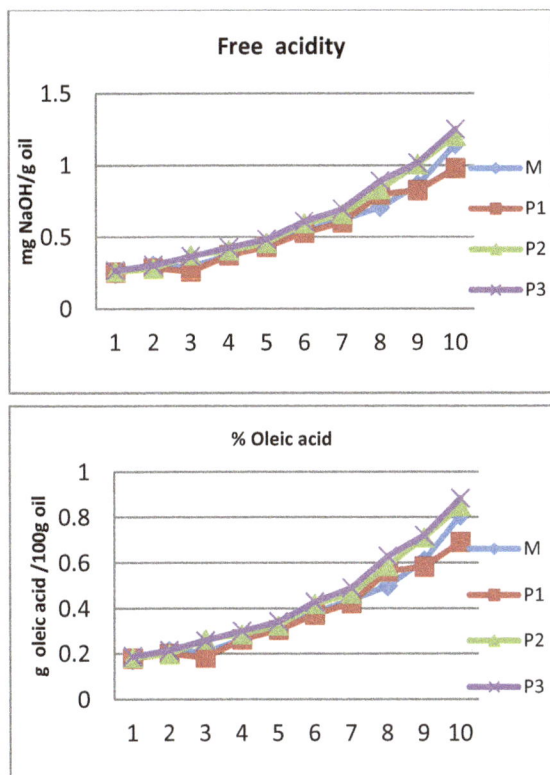

Figure 2. Free acidity and oleic acid percentage values of oil samples enriched with antioxidants from rosemary oil, used in 10 days frying process

Anisidine value

Figure 3. P- Anisidine values of oil samples enriched with antioxidants from rosemary oil, used in 10 days frying process

It can be concluded that an amount of 500 ppm antioxidant INOLENS 4 is more efficient than 1000 ppm antioxidant HT SyneROX in the accumulation of free fatty acids during the 10 days of the experiment, representing 68 h hours of heat treatment at a temperature of $180^0C \pm 1^0C$. The value of p - anisidine is a measure of the presence of certain oxidation by-products (primarily aldehydes) and has a good ability to discriminate between samples with different degrees of oxidation.

The results of the experiment reveals the fact that P3 sample with 1000 ppm antioxidant SyneROX HT had the smallest p-anisidine values throughout the experiment with a value of 16.61 at the end of first experimantal day and 43.38 in day 10, therefore p anisidine value in P3 incresed 2.83 times compared to day one. Higher values recorded P2 sample, which contained sunflower oil with added antioxidant INOLENS 1000 ppm as rosemary extract, at the end of the 10th day of the experiment (Figure 3).

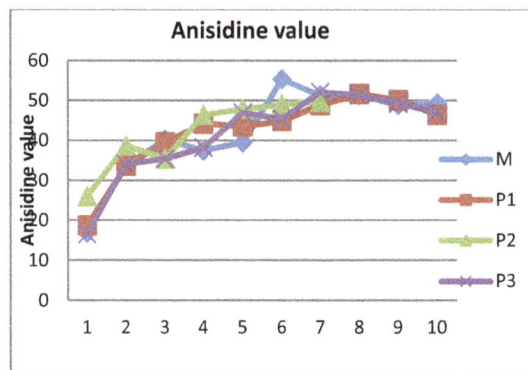

It was observed the fact that the frying process increased the extinction values at 232 nm and 270 nm, which indicates the formation of compounds such as conjugated dienes and trienes following the removal of double bonds during frying (Figure 4). These findings are consistent with results obtained by Al Kahtani in 1991. Most of frying oils that have a high percentage of polar compounds have a high content of diene and triene. Although these compounds may form polymers, there is a balance between the rate of conjugated diene formation and the rate at which these compounds form polymers during the frying process (Yoon et al., 1987).

In conclusion these two antioxidant extracts and different concentrations, both INOLENS 4 and SyneROX HT did not affect the development of both 232 nm and 270 nm extinction over a period of 10 days of the experiment, 68 h hours of heating treatment at a temperature of $180^0C \pm 1^0C$, except for the first two days in which the samples P1 and P2 have had lower extinction, the following days the differences were not significant.

Figure 5 shows the increase in percentage of total polar compounds (TPC) of the samples containing oils enriched with antioxidants in rosemary extracts that have been used in continuous frying processes. Of all the physical and chemical analysis, the content of TPC is one of the most objective and valid criteria for the evaluation of the deterioration of oils and fats used for frying processes. However, the standard method for the measurement of TPC by column chromatography on silica gel may be correct, but it is time-consuming and relatively expensive.

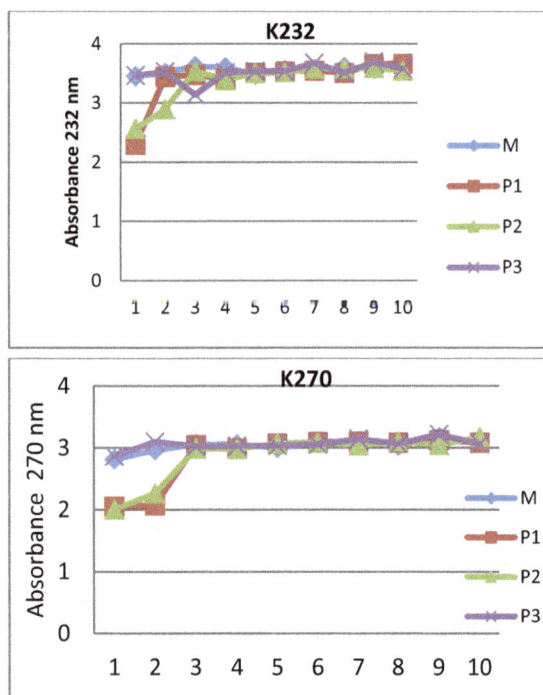

Figure 4. K232 and K270 values of oil samples enriched with antioxidants from rosemary oil, used in 10 days frying process

During the frying process, the oil decomposes into peroxides, acids and other radicals that are formed in the frying oil. This induced polarization of oil molecules. TPC increased in oils during frying experiment being correlated with frying time. Once frying time has progressed, it was observed that the growth rate in TPC was relatively lower in sunflower oil samples enriched 1000 ppm SyneROX HT and INOLENES 4 samples P3 and P2 respectively. After the first day of frying to $180^{0}C$ the TPC content of the control samples was 6.75%, TPC of P1 was 6%, 6.5% of P2 and 5.5% of P3. At the end of ten days, the TPC increased to 17% in M, 16% in P1, 15% in P2 and 15.25% in P3.

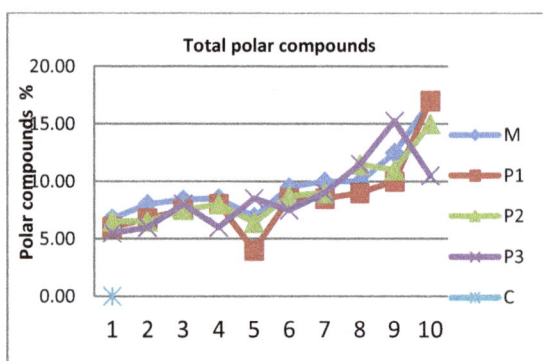

Figure 5. TPC values of oil samples enriched with antioxidants from rosemary oil, used in 10 days frying process (C refers to TPC value of fresh oil)

Although TPC content was the highest in the tenth day, these oils were acceptable after 68 h of the heat treatment conditions of experiment, given that many countries have established a maximum acceptable level for 25-27% TPC content.

After using Oxifrit kit for total oxidized compounds the results showed that all the samples were marked "good" (first colour - 1) on the scale of 4 colors of the kit even in the 9th day of the experiment; changes occurred on day 10 of the experiment the sample P3, sample containing sunflower oil with 1000 ppm antioxidant SyneROX HT which recorded a value of 2 (blue-green) meaning the oil is still good.

The results obtained from the use of the colorimetric test kit FRITEST, which is sensitive to carbonyl compounds, where the analysis consist of comparing the reacted sample mixture to a scale consisting of a choice of three colors from yellow to orange, showed that after 10 days of experiment or 68 hours of frying, the samples had a value of 2. A value of 2 or higher obtained from Fritest indicates that the oil should be replaced. An exception was exhibited by P1, sample containing sunflower oil with added antioxidant INOLENS 4, 500 ppm as rosemary extract, who recorded a value of 1, the color remains almost unchanged after ten days of experiment.

CONCLUSIONS

Both the refractive index and the percentage of solids in the oil samples taken during the frying experiment increased, the differences between the three samples of the oil enriched with antioxidant extract and the control oil sample were not significant. An amount of 500 ppm antioxidant INOLENS 4 is more effective than the 1000 ppm antioxidant SyneROX HT in the accumulation of free fatty acids during the 10 days of the experiment (representing 68 h hours of heat treatment at a temperature of $180^{0}C \pm 1^{0}C$). The amount of p-anisidine in P3 sample, 1000 ppm antioxidant SyneROX had the lowest values throughout the 10 days of frying.

TPC of oils used in frying increased during the experiment being correlated with frying time; there was relatively slower growth rate of TPC in sunflower oil samples enriched with 1000 ppm SyneROX HT and INOLENES 4 in samples P3 and P2. Although TPC content was highest in the tenth day, these oils were acceptable after 68 h of heat treatment, under the conditions of the experiment, given that many countries have established a maximum acceptable level for 25-27% of TPC content.

In conclusion, the value of p - anisidine, UV analysis of the lipid and the percentage of total polar compounds may be used as a complementary method to determination of free acidity level, to control the quality of vegetable oil during continuous frying processes. In addition, the combination of acid index value and some of these parameters provides additional information regarding the quality of oil that can be used to determine more accurate and efficient quality control methods for fried products.

REFERENCES

Al-Harbi M.M. and Al-Kahtani H.A., 1993. Chemical and biological evaluation of discarded frying palm oil from commercia lrestaurants. Food Chemistry, 48: p. 395-401.

Al-Kathani H.A., 1991. Survey of quality of used frying Oils from Restaurants. Journal of American OilChemists' Society, 68: p. 857-862.

Andrikopoulos N.K., Kalogeropoulos N., Falirea A., Barbagianni M.N., 2002. Performance of virgin olive oil and vegetable shortening during domestic deep-frying and pan-frying of potatoes. Intern. Journal of Food Science and Technology, 37: p. 177-190.

Basaga H., Tekkaya C. and Ackikel F., 1997. Antioxidative and free radical scavenging properties of rosemary extract. Food Sci. Technol. (London), 1997 30 105-8.

Blumenthal M.M., 1991. A new look at the chemistry and physics of deep fat frying. Food Technology. 45: p. 68-71, 94.

Bracco U., Loliger J. and Viret J.L., 1981. Production and use of natural antioxidants. J. Amer. Oil Chem. Soc., 1981 58 686-90.

Boskou D., 2003. Frying fats In: Chemical and functional properties of food lipids. Eds: Kolakowska A., Sikorski Z.E, CRC Press.

Cuvelier M.E., Richard H. and Berset C., 1996. Antioxidative activity and phenolic composition of pilot-plant and commercial extracts of sage and rosemary. J. Amer. Oil Chem. Soc., 1996 73 645-52.

Chammem N, Saoudi S, Sifaoui I, Sifi S, Person M, Abderraba M, Moussa F, Hamdi M, 2015, Improvement of vegetable oils quality in frying conditions by adding rosemary extract, Industrial Crops and Products, Volume 74, 15, p. 592-599.

Gertz C., 2000. Chemical and physical parameters as quality indicators of used frying fats. European Journal of Lipid Science and Technology 102: p. 566-572.

Gray J.I., 1978. Measurement of lipidoxidation, a review. Journal of American Oil Chemists' Society, 55: p. 539-546.

Kaufmann A., Ryser B., Suter B., 2001. Comparison of different methods to determine polar compounds in frying oils. Eur. Food Res. Technol., 213: p. 377-380.

Lalas S., and Dourtoglou V., 2003. Use of Rosemary Extract in Preventing Oxidation During Deep-Fat Frying of Potato Chips JAOCS 80, p. 579-583.

Melton S.L., Jafar S., Sykes D., Triglano M.K., 1994. Review of Stability Measurements for Frying Oils and Fried Food Flavor. Journal of American Oil Chemists' Society, 71: p. 1301-1308.

Mehta U. and Swinburn B., 2001. A review of factors affecting fat absorption in hot chips. Critical Reviews in Food Science and Nutrition, 41(2): p. 133-154.

Nakatani N. and Inatani R., 1984. Two antioxidative diterpenes from rosemary (Rosmarinus officinalis L.) and a revised structure for rosmanol. Agric. Biol. Chem., 1984 48 2081-5.

Nichita C., Neagu G., Cucu A., Vulturescu V., Berteşteanu S.V.G., 2016, Antioxidative properties of Plantago Lanceolata L. Extracts evaluated by chemiluminescence method. AgroLife Scientific Journal, Volume 5, Number 2, ISSN 2285-5718, p. 95-102.

Paul S. and Mital G.S., 1997. Regulating the use of degraded oil/fat in deep fat/oil food frying. Critical Reviews in Food Science and Nutrition, 37: p. 635-662.

Popa G., Cornea C.P., Luta G., Gherghina E., Israel-Roming F., Bubueanu C., Toma R., 2016, Antioxidant and antimicrobial properties of Laetiporus sulphurous (Bull.) Murrill. AgroLife Scientific Journal, Vol. 5, No. 1, ISSN 2285-5718, p. 168-173.

Shahidi F., Ambigaipalan P., 2015. Phenolics and polyphenolics in foods, beverages and spices: Antioxidant activity and health effects, Journal of Functional Foods, Vol. 18, Part B, p. 820-897.

Soriano J.M., Molto J.C., Manes J., 2002. Hazard Analysis and Critical control points in deep-fat frying. European Journal Lipid Science Technology, 104: p. 174-177.

Zayova E., Nikolova M., Dimitrova L., Petrova M. 2016, Comparative study of in vitro, ex vitro and in vivo propagated Salvia Hispanica (Chia) plants: morphometric analysis and antioxidant activity. AgroLife Scientific Journal, Vol. 5, No. 2, ISSN 2285-5718, p. 166-174.

Quiles J.L., Tortosa M.C.R., Gomez J.A., Huertasa J.R., Mataix J., 2002. Role of vitamin E and phenolic compounds in the antioxidant capacity, measured by ESR, of virgin olive, olive and sun flower oils after frying. Food Chemistry 76: p. 461-468.

Yoon S.H., Kim S.K., Kim K.H., Kwon W.T. and Teach Y.K., 1987. Evaluation of physicochemical changes in cooking oils during heating. Journal of American OilChemists' Society, 64: p. 870-873.

***IUPAC, 1992. International Union of Pure and Applied Chemistry-Standard Methods for the Analysis of Oils, Fats and Derivatives, IUPAC Standard Method 2.501: Determination of Peroxide Value, 7th ed., Pergamon, Oxford (UK).

11

INFLUENCE OF SOWING PERIOD AND FERTILIZATION ON THE NAKED OAT CROP GROWN IN THE ILFOV COUNTY

Doru Ioan MARIN, Ciprian BOLOHAN, Cristina Andreea OPREA, Leonard ILIE

University of Agronomic Sciences and Veterinary Medicine of Bucharest,
59 Mărăşti Blvd, District 1, 011464, Bucharest, Romania

Corresponding author email: dorumarin@yahoo.com

Abstract

Naked oat is a spring cereal crop with a good yield potential and high nutritional grain value owing to its high content of protein, fiber, fat and minerals. The trend of climate change by increasing temperatures and aridity has a negative influence upon plant growth and crop yield stability. Analyzing the evolution of climatic conditions for the Ilfov area (Southeastern Romania) in the past 10 years (2006-2015), we observed an increase in average temperature by 1.5^oC and a decrease in the annual rain autumn amount by 45.4 mm.
Research was conducted for the naked oats crop (GK Zalan variety) and the purpose was to analyze plant growth and yield potential by sowing the crop in autumn, as compared with the spring crops and the influence of mineral fertilization. All the analyzed parameters (panicles/sqm; number of panicle/plant; yield grain /panicle; thousand grains weight; hectolitre mass, grain yield) had maximum values for the autumn sowing period (October) and $N_{100}P_{50}$ fertilization level. Grain yield was 3,962 kg.ha^{-1} exceeding the spring crop yield by 55% (2,550 kg.ha^{-1}).

Key words: naked oat, Avena nuda L., sowing period, fertilizer, yield.

INTRODUCTION

Avena nuda L. (naked oat) is an oats variety in which the grains are naked after harvest.
Naked oat grains have a high nutritional value owing to their high content in protein, fiber and fat. The nutritional value of *Avena nuda* L. is higher compared to other cereal crops (Ahokas, 2005; Biel, 2014; Wood, 2007; Zarkadas, 1995), being classified as a functional food. Naked oat has a high content in β-glucan, favourably influencing the cholesterol content, as well as the functionality of the cardiovascular system and digestive system.
In temperate climate conditions naked oat is a spring cereal crop.
Naked oat has a good yield potential depending on the technology applied (Mut et al., 2016; Klimek-Kopyra et al., 2015).
The trend of climate change, i.e. increasing temperatures and aridity, has a negative influence upon plant growth and crop yield stability. According to NMA, in Romania between 1901 and 2008 average temperature increased by approx. 0.8°C, and rainfalls decreased quantitatively. These changes were more significant after 1970 for the Southern

and Eastern areas of the country. Forecasts for the next period reveal continous warming and significant reduction of rainfalls especially in summer (Romania's National Strategy on Climate Change 2013-2020).
Finding technical solutions to limit the impact of climate change on crops and food resources is a priority for agricultural research.

MATERIALS AND METHODS

Experiments were placed at the Experimental Field of Moara Domneasca Didactic Farm (Ilfov County) on a chromic luvisol with a moderately acidic reaction.
Research started by analyzing the evolution of climatic conditions in the past ten years (Table 1) relative to the average annual values of the area (Normal). During 2006-2015 an increase in average temperature of 1.5°C and a reduction in the annual rainfalls amount of 45.4 mm were observed.
In winter (December-February) the average temperature for 2015-2016 was 0.27°C, compared to -1.12°C normal values.
In these climatic conditions we considered appropriate a review of the technological elements (sowing date, culture density, variety,

dosage and type of fertilizer, soil tillage system, etc.), especially for some spring crops.

Research was conducted for the naked oat crop *Avena nuda* L. Biological material was represented by GK Zalan variety.

Table 1. Climatic condition of Moara Domnească, Ilfov

Month	Temperature (°C)			Rainfall (mm)		
	Normal	Average 2006-2015	2014-2015	Normal	Average 2006-2015	2014-2015
October	11.0	12.3	11.8	35.8	48.24	64.2
November	5.3	6.8	6	40.6	24.5	49.1
December	0.4	1.4	1.4	36.7	46.5	84.6
January	-3.0	-1.5	-1.5	30.0	36.0	33.4
February	-0.9	0.9	2	32.1	16.9	21.4
March	4.4	6.5	6.2	31.6	26.9	65.0
April	11.2	12.6	11.7	48.1	39.5	2.0
May	16.5	17.6	18.6	67.7	63.2	33.6
June	20.2	21.6	21.0	86.3	70.5	56.8
July	22.1	23.4	25.3	63.1	51.1	5.2
August	21.1	23.6	24.3	50.5	38.0	48.4
September	17.5	18.3	18.8	33.6	48.7	88.0
Average °C Sum -mm	10.5	11.96	12.1	556.1	509.87	551.7

The purpose of the research was to analyze plant growth and yield potential by sowing the naked oat crops in autumn, as compared with the spring crops.

Sowing was conducted in two periods: Period I - 18 October 2014 and Period II - 28 March 2015.

The effect of three fertilizations (N_0P_0, $N_{100}P_0$, $N_{100}P_{50}$) on plant development and crop yield was analyzed. Nitrogen fertilizer was applied in divided doses, 50% of the dose at sowing and 50% during vegetation and phosphorus fertilizer (as NP 20:20 complex fertilizer) was applied at sowing.

During the vegetation period observations and biometric measurements were conducted.

RESULTS AND DISCUSSIONS

Influence of sowing period and fertilization upon naked oats crop density
Naked oat crop in May (Table 2) at sowing presented a higher density in the variants fertilized with up to 12% ($N_{100}P_{50}$), compared to the unfertilized variants. At spring sowing

crop density was between 277 and 291 plants/sqm.

Table 2. *Avena nuda* L. crop density (plants/sqm) in spring, depending on sowing period and fertilization

Variants	Period I - October			Period II - March		
	No. of plants/sqm	%	Diff	No. of plants/sqm	%	Diff
N_0	228	100	Ct	277	100	Ct
N_{100}	245	107	17	291	105	14
$N_{100}P_{50}$	256	112	28	283	102	6

Influence of the sowing period and fertilization on the number of panicles
An important component of yield was the number of harvested panicles.

The data presented in Table 3 show that, depending on fertilization, the number of panicles/sqm was between 242 and 350 panicles/sqm for the autumn sowing period while for the spring sowing period was between 264 and 322 panicles/sqm.

$N_{100}P_0$ fertilization generated a 35% increase for the autumn sowing period and a 17% increase for the second sowing period, and nitrogen and phosphorus fertilization resulted in an increase of 45% in Period I and 22% in Period II.

Table 3. Panicles density to naked oat crop depending on sowing period and fertilization

Variants	Period I - October			Period II - March			PI com. PII %
	No. of pn/sqm	%	Diff	No. of pn/sqm	%	Diff	
N_0	242	100	Ct	264	100	Ct	22
N_{100}	327	135	85	310	117	46	17
$N_{100}P_{50}$	350	145	108	322	122	58	28

Phosphorus-based fertilizer had a higher effect to the oat sown in autumn, as the number of panicles was 9% higher.

The number of panicles formed by the productive twinning of oat plants (Table 4) was favourably influenced by fertilization in both sowing periods; maximum results were recorded for autumn sowing 1.33-1.37 panicles/plant. For the variant sown in March, there were 1.06-1.14 panicles/plant on average, the productive twinning being influenced more strongly by low rainfalls in April and May

(31% of the value of rainfalls in the area). Naked oat sown in autumn resulted in an increase in the number of panicles/plant by 11-25%, compared to the crop sown in March.

Table 4. Number of panicles/plant in naked oat crops depending on sowing period and fertilization

Variants	Period I - October			Period II - March			PI comp. PII %
	No pn/pl	%	Diff	Nr pn/pl	%	Diff	
N_0	1.06	100	Ct	0.95	100	Ct	111
N_{100}	1.33	125	0.27	1.06	111	0.11	125
$N_{100}P_{50}$	1.37	129	0.31	1.14	120	0.19	120

Figure 1. Naked oat plants development depending on sowing period (Left - Period I, Right -Period II) at 6/19/2015

Influence of the sowing period and fertilization upon grain yield

Sowing period and fertilization strongly influenced the nude oats yield level in 2015 (Table 5).

Grain yield had the highest level in the three variants of fertilization for the autumn sowing period. Fort the fertilizated variant $N_{100}P_{50}$ the recorded grain yield was of 3962 kg.ha^{-1}. The fertilizer resulted in yield increase of 1852 kg.ha^{-1} for nitrogen a dose of 100 kg.ha^{-1} and 2246 kg.ha^{-1} from the nitrogen and phosphorus. For the second sowing period (spring), the maximum yield level of 2550 kg.ha^{-1} was recorded for nitrogen and phosphorus fertilization, and the yield increase was 1145 kg.ha^{-1} for N_{100} and 1412 kg.ha^{-1} for $N_{100}P_{50}$ compared to the unfertilized variant.

Analyzing the sowing period effect on yield, it resulted that autumn sowing generated important yield increases of 365 kg.ha^{-1} for the

unfertilized variant, i.e. 1145 kg.ha^{-1} for nitrogen fertilization and 1412 kg.ha^{-1} when $N_{100}P_{50}$ surpassed the oats yield sown in Period II by 27-55%.

Table 5. Sowing period and fertilization influence on naked oats yield

Variants	Period I – October			Period II – March			Diff PI-PII kg.ha^{-1}	PI comp. PII %
	kg.ha^{-1}	%	Diff	kg.ha^{-1}	%	Diff		
N_0	1716	100	Ct	1351	100	Ct	365	127
N_{100}	3568	208	1852	2423	179	1072	1145	147
$N_{100}P_{50}$	3962	231	2246	2550	189	1199	1412	155
LSD (5%) = 217 kg								

Influence of the sowing period and fertilization on some yield parameters

The analysis of several yield parameters (yield/panicle - YGP, a thousand grains weight - TGW, hectolitre mass - HM see, Table 6) shows that they were favourably influenced by autumn sowing and fertilizer applied. Thousand grains weight per panicle increased depending on fertilization with 0.20-0.34 g, TGW with 0.3-1.1 g and HM with 0.5-1.3 kg.

Table 6. Sowing period and fertilization influence on naked oats yield parameters

Variants	Period I - October			Period II - March		
	YGP g (%)	TGW g (%)	HM kg (%)	YGP g (%)	TGW g (%)	HM kg (%)
N_0	0.71 (100)	22.3 (100)	58.1 (100)	0.51 (100)	22.0 (100)	57.6 (100)
N_{100}	1.09 (153)	23.8 (107)	59.6 (102)	0.78 (153)	23.0 (105)	58.9 (102)
$N_{100}P_{50}$	1.13 (159)	24.2 (108)	60.0 (103)	0.79 (155)	23.1 (105)	58.7 (102)
YGP-yield grain panicle; TGW-thousand grains weight; HM-hectolitre mass						

CONCLUSIONS

Considering the conditions of the agricultural year 2014-2015, the results obtained show that the cultivation of the naked oats variety GK Zalan as winter grain leads to substantial yield increases under the pedological and climatic conditions of the Ilfov region.

In spring plants had a quick start of the vegetative development, a vigorous development with a higher capitalization of fertilizers and soil water reserve.

All the analyzed parameters recorded the highest values for the autumn sowing period and $N_{100}P_{50}$ fertilization level.

Grain yield was 3962 kg.ha^{-1}, exceeding the spring crop yield by 55%. Yield increase was 2246 kg.ha^{-1}, compared to the unfertilized variant and to variant $N_{100}P_0$, 394 kg.ha^{-1}. $N_{100}P_0$ fertilization generated an yield increase of 1852 kg.ha^{-1} for the sowing Period I and of 1072 kg.ha^{-1} for Period II.

These results lead to further research for the elaboration of an alternative naked oat crop technology in the context of climate changes.

REFERENCES

Ahokas H., Heikkilä E., Alho M.L., 2005. Variation in the ratio of oat (*Avena*) protein fractions of interest in coelic grain diets. Genetic Resources and Crop Evolution, 52: p. 813-819.

Biel W., Jacyno E. and Kawęcka M., 2014. Chemical composition of hulled, dehulled and naked oat grains. South African Journal of Animal Science, 44(2) p. 189-197.

Klimek-Kopira A., Kulig B., Oleksy A., Zajac T., 2015. Agronomic performance of naked oat (*Avena nuda* L.) and faba bean intercropping. Chilean Journal of Agricultural Research, 72(2), Spril-June, p. 168-173.

Mut Z., Akay H., 2010. Effect of seed size and drought stress on germination and seedling growth of naked oat (*Avena sativa* L.). Bulgarian Journal of Agricultural Science, 16, No. 4, p. 459-467.

Mut Z., Erbas Kose O.D., Akay H., 2016. Grain yield and some quality trait of naked oat cultivars. Anadolu Tarim Bilimleri Dergisi/Anadolu Journal of Agricultural Sciences, 31, p. 96-104.

Klimek-Kopira A., Kulig B., Oleksy A., Zajac T., 2015. Agronomic performance of naked oat (*Avena nuda* L.) and faba bean intercropping. Chilean journal of Agricultural Research, 72(2), Spril-June, p. 168-173.

Peterson D.M., 1998. Malting oats: Effects on chemical composition of hulled and hull-less genotypes. Cereal Chem. 75, p. 230-234.

Wood P.J., 2007. Cereal ß-glucans in diet and health. J. Cereal Sci. 46, p. 230-238.

Zarkadas C.G., Yu Z., Burrows V.D., 1995. Protein quality of three new Canadian-developed naked oat cultivars using amino acid compositional data. Journal of agricultural and Food Chemistry, 43(2) p. 415-421.

***2012. MMP-Romania's National Strategy on Climate Change 2013-2020. http://www.mmediu.ro.

***http://www.meteoromania.ro.

12

INPUT USAGE AND PROBLEMS IN GREEN BEAN PRODUCTION: A CASE OF BURDUR PROVINCE, TURKEY

Mevlüt GÜL, Halil PARLAK

University of Süleyman Demirel, Agriculture Faculty, Department of Agricultural Economics, 32260 Isparta, Turkey

Corresponding author email: mevlutgul@yahoo.com

Abstract

The production of green beans increased by 6.5 times from 3,106 tons in 1991 to 20,199 tons in 2015 in Burdur. In this study, it was aimed to reveal the use of inputs and problems of farmers who produce green beans in Burdur province. The study was carried out in the central district of the province of Burdur, where the production of green beans was the most intensive. Stratified sampling method was applied and the number of samples according to this method was determined as 112 farmers. Data were obtained by the face-to-face survey method. Sales quantities of farmers in the region varied depending on the cultivation area and yield of green bean. It was found that the use of unit labour in the unit area was 251.72 MLU hours in the production of green beans. Machine power usage was calculated as 2.18 hours per decares. The amount of seed used was 8.20 kilograms. The most important criterion in the preference of green bean cultivation was the high possibility of irrigation and productivity in the region. Sale of green beans starts at the end of July and lasts until the 15th of September. The prices of green beans were largely determined by the companies. Farmers sell their products wholesaler-trader. It can be said that the cultivation of green bean cultivation will continue through the factors such as the suitability of climate conditions, irrigation opportunities in the region, high yields and inadequate supply. However, the presence of a single collection center for the sale of products in the region is perceived as a threat. It can be stated that the producer's activity in the market may be more prominent with the establishment of the producer association and the development of organizational awareness in green bean growers in the region.

Key words: green bean, input, labour, problems, Burdur, Turkey.

INTRODUCTION

Bean (*Phaseolus vulgaris* L.) is an important cultivar of the Leguminosae family, which has a very high nutritional value and is abundantly consumed throughout the world. Beans can be evaluated in various forms such as fresh vegetables, dried grains and canned foods (Bozoğlu, 1995). The bean plant, which originated in Central America, came to Anatolia 250 years ago and found a very wide spreading area (Şehirali, 1988).

In the world 1,543,335 hectares of area and 21,365,119 tons of fresh bean production was realized in 2013. According to the figures for 2013, the most important producer of green beans in the world is China with a share of 78%. Other important countries in the production of this product are Indonesia, Turkey, India, Thailand, Egypt, Spain and Italy (FAOSTAT, 2017). Turkey meets 3% of the world's bean production. According to the data of 2015, 640,836 tons of green beans were

produced in Turkey from 50,122 hectares. Parlak and Gül (2016) pointed out the changes in the green and dry bean market in the world and in Turkey. Parlak and Gül (2016) reported that the world green bean production rose 5 times with both in terms of green bean yield and planting area expansion, Turkey ranks sixth in the world's green beans production and sixth in yields.

In 1992, 535,950 decares of green bean cultivation area in Turkey decreased by 6% and it was realized as 501,218 numbered. In 1991, 436,000 tons of green bean production increased by 47% to 640,836 tons (Table 1). So while the fresh bean fields remained stable, production growth continued to increase due to increased yields.

Turkey's share in green bean harvested area is the highest in Samsun with a rate of 15.7%. Turkey has the highest share in fresh bean production with a rate of 18.52% in Samsun province. Samsun is followed by Tokat with 54,783 tons, Bursa with 53,890 tons, Antalya with 50,582 tons, Izmir with 42,586 tons, Mersin with 31,771 tons and Burdur with 20,199 tons. When considering the change of fresh bean cultivation area compared to 2000 in provinces, the highest increase was in Burdur (1.86 times), İzmir (1.62 times) and Samsun (1.42 times) respectively.

Table 1. Development of green bean cultivation area and production in Turkey

Year	Harvest area (decare)	Harvest area index (1992 = 100)	Production (ton)	Production index (1991 = 100)
1991	-	-	436,000	100
1992	535,950	100	450,000	103
1993	522,690	98	440,000	101
1994	540,420	101	435,000	100
1995	556,720	104	460,000	106
1996	548,570	102	455,000	104
1997	537,520	100	450,000	103
1998	546,100	102	455,000	104
1999	558,990	104	471,000	108
2000	584,170	109	514,000	118
2001	571,870	107	490,000	112
2002	570,980	107	515,000	118
2003	575,670	107	545,000	125
2004	562,710	105	582,000	133
2005	560,500	105	555,000	127
2006	537,824	100	563,763	129
2007	519,813	97	519,968	119
2008	530,200	99	563,056	129
2009	535,172	100	603,653	138
2010	531,340	99	587,967	135
2011	528,931	99	614,948	141
2012	528,506	99	621,036	142
2013	506,619	95	632,301	145
2014	501,767	94	638,469	146
2015	501,218	94	640.836	147

Source: TUİK, 2017

When the change in the production of green beans is compared with the year 2000, with a maximum increase of 4.27 times in the province of Burdur. Burdur was followed by Adana with 3.76 times increase, Bursa with 2.22 times increase and İzmir with 2.16 times increase and Tokat with 2.11 times increase.
The development of green bean harvest area and production in Burdur province is given in Table 2. When the harvest area of green beans in Burdur province is examined by years; the area of fresh bean harvest, which was 5,780 decares in 1992, increased by 85% in 2015 and reached 10,699.
Production increased by 6.5 times compared to 3,106 tons in 1991 and increased to 20,199 tons in 2015. The share of Burdur in Turkish production has increased in general from 1991

onwards. Burdur province received 0.71% of the green beans produced in Turkey in 1991 and this ratio increased to 3.15% in 2015.

Table 2. Development of green bean cultivation area and production in Burdur

Year	Harvest area (decare)	Harvest area index (1992 = 100)	Production (ton)	Production index (1991=100)	Production share in Turkey (%)
1991	-	-	3,106	100	0.71
1992	5,780	100	3,742	120	0.83
1993	5,820	101	4,100	132	0.93
1994	5,080	88	4,395	142	1.01
1995	5,230	90	4,275	138	0.93
1996	4,680	81	4,577	147	1.01
1997	5,000	87	4,315	139	0.96
1998	5,560	96	4,834	156	1.06
1999	4,850	84	4,574	147	0.97
2000	5,740	99	4,732	152	0.92
2001	5,280	91	5,158	166	1.05
2002	4,980	86	4,605	148	0.89
2003	5,450	94	5,360	173	0.98
2004	5,030	87	5,686	183	0.98
2005	4,940	85	7,966	256	1.44
2006	5,012	87	9,266	298	1.64
2007	4,706	81	4,907	158	0.94
2008	8,939	155	11,489	370	2.04
2009	10,344	179	17,313	557	2.87
2010	10,609	184	19,276	621	3.28
2011	11,855	205	21,213	683	3.45
2012	11,713	203	21,068	678	3.39
2013	11,968	207	21,708	699	3.43
2014	11,114	192	20,208	651	3.17
2015	10,699	185	20,199	650	3.15

Source: TUİK, 2017

When the harvest area of green beans was examined in the Burdur, the highest increase was realized in the Central district with an increase of 9.83 times. According to the year 2000, the production of green beans in Burdur was also realized in the Central district with a maximum of 54 times. The share of central province in Burdur production is 80.20% and it is in the first place.
Green beans in Turkey constitute 5.2% of vegetable production value and 1.4% of total plant production value with 1.3 billion TL production value in 2014. Turkey appears to be self-sufficient in the production of green beans. However, improvements in politics are important at the point of development of production.
As the research area, Burdur province was selected as the highest increase in harvest area and in production of green bean in Turkey

between 1991 and 2015. In Burdur, the production of green beans increased by 6.5 times from 1991 to 31,099 tons, reaching 20,199 tons in 2015. The aims of this study at this point were: (i) to examine the input use of green bean producing farms in Burdur, (ii) marketing activities, (iii) identifying problems and suggesting solutions.

Studies on the technical aspects of green bean production in Turkey have been made. The number of studies examining the production of green beans economically is limited. For this reason, the findings that would be obtained are thought to be useful to the farmers, researchers and related institutions working on the subject.

MATERIALS AND METHODS

Material

The main material of this study was obtained from the farmers who grown green beans in Burdur province. Secondary data related to the study were obtained from FAO, TUIK, Provincial and District Food, Agriculture and Livestock Departments. The data used in the research belonged to the production period of 2016.

Methods

Sampling Method

Sample volume was calculated as 112 farmers using Neyman stratified sampling method with considering 95% confidence interval and 5% error margin. Farmers in the production of green beans were divided into four groups according to their frequency distribution (first layer was 1.00-3.99 decares of green bean cultivation area, second layer was 4.00-7.99 decares green bean cultivation area, third layer was 8.00-15.99 decares of green bean planting area, layer 4 referred to 16.00 decares and above the green bean planting area).

According to this; there was interviewed with 24 farmers from the first layer, 13 from the second layer, 20 from the third and 55 farmers from the fourth layer. Farmers were selected randomly.

Method to use in obtaining data

The data needed for the analysis were obtained by face-to-face surveys of green bean farmers in Burdur. The questionnaire included the following questions: (i) age of the farmer,

education level, experience, cooperative membership, agriculture and non-agricultural income, (ii) population and family labour force, (iii) non-family labour force status, (iv) land-saving structure, (v) green bean marketing, (vi) support from the state of green bean producers, (vii) information on the amount of credit and credit utilization, and (viii) problems with green bean growing.

Evaluation of data

The data collected from the identified farmers were transferred to the computer, calculations were made and tables were created and these tables were interpreted using absolute and relative distributions.

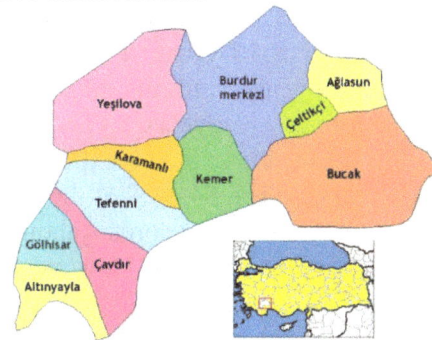

Figure 1. Research area

RESULTS AND DISCUSSIONS

As a result of the farmer-level survey, 21.4% of the green bean producers interviewed were in the first group, 11.6% in the second group, 17.9% in the third group and 49.1% in the fourth group (Table 3).

Table 3. Distribution of green bean producers interviewed

Farmer width groups	Frequency (N)	Rate (%)
I (1.00-3.99 decares)	24	21.4
II (4.00-7.99 decares)	13	11.6
III (8.00-15.99 decares)	20	17.9
IV (16.00 decares and more)	55	49.1
Total	112	100.0

1 decares equal 0.1 hectares

Socio-economic indicators of farmers producing green beans in the study area were given in Table 4. The average age of farmers producing green beans in the region was 42.57 years. Educational levels of farmers were 5.46 years. The education level of the first group farmers was 5.13 years, the second group was 5.23 years, the third group was 5.30 years and the fourth group was 5.71 years. Therefore, the education level of the farmers was above the elementary school level and it was close to the average level of education in Turkey. The household size of the green bean producers in the region was 4.22 persons. This value is 3 persons for Burdur province (TUİK, 2017). Therefore, the household size of green bean producers was above the value of Burdur province.

In the farmers' group interviewed, the fourth group had the highest number of experiments in plant production with 16.33 years. The first group of farmers had been dealing with green bean agriculture for about 14 years. The second group farmers were dealing with green bean production for about 10 years, the third group farmers about 6 years, and the fourth group farmers for about 13 years. The farmers average more than 11 years of experience in green bean growing. The 54.46% of the farmers stated that they keep records about the production of green beans. However, this record was in an unregulated system. Therefore, the record keeping in farmers was low. As a consequence, farmers were less likely to be able to follow developments related to green bean production activities and to plan and to see their net profits.

In the surveyed farmer groups, the fourth group with 30.91% was the most engaged in non-agricultural work. This value was 17.86% in farmers' average.

Therefore, the agricultural activity in the income of the interviewed farmers had a significant share. It was important to produce green beans so that they could continue their lives.

When some of the ownership indicators of the farmers are examined; Computer ownership was found to be as low as 10.71%. As a matter of fact, this rate was 22.9% in Turkey.

Internet ownership was also in the fourth group of farmers with only 3.64%. In Turkey, the internet access rate was 76.3% and the internet access rate was 93.7% on an individual basis (TUİK, 2017).

About 99.11% of the farmers in the area had ownership of mobile phone, 55.36% of them owned the car, and 63.39% owned the credit card.

The 66.07% of green bean growers were borrowed. The highest indebtedness was in the first and fourth group of farmers. Survey ownership of social security was high. This assurance was in 100% of the first, second, third group farmers and 87.27% of the fourth group farmers.

Table 4. Some socio-economic indicators in the production of green beans

Features	Farm groups				Average
	I	II	III	IV	
Farmer age (years)	45.54	57.23	37.75	39.56	42.57
Farmers' education level (year)	5.13	5.23	5.30	5.71	5.46
Household size (person)	3.75	3.85	3.50	4.78	4.22
Farmer's experience in plant production (years)	15.21	12.23	11.10	16.33	14.68
Farmer's experience in green bean production (years)	13.67	9.69	6.35	12.09	11.13
Farmer keeping records of operation (%)	45.83	76.92	25.00	70.91	58.04
Doing non-agricultural business (%)	4.17	0.00	10.00	30.91	17.86
Ownership of computers (%)	0.00	7.69	5.00	18.18	10.71
Internet ownership rate (%)	0.00	0.00	0.00	3.64	1.79
Ownership of mobile phones (%)	100.00	100.00	95.00	100.00	99.11
Ownership rate of cars (%)	95.83	15.38	25.00	58.18	55.36
Credit card ownership rate (%)	54.17	92.31	90.00	50.91	63.39
Debt rate (%)	100.00	23.08	25.00	76.36	66.07
With social security (%)	100.00	100.00	100.00	87.27	93.75
Ratio of farmers engaged in livestock (%)	0.00	7.69	90.00	25.45	29.46
Own land (%)	100.00	100.00	85.71	62.90	68.72
Rented land (%)	0.00	0.00	14.29	37.10	31.28
Agricultural membership (%)	70.83	69.23	55.00	67.27	66.07
Credit users (%)	4.17	7.69	15.00	7.27	8.04
A tendency to continue to produce green beans *	4.54	4.54	4.70	4.29	4.45
Knowledge of green beans **	3.79	3.23	3.20	4.25	3.85
Satisfaction with green bean production **	3.00	4.23	3.50	3.31	3.38

*: 1 = Absolutely not thinking; 2 = Does not think; 3 = Undecided; 4 = Thinking; 5 = Definitely thinking

**: 1 = Very low; 2 = Low; 3 = Medium; 4 = High; 5 = Very high

It was the third group with 90% of the farmer groups engaged in livestock. These were usually dairy cattle breeding. The first group of farmers did not have livestock production activity.

The 68.72% of the total land of farmers producing green beans was owned and 31.28% was rented. The green bean area was increased with the rental land ratio was increased.

When producers were investigating their tendency to continue to produce green beans; it was determined that the farmers thought to continue. The tendency to continue was stronger in the first, second and third group farmers. Farmers' knowledge of green bean production was also generally high. The level of knowledge was higher in the fourth group of farmers. The level of satisfaction with the production of green beans was moderate. The highest level of satisfaction with green bean production was found in the second group of farmers (Table 4).

The 4.17% of the first group farms, 7.69% of the second group farms, 15.00% of the third group farms and 7.27% of the fourth group farms were using agricultural credit. In the region average 8.04% of the farmers who produce green beans used the loan. Therefore, the use of credit in farms that produce green beans was low (Table 4).

It was found that 66.07% of the farmers were members of any agricultural organization. These organizations were Agriculture Chamber, Agricultural Credit Cooperative, Pankobirlik and Breeding Cattle Breeders' Union (Table 4).

Soil processing in the production of green beans takes place in April. About 47.32% of the farmers carried out this process with their equipment. Green bean planting usually takes place in May (86.61%). The vast majority of the farmers were renting machinery (76.79%) during sowing. Drip irrigation method was used as irrigation method. Irrigation was carried out in May-August period. Animal manure was used by 88.39% of farmers. Leaf fertilization was done by 97.32% of the farmers.

In the production of green beans, the amount of seeds used during the sowing was examined. The first group used 8.86 kg per decares, 7.29 kg for the second group of farmers, and 7.90 kg for the third group of farmers and 8.26 kg of seeds were used in the fourth group. The highest seed usage per decare was in the first group (8.86 kg per decares) (Table 5). In green bean growing, 7-10 kg of seeds are recommended for use in sowing (MEGEP,

2008). Therefore, the amount of seed used in the research area was within the recommended limits.

Table 5 also showed the cost of seedlings of green bean producers. The farmers in the first group had higher seed costs (357.08 TL per decares). It was determined that the average cost of seedlings was 326.30 TRL per decares.

Table 5. Green bean seeds amount used of the farmers

Farmer width groups	Seed amount (kg per decares)	Seed cost (TRL per decares)
I	8.86	357.08
II	7.29	287.85
III	7.90	316.80
IV	8.26	325.42
Total/average	8.20	326.30

1 Euro equal 3.343611 TRL (Turkish Liras)

The 41.7% of the farms producing green beans in the region used their machine power in production. About 58.93% of the farmers provided the machine power by renting for the production of green beans. While 92.31% of the second group of farmers used their own machine power, 80% of the third group farmers had used the machine power by renting (Table 6).

Table 6. Machinery renting of the farmers

Farmer width groups	Owned (%)	Rented (%)
I	25.00	75.00
II	92.31	7.69
III	20.00	80.00
IV	43.64	56.36
Total/average	41.07	58.93

It was found that the green bean growers at the study site used and preferred their inputs according to their own knowledge and experience. The opinions of the dealers were also "absolutely important" in green bean farming. It was determined that farmers' neighbours and relatives and the technical staff in the Provincial Directorate of Agriculture opinions on the input used was also partially complied. Written sources were the least used information source (Table 7).

Labour employed in the green bean production was calculated in unit of male labour unit (MLU: Male labour unit. Here, male aged 15-49 was taken as =1, female aged 15-49 was taken as =0.75, male aged >50 was taken as=0.75, female aged>50 was taken as 0.50 and child aged 7-14 was taken as =0.50 male labour unit) (Erkuş et al., 1995).

Labour force (MLU as hours) and machine power used in the production of green beans were shown in Table 8.

In the first group of farmers, the use of the family labour force per decares (351.44 MLU hours) was higher than the other groups. The fourth group farmers (199.43 hours) used more non-family labour than the other groups of farmers.

Table 7. Importance of information sources on the input used

Information sources	Farm width groups*				Average
	I	II	III	IV	
According to your own knowledge and experience	4.5	5.0	4.2	4.5	4.5
Dealer recommendations	4.5	5.0	4.1	4.5	4.5
Recommendations of technical staff in Provincial Directorate of Agriculture	3.9	4.2	2.0	2.7	3.0
Neighbours and relatives recommendation	3.3	2.5	1.8	2.8	2.7
Buyer recommendation (trader)	3.1	1.2	1.9	2.6	2.4
Books, magazines, newspapers, brochures, etc.	2.0	4.1	1.8	1.3	1.9

*: Likert Scale: 1 = Definitely no; 2 = No; 3 = Partially; 4 = Yes; 5 = Absolutely yes

As the amount of green bean land increased, the amount of non-family labour force increased accordingly.

As the area of green beans planting in the same way decreased, the use of family labour increased and the use of non-family labour decreased.

In Burdur province, the average labour used for the production of fresh beans was calculated as 251.72 MLU hours. Again in this direction, it was determined that the fourth group (2.00 hours) used the least amount of machine power. The fourth and third group farmers was near the collection center which is the market for green beans, so they used less machine power and also transport cost was less than the other farmers.

In the average farmers, the use of machinery in green bean production was calculated as 2.18 hours (Table 8).

Figure 2 showed the SWOT analysis results of the green bean production. This analysis of the sector was carried out using information obtained from 112 producers, brokers, traders, authorizing firms and relevant experts.

Table 8. Usage of labour and machine power in the green beans production

	Farm width groups*				Average
	I	II	III	IV	
Family labour force (hours per decares) (MLU)	287.74	134.76	47.39	57.21	68.19
Non-family labour force (hours per decares) (MLU)	63.70	100.60	147.22	199.43	183.54
Total labour force (hours per decares) (MLU)	351.44	235.36	194.61	256.64	251.72
Machine power usage (hour per decares)	3.57	4.26	2.25	2.00	2.18
Family labour force (%)	81.88	57.26	24.35	22.29	27.09
Non-family labour force (%)	18.12	42.74	75.65	77.71	72.91
Total labour force (%)	100.00	100.00	100.00	100.00	100.00

MLU: Man Labour Day--hours

The strongest aspects of fresh bean production in the region were that the climate of the research area is appropriate, the yield is high compared to other crops, the water required for green bean production is found and the product is easy to sell.

The lack of knowledge of green bean production, high labour use, labour cost, inability to find a market for the product in some cases and lack of producer cooperatives were the weaknesses of green bean production.

Opportunities in the green bean production in the region where the work is done can be defined as provide employment for the people, an important source of income, and the sale in cash.

The increase in the number of producers in the region, the inability to obtain fraud/sales amounts, the climate change, the lack of support/inadequacy were expressed as threats to the green beans production (Figure 2).

Problems in the green bean production

Problems related to production, marketing and financing (credit) in the research area were also examined. It was determined that the most problem was related to marketing and production. There was no problem with credit in the region. Problems related to production and marketing were listed below.

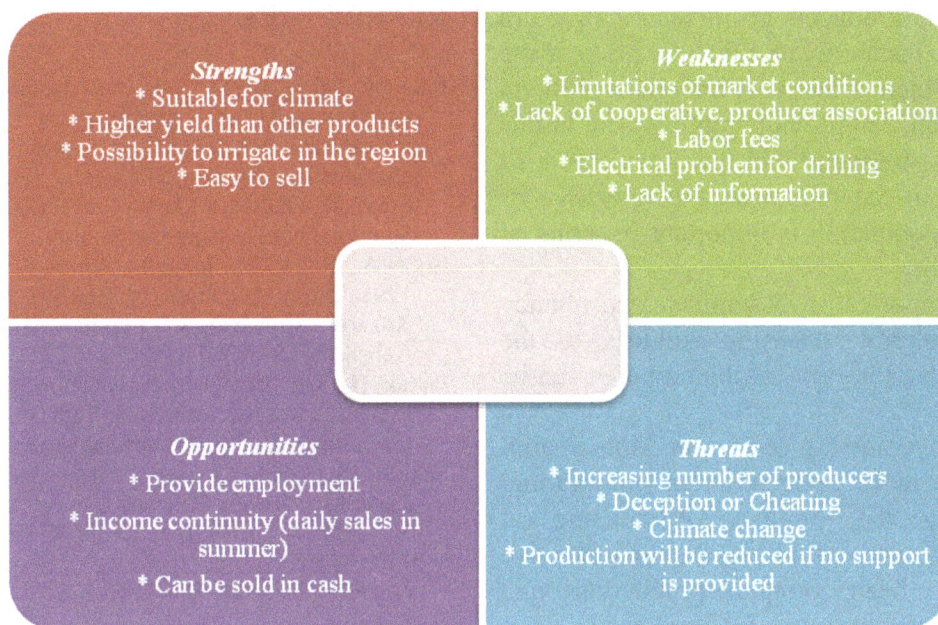

Figure 2. SWOT analysis of green bean production

Problems with production:

-Bourgondia seed which is a widely used green bean variety in the region, is expensive,

-Lack of information about the producers' struggle against disease and harmfulness,

-The lack of information on the green bean cultivation and sometimes the hail reduces crop yield,

-The fact that producers cannot get electricity subscription to the drilling wells used in the irrigation,

-The inputs used in production are expensive,

-The lack of consultancy and agricultural extension activities in the region.

Problems with marketing:

-The price of green beans is low and the price is unstable,

-They cannot get the money (deceit) after the farmers sold on account their products to merchants

-At the time of sale, the number of buyers in the collection center is low,

-The determination of the price of green beans by mutual agreement among traders,

-When a village outside the collection center, traders lower the price.

CONCLUSIONS

The average age of farmers interviewed in the region was 42.57 years. The household size was 4.22 persons. The farmers' experience in producing green beans was 11.13 years. The 68.72% of the land was own land. Farmers generally rented land for the production of green beans.

Fresh bean producers were using their own knowledge and experience to supply the inputs they used. The views of agricultural chemical dealers were also "absolutely important" in green bean growing. One of the most important contributions of green bean growing is the used labour force. Green bean provides employment for over 30 days in one production season per decare. Small-scale enterprises were able to evaluate more of the family workforce. The need for non-family labour is increase with the scale of planted area.

The 8.20 kilograms of seed were used in the region. The maximum used of seeds was in the first group of farmers (8.86 kg) and the second group of farmers (7.90 kg) was the least seeded.

The majority of the interviewed farmers were the only green bean source of livelihood. Therefore, the low price of fresh beans, instability is one of the most important problems.

According to the information obtained, some farmers' green beans production were the sole source of income. Therefore, it is important to establish a cooperative or producer association for green beans in the region. The fact that the producers go to such a structure against the instability and the low price of the product, which is the most important problem in the region, can provide improvement in this case. It can be stated that the organization of producers in the region as a cooperative or producer association is important in terms of price stability.

It is also important to disseminate consultancy and agricultural extension services in the region. Thus, the yield of the unit area can be increased.

Burdur Governorship or Burdur Municipality may be more effective in promoting/ announcing the product. Advertising at various fairs and exhibitions can be a positive on green bean producer/production and therefore positive results for Burdur.

Improvements in input prices used in the green bean production can increase income by the farmer.

Improvement of the collection center located in the district can provide more participations of buyers and sellers in this market.

ACKNOWLEDGEMENTS

This study is a part of the postgraduate thesis. This study was supported under the scope of the Project number BAP 4767-YL1-16 from the Süleyman Demirel University Department of Scientific Research Projects.

REFERENCES

Bozoğlu H., 1995. An investigation on the determination of genotype x environment interactions and heritability of some agronomical characters in dry bean (*Phaseolus vulgaris* L.) (In Turkish). Ondokuz Mayıs University, Institute of Science, PhD Thesis, 99 p., Samsun.

Erkuş A., Bülbül M., Kiral T., Açil A.F., Demirci R., 1995. Tarım ekonomisi (In Turkish). Ankara: T.C. Ankara Üniversitesi Eğitim, Araştırma ve Geliştirme Vakfı Yayınları.

Parlak H., Gül M., 2016. Production of beans and external trade development in Turkey and the world (in Turkish). Bahçe, 45 (Special issue): p. 423-434.

Şehirali S., 1988. Yemeklik dane baklagiller (In Turkish). Ankara Üniversitesi Ziraat Fakültesi Yayınları, No: 1089. Ders Kitabı, 435 p., Ankara.

***FAOSTAT, 2017. Food and Agriculture Organization of the United Nations Statistical Data. (Internet address: http://www.fao.org/faostat/en/#home, accessed on (02.01.2017).

***MEGEP, 2008. Bahçecilik, Fasulye Yetiştiriciliği (In Turkish). Mesleki Eğitim ve Öğretim Sisteminin Güçlendirilmesi Projesi, 11 p., Ankara.

***TUİK, 2017. TUİK (Turkish Statistical Institute) Statistical Data. (Internet address: http://www.tuik.org.tr, accessed on (20.01.2017).

13

PARADIGM SHIFT FROM NON-BT TO BT COTTON AND FACTORS CONDUCING BT COTTON PRODUCTION IN A SOUTHERN PUNJAB'S DISTRICT OF PAKISTAN

Muhammad MUDDASSIR[1], Muhammad SHAHID[2], Ahmed Awad Talb ALTALB[3], Syed Muhammad Waqar AHSAN[2], Muhammad MUBUSHAR[1], Muhammad Abubakar ZIA[1], Mehmood Ali NOOR[4,*]

[1]Department of Agricultural Extension and Rural Society, College of Food and Agricultural Sciences, King Saud University, Riyadh 11451, Kingdom of Saudi Arabia
[2]Institute of Agricultural Extension and Rural Development, University of Agriculture, Faisalabad-38040, Pakistan
[3]Department of Agricultural Extension and Technology Transfer, Faculty of Agriculture and Forestry, University of Mosul, Mosul, Iraq
[4]Institute of Crop Science, Chinese Academy of Agricultural Sciences, Key Laboratory of Crop Physiology and Ecology, Ministry of Agriculture, Beijing 100081, China

Corresponding author email: mehmood2017@gmail.com

Abstract

Pakistan is one of the developing countries and major portion of its economy depends on agriculture. The study aim was to identify the different factors affecting the adoption of Bt cotton and analysis of paradigm shift from Non-Bt to Bt cotton in tehsil Jatoi of Muzaffargarh district, which provided a guideline for extension organizations to develop better strategies in future for effective extension work towards Bt cotton production. The data were collected from 120 Bt cotton growers through random sampling technique. The data were analyzed through Statistical Package for Social Sciences (SPSS). Descriptive statistics such as frequencies, percentages, means, standard deviations and rank order were used for interpretation of the data. The study revealed that maximum respondents were belonging to the middle age group, agriculture and livestock farming were their source of income and their maximum cultivation was under Non-Bt cotton. Cotton growers were highly dependent on pesticide companies for agricultural information. The higher crop yield was the major factor which shifted the farmers to grow Bt cotton. Unapproved Bt, high fertilizer requirement and non-availability of seed were the threatening factors being faced in Bt cotton cultivation, ranked at medium scale. Opportunities of less use of pesticides, increase in production and net annual income, reduction in health hazards, less cost of production and availability of certified seeds was generated by the cultivation of Bt cotton, was recorded at medium scale.

Key words: cotton, Bt and Non-Bt, paradigm shift, adoption, factors.

INTRODUCTION

Pakistan is one of the developing countries and major portion of its economy depends on agriculture. About 70.5% of the population of Pakistan are involved directly or indirectly in the agriculture industry. It contributes almost 21% to GDP of Pakistan. Cotton accounts for 8.6% of the value added in agriculture and about 1.8% to GDP. The crop was sown on the area of 3106 thousand hectares during growing season of 2012, 10.1% more than previous year (2820 thousand hectares) (Govt. of Pakistan, 2013). The main cotton growing districts in the Punjab province are Muzzaffargarh, Lodhran, Rahim Yar Khan, Vehari, Bahawlanagar, Multan, Rajanpur, Bahawalpur and Dera Ghazi Khan. Its production was 5% less than the target of 13.36 million bales mainly due to the shortage of irrigation water, less use of DAP in the cotton crop, attack of Cotton Leaf Curl Virus (CLCV), mealy bug and white fly on the crop and last picking of it was affected due to higher prices of wheat announced by the government (Govt. of Pakistan, 2013). Pakistan is the largest exporter of cotton material such as (cotton fabric, cotton yarn and other items, manufactured of cotton) to the USA and other countries of the world. The statistical analysis showed that cotton has importance in Pakistan

Economy (Shafiq-ur-Rehman, 2009).

Punjab and Sindh are two major cotton producing provinces in the country. The share of Punjab in the production is about 80% and the contribution of Sindh is about 20%. The share of other provinces like Baluchistan and Khyber Pakhtunkhwa is very low as compared to Punjab and Sindh provinces (Govt. of Pakistan, 2013). Biotechnology is playing a vital role in the cotton industry, the Bt technology is providing the varieties of high yield (Gandhi and Namboodiri, 2006). Bt cotton showed some problems in the local environment, to solve these issues government start Bt breeding programs. These breeding programs conducted various trials by sowing different varieties of Bt and non-Bt cotton on experimental basis, with different variations. Resultantly, they developed variety of seed Bt cotton, which could be survived at high temperature and low water with high yield. It has been noticed that these locally developed varieties according to local environment showed better results, and suggested to cotton growers to grow these locally developed Bt varieties (Mahmood and Farooq, 2011).

The Bt cotton was adopted in Pakistan mainly due to requiring less number of sprays and high productivity. The average yield of Bt cotton is 23-28 mounds per acre as compared to conventional cotton, which is 17-20 mounds per acre (Abid, 2010). The small farmer reported the main reason of the adoption of Bt cotton was high yield. When they analyzed the data, it was revealed that the adoption of Bt cotton was due to three main reasons, one it reduced the pesticide application, second it increased the seed cost and the third main reason of the adoption of Bt cotton was its high yield (Gouse et al., 2002). Subramanian and Qaim (2008) studied the impact of Bt cotton on poor households in rural India. They focused on insect resistant *Bacillus thuringiensis* (Bt) crops, especially Bt cotton. Bt technology had been adopted by a large number of farmers in the world. Data were collected from Kanzara village which was located in Akola district of Maharashtra the state with the largest area under cotton in India. They used a microeconomic modeling approach and comprehensive household survey data. Study showed that the farmers got a number of

benefits from Bt cotton in which some were insecticide savings, higher yields through reduced crop losses, and net revenue gains, in spite of higher seed prices. This study also showed that Bt cotton entails positive direct and indirect welfare effects in the rural economy. The study concluded that Bt cotton contributes to poverty reduction and rural development.

Cotton production in Pakistan faces a rising occurrence of diseases and insect's attack. In Pakistan agriculture is particularly susceptible to pest invasion and climatic extremes because of their capacity to stand any form of financial risks (Farooqi, 2010; Ahmad et al., 2016; Bakhtavar et al., 2015). While on the other side the cultivation of Bt cotton at a commercial level have gained importance, in terms of reduced pesticide use and costs and higher yields (Qaim and Matuschke, 2005). There are many extension organizations working in Pakistan for the introduction of Bt cotton among the farmers. However, despite all efforts, farmers have not been adopted to Bt cotton as desired by the extension organizations. A conventional cotton profit is reduced because of pesticide cost and insect attack losses. While in case of Bt cotton the pesticide costs and losses by pest attack were reduced. Keeping in view all these facts, the present study was designed for analysis of paradigm shift from Non-Bt to Bt cotton in tehsil (local administrative division) Jatoi. The objective of the study was to identify the different factors affecting the adoption of Bt cotton, which provides a guideline to extension organizations to develop better strategies in future for effective extension work towards Bt cotton production.

MATERIALS AND METHODS

The district Muzzafargarh (Punjab province) is about 34 km away from Multan cross, the river Chenab on its' east. It comprises of four tehsils namely Muzaffargarh, Ali pur, Kotaddu and Jatoi. Most of the area in tehsil Jatoi is canal irrigated. The climatic conditions of this area are suitable for successful in cotton cultivation. The basic objective of this research project was to conduct an analysis of paradigm shift from Non-Bt to Bt cotton in tehsil Jatoi. All the Bt and Non-Bt cotton farmers of this area were considered as the population of the study. The sample of the population was limited to 120 Bt cotton growers selected through random sampling technique. Four union councils (UCs) were selected randomly from the total 16 rural union councils. From each randomly selected rural union council, two villages were selected randomly. From each selected village, 15 Bt cotton growers were selected randomly. So, the total sample size of the respondents was consisted of 120 and the study was performed in the year 2014. The questionnaire formulated in Urdu language according to the requirement of the research objectives. But at the time of interview local language was used to ask the question from the respondents. The data were analyzed with the help of Statistical Package for Social Sciences (SPSS). Descriptive statistical such as frequencies, percentages, means, standard deviations and rank order were used for interpretation of the data. In order to know the relative ranking of various factors, their weighted scores were calculated by multiplying the score value allotted to each category of the scale with frequency count. Means were calculated as sum of values divided by number of observations. Then, factors were ranked taking their mean value into consideration.

RESULTS AND DISCUSSIONS

Survey results showed that less than half (47.5%) of the respondents were middle aged, followed by old (25%) and 27.5% respondents were young (Table 1). It means that nearly about half of the respondents related to the middle aged category. These results are more or less similar to those of Gangil and Dabos (2005) who reported that 44.5% of the respondents fell middle age (35-50 years) category, whereas 25% and 31.8% belonged to young and old age categories respectively. Many research studies have confirmed that education plays a vital role in the adoption process because it is easy to understand and getting acquired information by educated person than the illiterate one (Muro and Burchi, 2007; Fiaz et al., 2016). Table 1 also indicated that 30.0% of the respondents were primary to middle educated, and 26.7% of them were

illiterate. Among the literate respondents, 23.3% of them who had primary level. However, 20.0% of the respondents were up to secondary and higher secondary, respectively. An analysis of the above data signified that quite a good number (30.0%) of the respondents were educated (Table 1). These results were also more or less similar to that of Arshad (2007) who found that 58% of the respondents were illiterate, 14% of them had an education level up to primary and 11% of the respondents had up to the middle level of education. While, 8% and 9% of the respondents were matriculation and above matriculation.

Table 1. Distribution of respondents according to their age and education

Age (Years)	Respondents %	Education	Respondents %
Up to 35	27.5	Illiterate	26.7
36-50	47.5	Up to Primary	23.3
Above 50	25.0	Primary-Middle	30.0
Total	100.0	Matriculation and above	20.0
		Total	100.0
Source of income and area under cotton crop			
Source of income	Respondents %	Area under cotton cultivation (acres)	Respondents %
Crop farming	10	Up to 12.5	51.7
Crop and livestock farming	90	13-25	28.3
Total	100	Above 25	20.0
		Total	100.0
Total	100.0		

Results also showed that 51.7% of the respondents had up to 12.5 acres area under cotton, whereas 28.3% of the respondents having 13-25 acres under cotton cultivation, and 20.0% of the respondents were cultivating more than 25 acres of cotton (Table 1). According to the source of income the results showed that a large majority of the respondents (90%) had their source of income mainly from crop and livestock farming, followed by 10% of the respondents having crop farming as their sources of income. A small number of farmers were dependent on agriculture and service as their sources of income.

In Table 2 results showed that 42.5% of the respondents cultivated the Bt cotton and 75% were with Non-Bt cotton under the area of up to 5 acres, 30.0% of the respondent were Bt cotton growers and only 8.3% of the respondents were growing Non-Bt cotton under the area of 5 to 10 acres. Whereas, 27.5% of the respondents growing the Bt cotton and 15.8% of the respondents were growing Non-Bt cotton under the area of more than 10 acres.

Table 2. The area cultivated under Bt and on Bt cotton

Area	BT Cotton	Non-BT Cotton
(Acres)	Respondents (%)	Respondents (%)
Up to 5	42.5	75.8
5.1-10	30.0	8.3
Above 10	27.5	15.8
Total	100.0	100.0

Table 3 describes the distribution of respondent according to the source of information. Data showed that the level of dependency on pesticide companies were high (37.5%) to medium (35.8%) and ranks first with mean and standard deviation of 3.31 and 0.90, respectively. Only 10% respondents showed very high level of confidence on pesticide companies regarding source of information and 10% respondents out of 120 respondents didn't show their interest. Respondents that had got information from their neighbors or fellow farmers ranked 2nd in rank order with the mean 3.31 followed by progressive farmers, agricultural department (Extension) and pesticide dealers, that were ranked 3rd, 4th and 5th with the means 1.93, 2.72 and 2.37, respectively.

Table 3. Distribution of the respondents according to their sources of information

Source of information	Level of dependency (%)						Mean	Std. Deviation	Ranks order
	Very low	Low	Medium	High	Very High	No response			
Pesticide companies	0.0	6.7	35.8	37.5	10.0	10.0	3.31	0.90	1
Fellow farmers	3.3	10.8	39.2	32.5	6.7	7.5	3.31	0.90	2
Progressive farmers	6.7	24.2	30.0	22.5	13.3	3.3	1.93	1.07	3
Agriculture Department	6.7	10.8	25.8	3.3	3.3	50.0	2.72	1.01	4
Pesticide dealers	22.5	31.7	14.2	15.0	3.3	13.3	2.37	1.16	5
NGOs	15.0	33.3	24.2	3.3	0.0	24.2	2.21	0.81	6
Newspapers	22.5	25.0	13.3	2.5	3.3	33.3	1.59	0.70	7
Television	40.8	25.0	17.5	3.3	3.3	10.0	1.82	0.97	8
Fertilizer companies	15.0	37.5	3.3	3.3	0.0	40.8	2.72	1.01	9
Radio	35.8	30.0	9.2	2.5	2.5	20.0	1.92	0.73	10
Fertilizers dealers	19.2	40.8	0.0	3.3	0.0	36.7	2.37	1.16	11
Internet	14.2	23.3	1.7	1.7	0.0	59.2	1.78	0.71	12
Mobile	44.2	28.3	10.0	0.0	0.0	17.5	1.80	0.69	13

Respondents did not fully dependent on a single source of information regarding Bt cotton technology, but they got it from multiple sources. High (32.5%) to medium (39.2%) level of dependency has been found in fellow farmers, but in case of progressive farmers medium (30.0%) to low (24.2%) dependency level was found. This might be due to unavailability of approach for progressive farmers to small or poor farmers. Agricultural department (Extension) showed very poor performance in delivering Bt cotton technology information. Only 25.8% respondents were showed a medium level of dependency on agriculture department and 50% respondents

were refused to answer the question. This might be due to poor or lack of knowledge about Bt cotton technology information delivered by the extension agents. The need is to give the training on Bt cotton technology to extension agents that should help the farmers. Respondents also showed low (31.7%) to very low (22.5%) dependency levels on pesticide dealers for Bt cotton technological information and ranked at 5th on rank order with 2.37 mean and 1.16 standard deviation values. While electronic and print media received lowest number on rank order that indicates respondents did not had access to these technologies. This might be due to the poor financial situation of the respondents that kept them away from source of advanced information (Muddassir et al., 2016).

Table 4 described the respondent's reason for shifting from Non-Bt cotton to Bt cotton. The maximum respondents shifted from Non-Bt to Bt cotton variety for their higher yields. High yields ranked 1st on rank order. The dependency level was found high (49.2%) and tends toward medium (27.5%) with 3.43 as mean and 0.92 as standard deviation, respectively. The optimum pesticide use was ranked 2nd and respondents showed high (45.8%) level of dependency with mean and

standard deviation of 3.39 and 0.92, respectively, followed by insect/pest control and resistivity against bollworm in Bt cotton than Non-Bt. Respondents also showed medium (58.3%) level dependency on plant height and it ranked 5th with mean and standard deviation of 3.20 and 0.88, respectively. Other factors such as less irrigation, multiple harvest, growth rate and high potential ranked 15th, 14th, 13th and 12th with means 2.84, 2.89, 2.90 and 2.99, respectively. The level of dependency was also in medium categories, indicated that these factors did not seriously contribute to shift from Non-Bt to Bt cotton.

Table 4. The distribution of the respondents according to factors which shifted them to grow Bt cotton instead of non Bt cotton

| Factors | Level of dependency (%) | | | | | | Mean | Std. Deviation | Rank Oder |
	Very low	Low	Medium	High	Very High	No response			
High yields	3.3	13.3	27.5	49.2	6.7	0.0	3.43	.92	1
Optimum pesticide use	3.3	13.3	30.8	45.8	6.7	0.0	3.39	.92	2
Insect/ pests control	3.3	16.7	32.5	37.5	10.0	0.0	3.34	.98	3
Resistance variety	3.3	16.7	35.0	35.0	10.0	0.0	3.32	.98	4
Plant height	3.3	10.8	58.3	17.5	10.0	0.0	3.20	.88	5
Good quality of seed	5.0	9.2	56.7	25.0	4.2	0.0	3.17	.78	6
Low cost of inputs	4.2	16.7	40.8	34.2	4.2	0.0	3.17	.88	7
Farm income raise	6.7	17.5	35.8	36.7	3.3	0.0	3.13	.97	8
Number of boll	3.3	10.0	54.2	25.8	3.3	3.3	3.12	.81	9
Production increase	4.2	22.5	40.0	28.3	5.0	0.0	3.07	.90	10
More profit	7.5	20.0	39.2	30.0	3.3	0.0	3.02	.97	11
High potential	3.3	25.0	47.5	17.5	6.7	0.0	2.99	.91	12
Growth rates	7.5	20.0	50.8	18.3	3.3	0.0	2.90	.90	13
Multiple harvesting	6.7	20.8	55.8	10.0	6.7	0.0	2.89	.91	14
Less irrigation requirement	3.3	20.0	62.5	7.5	3.3	3.3	2.84	.74	15

It is evident from the data given in Table 5, that high seed rate, non-availability of seed and high requirements of fertilizer were ranked 1st, 2nd and 3rd on the basis of threats. They fell between high and very high category but inclined towards high category with mean values of 2.99, 2.98 and 2.97, respectively on the basis of threats. Un-approved Bt cotton

variety, save anti-law methods and lack of knowledge about Bt technology were in between low and medium category but inclined towards low category having mean values of 2.92 and 2.76 and 2.57, respectively. Climatic variations (temperature, humidity and wind) inclined towards low category with mean values of 2.35, respectively.

Table 5. Distribution of the respondents according to the threats which they are facing in Bt cotton cultivation

Threats	Very low	Low	Medium	High	Very High	No response	Mean	Std. Deviation	Rank order
High seed rate	3.3	15.0	33.3	14.2	3.3	30.8	2.99	0.90	1
Non availability of seed	0.0	21.7	46.7	13.3	3.3	15.0	2.98	0.76	2
High fertilizer requirements	5.0	20.8	50.8	20.0	3.3	0.0	2.97	0.83	3
Un-approved Bt verities	4.2	17.5	54.2	10.8	3.3	10.0	2.92	0.79	4
Save anti-law methods	14.2	20.8	43.3	18.3	3.3	0.0	2.76	1.02	5
Lack of Bt technology	6.7	10.0	15.0	6.7	0.0	61.7	2.57	0.98	6
Climatic variations	14.2	35.8	40.0	3.3	0.0	6.7	2.35	0.78	7

As Bt cotton comprises of endogenous insecticidal protein to kill insect, so it reduced the number of sprays and cost of production, that ultimately alleviating the poverty of the rural people. It is evident from the data given in Table 6 that reduction in pest infestation and increase in production were ranked 1st and 2nd on the basis of opportunities and fell between high and very high category but inclined towards high category with mean values of 3.34 and 3.20, respectively on the basis of opportunities, reduction in health hazards and increase in net annual income were in between low and medium category, but inclined towards low category having mean values of 3.08 and 3.00, respectively. Reduction in pesticide applications and less cost of production inclined towards low category with mean values of 2.99 and 2.98, respectively. On the basis of opportunities, availability of certified seeds' mean value of 2.24 showed that it inclined towards very low category.

Table 6. Distribution of the respondents according to the opportunities in Bt cotton cultivation

Opportunities	Very low	Low	Medium	High	Very High	No response	Mean	Std. Deviation	Rank order
Reduced pesticides applications	0.0	27.5	42.5	26.7	0.0	3.3	3.34	0.92	1
Increase in production	5.0	10.0	53.3	23.3	8.3	0.0	3.20	0.86	2
Reduction in health hazards	4.2	23.3	40.8	25.0	6.7	0.0	3.08	0.95	3
Increase in net annual income	0.0	16.7	70.0	9.2	4.2	0.0	3.00	0.64	4
Less cost of production	3.3	16.7	61.7	15.0	3.3	0.0	2.98	0.77	6
Availability of certified seed	3.3	14.2	22.5	11.7	6.7	50.0	2.24	0.96	7

CONCLUSIONS

The study aim was to identify the different factors affecting the adoption of Bt cotton and analysis of Paradigm shift from Non Bt to Bt cotton in Tehsil Jatoi. It was concluded that cotton growers were highly dependent on pesticide companies for obtaining agricultural information. The higher crop yield was the major factor which shifted the farmers to grow Bt cotton. Unapproved Bt, high fertilizer requirement and non-availability of seed were the threatening factors facing in Bt cotton cultivation. Opportunities of less use of pesticides, increase in production and net annual income, reduction in health hazards, less cost of production and availability of certified seeds were found to be the main shifting drivers of Bt cotton. Results suggested that extension organizations should develop better strategies in future for effective extension services towards the Bt cotton production.

REFERENCES

Abid M., 2010. Productivity difference of Bt cotton on different farm sizes Unpublished thesis of M.sc (Hons) Agricultural Economics, University of Agriculture Faisalabad.

Ahmad W., Noor M.A., Afzal I., Bakhtavar M.A., Nawaz M.M., Sun X., Zhou B., Ma W., Zhao M., 2016. Improvement of Sorghum Crop through Exogenous Application of Natural Growth-Promoting Substances under a Changing Climate. Sustainability. 8, 1330; http://dx.doi.org/10.3390/su8121330.

Arshad M., Sohail A., Tayab A., Hafeez F., 2007. Factors influencing the adoption of Bt in Pakistan. Journal of Agriculture and Social Sciences, 3 (8): p. 121-126.

Bakhtavar M.A., Afzal I., Basra S.M.A., Ahmad A.H., Noor M.A., 2015. Physiological Strategies to Improve the Performance of Spring Maize (Zea mays L.) Planted under Early and Optimum Sowing Conditions. PLoS ONE 10: e0124441. doi:10.1371/journal.pone.0124441.

De Muro P., Burchi F., 2007. Education for Rural People and Food Security a Cross Country Analysis. Food and Agriculture Organization of the United Nations, Rome. ftp://ftp.fao.org/docrep/fao/010/a1434e/a1434e.pdf

Farooqi Z., 2010. Comparative performance of Bt cotton: economic and environmental aspects. Unpublished M.Sc. (Hons.) Thesis, Department of Environmental and Resource Economics, University of Agriculture, Faisalabad, Pakistan.

Fiaz S., Mobeen N., Noor M.A., Muddassir M., Mubushar M., 2016. Implications of Irrigation Water Crisis on Socioeconomic Condition of Farmers in Faisalabad District, Punjab, Pakistan. Asian Journal of Agricultural Extension, Economics & Sociology, 10(1): p. 1-7.

Gandhi V.P., Namboodiri N.V., 2006. The adoption and economics of Bt cotton in India. Published by Indian institute of management Ahmadabad.

Gangil D., Dabos Y.P.S., 2005. Effect of socio-economic variables on the level of knowledge and training needs of livestock. Kurukshetra, 53(4): p. 11-15.

Gouse M., Kirsten J.F., Jenkins L., 2002. Bt Cotton in South Africa: Ado ption and the impact on farm incomes amongst small-scale and large scale farmers Working paper. 2002-15.

Mahmood A., Farooq J., 2011. Introgression of desirable characters for growing cotton in Pakistan. Cotton research institute, Ayub agriculture research Faisalabad, Pakistan.

Muddassir M., Jalip M.W., Noor M.A., Zia M.A., Aldosri F.O., Zuhaibe AuH., Fiaz S., Mubushar M., Zafar M.M., 2016. Farmers' Perception of Factors Hampering Maize Yield in Rain-fed Region of Pind Dadan Khan, Pakistan. Journal of Agricultural Extension, 20(2): 1-17. http://dx.doi.org/10.4314/jae.v20i2.1

Qaim M., Matuschke I., 2005. Impacts of Genetically Modified crops in developing countries: A survey, Quarterly Journal of International Agriculture, 4(15): p. 207-227.

Shafiq-ur-Rehman M., 2009. Agriculture biotechnology annual: Islamabad, Pakistan (Gain Report No. PK9012). Washington, DC: USDA Foreign Agriculture Service, Global Agricultural Information Network (GAIN).

Subramanian A., Qaim M., 2008. The Impact of Bt Cotton on Poor Households in Rural India. Nature Precedings: hdl; 10101/npre, p. 1-23.

***Govt. of Pakistan. 2013. Pak Economic Survey 2011-12. Ministry of Finance, Economic Advisor's Wing, Islamabad, Pakistan.

14

PLASTICITY AND ADAPTABILITY OF TUNCELI GARLIC (*Allium tuncelianum* KOLLMAN) UNDER SEMIARID ECO-LOGICAL CONDITIONS OF SOUTH-EAST ANATOLIA

Süleyman KIZIL[1], Khalid Mahmood KHAWAR[2]

[1]Dicle University, Faculty of Agriculture, Department of Field Crops, 21280 Diyarbakir, Turkey
[2]Ankara University, Faculty of Agriculture, Department of Field Crops, 06100 Ankara, Turkey

Corresponding author email: suleymankizil@gmail.com

Abstract

Turkey, with important cultural heritage and rich history and enormous plant diversityhas poor agricultural practices that are making difficult to conserve many of the endemic local plant taxon. The soil conservation practices are not sufficient as the farming practices are more often sowing of soil depleting rather conserving. Allium tuncelianum (Kollman) Ozhatay, Matthew & S iranecivernacular "Tuncelisarımsağı" is an endemic specie of garlic native to the Eastern Anatolian province Tunceli, where temperate climate is dominant and the people in general and farmers in particular are not well aware of good farming practices. There is need to develop and introduce more new practical propagation and multiplication approaches for its conservation at naturalhabitat and outside without endangering surrounding environment. This study reports effect of four planting densities andintra-row spacingon some agronomical characteristics of A. tuncelianum yield and some agronomic characteristics. The study was carried under warm semi-arid ecological climatic conditions of Southeast Anatolia ensuring minimum soil depletion effects on the environment. Yield components like plant height, leaf length, leaf width, bulb diameter, flower inflorescence, bulb weight and number of scales onbulb changedsignificantly with range of 101.3-115.8 cm, 37.2-40.4 cm, 2.55-1.61 cm, 3.47-3.85 cm, 8.90-8.87 cm, 36.0-48.1 g and 1.67-1.71 respectively. These values did not show a significant difference with the yield component values at original habitat of the plant at Tunceli. The results of the study are very encouraging and suggest that the plant has large and increased plasticity with easy tolerance and adjustment for differences in climatic without significant loss in yield.

Key words: agronomy, bulb size, bulb yield, cultivation, Tunceli garlic, wild plants.

INTRODUCTION

Turkey lies on cross road to Asia, Africa and Europe and has very important cultural heritage and rich history that is spread all over. This transition has allowed Turkey to develop enormous plant diversity as well with more than 12.057 taxons growing on its soils. These taxons include many of the endemic plant species as well. However, large numberof farmers arenot well aware of good agricultural practices that are making very difficult to conserve both soiland endemic land races of plants (Firat and Tan, 1995). The locally employed soil conservation practices are not adequate, most often and the farming practises are leading to soil depletions on huge scale (Gunal et al., 2015).

No soil protection and poor reclamation work could lead to loss of locally available germplasm to huge scale. As mostof the cultivated land have steep mountainous slopes, farming practices are unsuitable for cultivation. Turkey needs in time, approaches to protect both its soil and fast eroding local germplasm through organised and systematic scientific studies.

The Fertile Crescenten compassing western Iran, Jordan, Syria, Palestine, Israel, Iraq, Lebanon, and southeast Turkey is an area of mega diversity and centre of origin and domestication of many important food, pasture, rangeland, feed, horticultural and medicinal temperate zone plant species (Harlan, 1992). Local land resources are fragile andemphasize importance of maintenance and conservationofbiodiversity without loss of productivity (Held and Cummings, 2014).

The genus *Allium* comprises around 750 species (Stearn, 1992) and issubdivided into 15 subgenera and 72 sections (Friesen et al., 2006; Hirschegger et al., 2010). It is a group of petaloid monocotyledonous genus with bulbs enclosed in fibrous or membranous tunics, which are often or almost tepal free, grow in a subgynobasic way and most of them producecysteine sulphoxides. It has center of diversity located in southwest & Central Asia and North America (Friesen et al., 2006). Many *Allium* species are grown for diverse purposes including their use in foods, in pharmaceutics for their medicinal characteristics (Fritsch and Friesen, 2002) and ornamental purpose that are popular with gardeners (Block, 2010).

A. tuncelianum Kollmanvernacular name "Tuncelisarimsağı" (Kizil et al., 2014) is an endemic garlic specienative to Munzur Mountains (2359 m) largely Ovacik, district of Tunceli province lying between Karasu and Murat dells and limited region lying in between Erzurum and Sivas provinces. It grows competing with other plants in natural environment at its habitat (Figure 1) on soil rich in metamorphic sedimentary volcanic and intrusive rocks (Baktir, 2005). The climate of the region is highly influenced by acold temperate continental climate/Mediterranean continental climate (Dsa) as described under köppen climate classifications (Anonymous, 2016 a, b, c) with extreme winter temperatures and heavy snowfalls. The climate of the area varies from warm to hot and dry during summer. Precipitation in the Munzur Valley and Ovacik district of Tunceli province is variable and ranges between 600-1.000 mm annually, with very little precipitation during summer months. The soils of region areclay loam, sandy clay loam and clay, with pH of 7.0 andwith 1-2% organic matter (TVICOM, 2005).

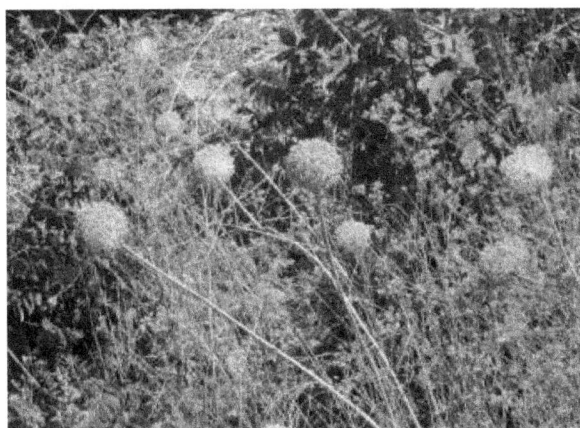

Figure 1. *Allium tuncellianum* plants growing on metamorphic sedimentary volcanic and intrusive rocks of Munzur Mountains competing with other plants in natural environment

A. tuncelianum contain 1-2 cloves that are locallyused in cuisine for culinary purposeand the cloves could be stored for a long time at 18-20°C. Appropriate agronomic techniques for its culture are still tobe developed. It is locally collected from wild regions for export tobig cities including Istanbul and abroad (European countries); that has been abode to a large number of immigrants from Turkey. Wild populations of *A. tuncelianum* are at a risk of extinction in Turkey, because ofcompetition among neighbouring species, land erosion, human activities like their irrational collection from wild for use in pharmaceutics, and their use for culinary purpose or salad (Kiralan et al., 2013; Aasim, 2015). Increased human activities like construction of houses, highways, cutting of forests, prairies and wilds for farming activities etc. have resulted in rapid or subtle un noticed habitat destruction of *A. tuncelianum*. These activities have also contributed to increased pollution in the area. Cumulative effects of all activities are gradually changing the life style and structure of local people living around the habitat of Tunceli garlic (Pers observations).

Commercially, *Allium* species are propagated either by seeds, cloves or bulbs (De Hertogh and Le Nard, 1993). All plants have different requirements for growth and development. Correct type of soil, planting depth and inter plant space influences uptake of water, nutrition, air, light by plants for growth (Amjad and Ahmad, 2012). However, very limited information is available for successful agronomic culture of *A. tuncelianum* (Kizil and Khawar, 2015). Thus, development of an agronomic technique to conserve and systematically grow Tunceli garlic for obtaining of uniform growth with high yield would be very desirable. It will help to meet ever rising demand of Tunceli garlic for local consumption in foods and industries.

Inthis regard, the study was conducted to determine effects of planting densities on some agronomic characteristics of *A. tuncelianum* under cold semi-arid climatic conditions of the South-Eastern Anatolian plains (BSk according to Köppen classification systems), compared to its habitat at Tunceli province in Eastern Anatolia (Dsa- according to köppen climate classification system), where it grows sparsely under cold temperate climatic conditions (BSk - according to Köppen classification system) on soils rich in metamorphic, volcanic, and sedimentaryrocks (Anonymous, 2016 a, b, c).

MATERIALS AND METHODS

Field studies were conducted under Diyarbakır ecological conditions at the Department of Field Crops, Faculty of Agriculture (latitude 37° 53' N and longitude 40° 16' E, 680 m), Dicle University during 2011-2012 and 2012-2013 growing seasons using bulbs of *A. tuncelianum* purchased from a local producer at Ovacik District of Tunceli, provinceTurkey.

The soil of the experimental area was sandy-loam which was deficient in organic matter (0.41%), available phosphorus (0.17%) and potassium (7.98%), with pH of 8.08. Climatic conditions in the experimental years, with mean temperature, relative humidity and total precipitation from September to July for 2011-12 was 15.4°C, 55.9%, and 625.7 mm, for 2012-13 15.0°C, 54.6%, and 431.6 mm, respectively. Long term with September to July, mean temperature, relative humidity % and total precipitation were 15.3°C, 55.6% and 495.4 mm, respectively (Table 1). Experimental fields were ploughed to achieve uniform texture and structure and watered three

days before planting. The Tunceli garlics were planted in the soil after field capacity of 0.34 bar of suction pressure was established after downward drainage of excess water at the time of planting. No fertilizer was applied before planting or during growth of Tunceli garlic. Planting was done with row spacing of 45 and 70 cm and plant spacing of 10 and 20 cm. The experimental design was a randomized complete block design with three replications for each experimental year with 48 bulbs for 20 cm and 63 bulbs for 10 cm spacing in each plot. Plots size was kept 5.4 m^2 (1.8 m × 3 m) and 8.4 m^2 (2.8 × 3 m) in each of the experiment. Hand planted bulbs at a depth of 5-8 cm had diameter of 2 to 4 cm. The plots were weeded when required. The plots were harvested manually on 5th June, 2012 for the first year and 18th June, 2013 for the second year. Plant height, plant stem diameter, leaf length, leaf width, leaf-less stem length, flower table diameter, bulb diameter, bulb circumference, bulb weight and number of bulbilsper bulb were investigated in the study.

Data obtained in the study were analysed statistically, using MSTAT - C (Michigan State University) computer program, and the means were grouped, using LSD test or t test at 0.05 level of significance.

Table 1. Means of temperature, humidity and precipitation at the site of experimentation for long years, 2012 and 2013 years

Months	Mean temperature (°C)			Humidity (%)			Precipitation (mm)		
	Long Years	2012	2013	Long Years	2012	2013	Long Years	2012	2013
January	2.7	2.4	2.7	78.9	84.5	83.8	70.3	78.3	82.2
February	4.5	2.0	6.1	76.8	68.2	82.3	53.7	74.4	85.2
March	9.0	5.2	9.5	67.4	58.6	62.7	61.1	44.0	19.8
April	14.1	15.2	14.5	66.0	58.4	63.6	64.3	26.2	39.4
May	19.2	19.6	19.0	56.4	58.2	61.7	57.9	41.0	98.0
June	26.1	27.6	26.7	31.7	28.0	27.6	10.6	7.0	2.8
July	30.9	31.2	31.2	22.9	21.1	19.4	0.5	1.6	0
August	30.1	31.0	30.4	20.8	20.9	19.0	0.0	0	0
September	25.0	26.1	24.5	29.7	23.3	25.0	6.5	1.8	0.0
October	17.8	18.5	17.0	47.6	55.1	28.1	50.1	107.4	0.0
November	9.8	12.0	11.4	65.2	77.3	68.8	53.0	83.2	53.8
December	4.5	5.1	-3.4	78.9	85.4	83.6	67.4	160.8	50.4
Mean	**16.1**	**16.3**	**15.8**	**53.5**	**53.3**	**52.1**	**495.2**	**625.7**	**431.6**

Source: State Meteorology Institute (Diyarbakir, Turkey)

RESULTS AND DISCUSSIONS

The effect of temperature on *A. tuncelianum* bulb germination and sprouting in the fields

were very clear. Allium species, depending on their origin, divided into four types. *A. tuncelianum* germinate over a wide temperature range (5-25 °C) (De Hertogh and Le Nard, 1993).

Statistical analysis in Table 2 indicated that there was no significant interaction effect between plant densities and years, while main effects of years significantly influenced plant height, stem diameter, leaf length, leaf width and plant densities that influenced leaf width.

Table 2. Results of analysis of variance and F values of the investigated characteristicsof *A. tuncelianum*

Source of Variance	Plant height (cm)	Stem diameter (cm)	Leaf Length (cm)	Leaf width (cm)	Scape length (cm)	Flower diameter (cm)	Bulb diameter (cm)	Bulb circumference (cm)	Bulb weight (cm)	Number of scale bulb
Years	101.24**	8.58*	9.41*	69.41**	39.44**	0.02	2.09	0.88	38.58**	19.79*
Plant densities	2.35^{ns}	0.78	2.16	10.68**	3.43	6.25**	9.97**	2.96	8.38**	2.01
Interaction	1.92	0.30	2.08	2.47	2.16	5.08*	7.12**	2.99	5.11*	2.36

*Significant at 0.05 probability level; ** Significant at 0.01 probability level; ns non-significant

Figure 2. *Allium tuncellianum* presents healthy plants in the fields showing no signs of stress in the new semi-arid climate of Diyarbakır

All bulbs developed healthy plants in the fields showing no signs of stress in the new semi-arid climate of Diyarbakır (Figure 2). Mean plant height for first experimental year was 101.3 cm, while it remained 115.8 cm during second year. This showed that more precipitation at active growing periods of *A. tuncelianum* during 2nd year of growth influenced plant height positively. A normal vertical crop with no lodging or bending of plants was noted during experimentation avoiding loss of yield showing plasticity of plants with the new environment.

Mean values of plant stem diameter duringfirst year was recorded as 1.26 cm and second year as 1.08 cm, mean values of leaf length was recorded as 37.2 cm during first year and 40.4 cm during second year, and mean values of leaf width were recorded as 2.55 cm during first year and 1.61 cm during second year (Table 3). Increase and decrease in plant density did not influence plant height, stem diameter, leaf length, scape length, bulb circumference and number of scales statistically.

Table 3. Mean values of plant height, plant stem diameter, leaf length and leaf width obtained from different plant densities of *A. tuncelianum*

Planting Densities (cm)	Plant height (cm)			Stem diameter (cm)			Leaf length (cm)			Leaf width (cm)		
	2011-12	2012-13	Mean	2011-12	2012-13	Mean	2011-12	2012-13	Mean	2011-12	2012-13	Mean
45 × 10	100.0	112.6	106.3	1.21	1.00	1.11	38.7	38.9	38.8	2.35	1.37	1.86 C
45 × 20	106.6	114.9	110.8	1.35	1.13	1.24	39.7	41.7	40.7	2.88	1.83	2.36 A
70 × 10	102.7	120.1	111.5	1.26	1.07	1.16	34.0	40.7	37.3	2.26	1.63	1.95 BC
70 × 20	96.0	115.8	106.0	1.20	1.13	1.17	36.5	40.4	38.5	2.71	1.60	2.15 AB
Mean	101.3 A	115.8 B		1.26A	1.08B		37.2 B	40.4 A		2.55 A	1.61 B	
LSD (0.05)	Years: 9.57			ns			Years: 2.15			Years: 0.20: Plant density: 0.21		

Means within a column followed by the same letter are not significantly different according LSD test at $p \leq 0.05$.

Leaf length andwidth of *A. tuncelianum* are important for improving the bulb; as they affect accumulation of carbohydrate and other essentials necessary for plant growth. Moreover, in the experiment, second year mean value (40.4 cm) was higher compared tothe first year (37.2 cm) values. Conversely, leaf width recorded during first year is higher compared to that recorded during second year (Table 3). Increased precipitation inactive growth periods during first year promoted vegetative growth especially the leaf width compared to leaf length during first year. Low moisture profile during 2nd year promoted elongation in leaves, a cause of moisture stress and reduction in leaf width. This resulted in low leaf area during 2nd year. It is assumed that increased leaf area during first year along with increased width resulted in interception of more solar energy for photosynthesis had positive bearing on all growth parameters according to the findings of Richards (2000).

The differences between scapes (is a long leafless internode forming the basal part or the whole of a peduncle in garlic) lengths were statistically significant. A significant interaction was noted between planting densities and interaction of year × planting densities for flower table diameter and bulb diameter. Mean values of scape length were determined as 76.9 cm during first experimental year and it was measured 84.6 cm during second year (Table 4).

Table 4. Mean values of stem length, flower table diameter and bulb diameter obtained from different plant densities of *A. tuncelianum*

Planting Densities (cm)	Scape length (cm)			Flower table diameter (cm)			Bulb diameter (cm)		
	2011-12	2012-13	Mean	2011-12	2012-13	Mean	2011-12	2012-13	Mean
45 × 10	77.6	83.5	80.6	8.67 bcd	8.60 cd	8.63	3.73 a	3.73 ab	3.79
45 × 20	83.6	85.1	84.4	8.27 d	8.97 bc	8.62	4.28 a	4.00 a	4.14
70 × 10	75.1	88.5	81.8	9.49 a	8.97 bc	9.23	2.60 c	3.77 ab	3.18
70 × 20	71.3	81.6	76.5	9.15 ab	8.83 bc	8.99	3.27 b	3.90 a	3.58
Ort.	76.9 B	84.6 A		8.90	8.87		3.47	3.85	
LSD (0.05)	Years: 3.59			İnt.: 0.52			Int.: 0.55		

Means within a column followed by the same letter are not significantly different according LSD test at $P \le 0.05$.

It was assumed that these differences were due to bulb diameter at planting and differences of precipitation received during experimental years. In respect to flowering table diameter, the maximum table diameter of 9.49 cm was obtained from plants in the scheme of 70 × 10 cm plant spacing during first experimental year, while the minimum table diameter of 8.60 cm was noted on 45 × 10 cm planting density during second year. No statistical difference was noted between years for table diameter. In the experiment, mean values of bulb diameter changed between 2.60 cm and 4.28 cm. The maximum bulb diameter was obtained from 45 × 20 cm plant density during first year, while the minimum bulb diameter was obtained from 70 × 10 cm density during first year of experiment (Table 4).

The higher yield and better control of over or under sized bulbs could be obtained if plants are grown at optimum (45 × 20 cm plant spacing) density. Bulb neck diameter, mean bulb weight and plant height decreased as population density of the plants increased. Total bulb yield could be increased as population density increases and depends on plant species (Kantona et al., 2003).

Only 10% of the whole Tunceli region (the natural habitat of Tunceli garlic) is arable with approximately 9% allocated to wheat, barley and rest to industrial crops like cotton, tobacco and sugar beets, small crops and vegetables. Moreover, rapid urbanization and rural depopulation; mainly due to economic reasons marketing problem and poor cropping pattern, lack of land and poor agronomic practices used by farmers are major problems at habitat. This has led local people towards scattered animal farming. Primarily, owing to these reasons, no agronomic trial has been conducted for Tunceli garlic at habitat or other places forquality bulb production.

Table 5. Mean values of bulb circumference, bulb weight and number of scale bulb obtained from different plant densities of *A. tuncelianum*

Planting Densities (cm)	Bulb circumference (cm)			Bulb weight (g)			Number of scale bulb		
	2011-12	2012-13	Mean	2011-12	2012-13	Mean	2011-12	2012-13	Mean
45 × 10	14.52	14.07	14.52	35.7 bc	37.2 bc	36.5	2.53	1.27	1.90
45 × 20	17.13	17.10	17.13	39.2 b	51.6 a	45.4	2.40	2.00	2.20
70 × 10	25.63	16.20	20.92	30.2 c	49.1 a	39.6	2.67	1.87	2.27
70 × 20	12.77	17.53	15.15	39.0 b	54.5 a	46.8	3.07	1.70	2.38
Mean	17.51	16.22		36.0		48.1	2.67 A*	1.71 B	
LSD (0.05)	ns			Int.: 7.30			Years: 1.51		

Means within a column followed by the same letter are not significantly different according LSD test at $P \le 0.05$.

Bulb circumference was determined as 17.51 cm duringfirst year and 16.22 cm during the second year. Although average temperature during both years of study was almost similar, a reduced precipitation was noted during 2nd year of experiment (Table 5). This had influenced the bulb circumference. The bulbs failed to gather necessary nutrients for their growth, affecting growth of leaves with poor photosynthesis due to reduced accumulation of carbohydrates. This showed that Tunceli garlic circumference was clearly affected by amount of precipitation that supposedly affected regulation and accumulation of starch/sugar ratio and bulb circumference. The results are according to Miko et al. (2000), findings, stating that reduced soil moisture at high temperatures induced reduction of 60% yield. It may be mentioned that garlic has a relatively shallow root system. Among plant densities, bulb length mean values changed between 14.07 and 17.53 cm. The maximum bulb circumference was obtained using widest spacing. So many studies have revealed correlation between sugars content of bulbs and storage temperature in the field and *in vitro* conditions (Salama et al., 1990; Iraqi et al., 2005). The storage of garlic bulbs in cold environment eliminates dormancy and stimulates sprouting.

An investigation of biochemical variation correlated to carbohydrate metabolism is of importance, due tochanges of chlorophyll, carbohydrate contents, amylase and invertase enzymes activity during stratification. Kahsay et al. (2013) observed thatthe narrowest intra-row spacing results indecreased bulb length.When intra-row spacing is increased from 5 to 10 cm, bulb length also increased from 4.1 to 4.6 cm (Table 5). This can be explained by intra row spacing or distribution, structure, and abundance of coexisting plants in population. Interaction of these determine the growth behaviour of plants, including productivity. The pattern of this competition in fluences availability of nutrients and moisture needed for growth. This helps in increased storage of carbohydrates in bulbs in the form of sucrose and more complex oligosaccharides (mainly fructans with various degrees of polymerisation) with improved bulb diameter/circumference/weight. This clearly indicate that competition for nutrients in soil is increased in narrow intra row spacing and decreased in wide intra row spacing. Moreover, their availability to plants was more during first year with more precipitation. Ahmed et al. (2007) observed that increased moisture level in soil at irrigation interval of 3 days had positive effects on number of leaves per plant, plant height at maturity, bulb yield, bulb weight. An increase in irrigation intervals decreased soil moisture content and had negative effects on growth, yield and other components of garlic development.

The maximum bulb weight was obtained from plant densities of 70×20 cm intra row spacing, while the minimum bulb weight was observed on 70×10 cm intra row spacing during first year (Table 5). Bulb weight was affected by density of plant population and had relationship with the percentage light interception by the *A. tuncelianum* leaf canopy (Addai and Scott, 2011). The importance of optimisation results in two advantages of avoiding competition among plants and allows sufficient space for growthby fetching optimum amount of water, nutrients and light for efficient growth of Tunceli garlic (Awas et al., 2010). The results are in line with Kantona et al. (2003), whonoticedonion yield increase from 17.4 to 39.5 tha^{-1} when plant population per square meter was increased from 50 to 150. Conversely, Rekowska and Skupien (2007) reported higher yield of bulbs and green leaves of garlic from narrowintra-row spacing.

Moreover, *A. tuncelianum* develops as a single cloveand do not constitute any bulbs except over riding bulbilson the cloves that acts as a potential propagules. These are capable to grow into a complete plant subject to cultural practices. Number of overriding bulbils changed depending on bulb size. Three of four over riding bulbils were noted onlarger bulbs, whereas, one overriding bulbil was induced on smaller bulbs in general. Maximum number of overriding bulbils was determined as 2.67 during first year and 1.71 during the second year (Table 5). Bulbs planted in any sowing pattern induced large number of viable seeds (Figure 3). Again induction of larger bulbils during first year with more precipitation seemed to be influenced by the soil moisture regime during first and 2nd years.

There are no reported scientific studies about agronomic yield in natural habitat of Tunceli garlic. Crude estimated yield at its habitat has range of 6500-9590 kg.ha^{-1} (Tunceli Sarımsagı, 2016). However, the results of the study suggests that *A. tuncelianum* has great plasticity and could be successfully grown for economic and commercial production by planting outside cold and temperate climate of its natural habitat to hot semi-arid ecological conditions successfully. This suggests that the plant has large plasticity to changes in ecological conditions with more likelihood tosurvivein other novel environmental and climatic conditions. It is well known that plasticity and level of tolerance to changing environmental conditions will influence in future natural selection and effect diversification among plant species (Sultan, 2004; Scheiner, 1998) that promote adaptive diversity (Sultan, 2004; Pigliucci and Murren, 2003). However, it should be known that besides phenotypic plasticity, genetic adaptationhave definite role in persistence of plant species under new environment (Richter et al., 2012) and thereby conservation of new plant species.

Figure 3. Seeds harvested from growing plants under semi-arid conditions of Diyarbakır

CONCLUSIONS

It follows from the above discussion that Tunceli garlic could adjust itself in semiarid environment that offers new avenues for multiplication adaptability and improvement through breeding. It will be desired to carry out adaptation studies at other places and climates for better comparison and understanding of the plant. Designing of experiments relating to fertilization and irrigation could further help in improving Tunceli garlic yield under semi-arid conditions.

ACKNOWLEDGEMENTS

This work was supported by a grant (Project number: 110 O 703) from the Scientific and Technical Research Council of Turkey (TUBITAK).

REFERENCES

Aasim M., 2015. Adventitous bulblet regeneration of endemic Ovacik garlic (*Allium tuncelianum* Kollman, Ozhatay, Mathew, Sıraneci) using wintered half clove explant. Rom Biotech Lett 20(5): p. 10845-10851.

Addai I.K., Scott P., 2011. Influence of bulb sizes at planting on growth and development of the common hyacinth and the lily. Agriculture and Biology Journal of North America 2(2): p. 298-314.

Ahmed H.G., Magaji M.D., Yakutu A.I., Aliyu L., Singh A., 2007. Response of Garlic (*Allium sativum* L.) to Irrigation Interval and Clove Size in Semi-Arid, Nigeria. Journal of Plant Sciences 2: p. 202-208.

Amjad A., Ahmad I., 2012. Optimizing Plant Density, Planting Depth and Postharvest Preservatives for *Lilium longifolium*. Journal of Ornamental and Horticultural Plants 2(1): p. 13-20.

Anonymous, 2016a. http://www.gps-latitude-longitude. com/gps-coordinates-of-tunceli (Accessed 20 December 2016).

Anonymous, 2016b. http://www.gps-latitude-longitude. com/gps-coordinates-of-erzurum (Accessed 20 December 2016).

Anonymous, 2016c. http://www.gps-latitude-longitude.com/gps-coordinates-of-erzincan (Accessed 20 December 2016).

Awas G., Abdisa T., Tolesa K., Chali A., 2010. Effect of intra-row spacing on yield of three onion (*Allium cepa* L.) varieties at Adami Tulu agricultural research center (mid rift valley of Ethiopia). Journal of Horticulture and Forestry 2(1), p. 07-011.

Baktir I., 2005. *In vitro* micropropagation of *Allium tuncelianum*. In: Proceedings of the GAP IV. Agriculture Congress, p. 206-208 (in Turkish).

Block E., 2010. Garlic and Other Alliums.Allium Botany and Cultivation, Ancient and Modern. Royal Society of Chemistry 11: p. 4-6.

De Hertogh A., Le Nard M., 1993. The physiology of flower bulbs. Elsevier Science Publ., Amsterdam.

Firat A.E., Tan A., 1995. Turkey maintains pivotal role in global genetic resources. Diversity, 11: p. 61-63.

Friesen N., Fritsch R.M., Blattner F.R., 2006. Phylogeny and new intrageneric classification of *Allium* L. (Alliaceae) based on nuclear ribosomal DNA ITS sequences. Aliso-Rancho Santa Ana Botanic Garden, 22: p. 372-395.

Fritsch R., Friesen N., 2002. Evolution, domestication and taxonomy. In H. D. Rabinowitch and L. Currah [eds.], Allium crop science: recent advances. CABI Publishing, Wallingford, Oxfordshire, UK, p. 5-30.

Gunal H., Korucu T., Birkas M., Ozgoz E., Halbac-Cotoara-Zamfir R., 2015. Threats to Sustainability of

Soil Functions in Central and Southeast Europe. Sustainability 7: 2, p. 2161-2188.

Harlan J.R., 1992. Crops and man. 2nd ed. American Society of Agronomy, Madison, WI.

Held C.C., Cummings J.T., 2014. Middle East patterns: places, peoples and politics. Westview Press (Sixth ed.). http://samples.sainsburysebooks.co.uk/97808133 48780_sample_390419.pdf.

Hirschegger P., Jernej J., Trontelj P., Bohanec B., 2010. Origins of *Allium ampeloprasum* horticultural groups and a molecular phylogeny of the section Allium (Allium: Alliaceae). MolPhylogenetEvol 54(2): p. 488-497.

Iraqi D., Quy Le V., Lamhamedi M.S., Tremblay F.M., 2005. Sucrose utilization during somatic embryo development in black spruce: involvement of apoplasticinvertase in the tissue and of extracellular invertase in the medium. J. Plant Physiol. 162: p. 115-124.

Kahsay Y., Belew D., Abay F., 2013. Effect of intra-row spacing on yield and quality of some onion varieties (*Allium cepa* L.) at Aksum, Northern Ethiopia. African Journal of Plant Science 7(12), p. 613-622.

Kantona R.A.L., Abbeyb L., Hillac R.G., Tabil, M.A., Jane N.D., 2003. Density affects plant development and yield of bulb onion (*Allium cepa* L.) in Northern Ghana. J. Veg. Crop Prod. 8(2): p. 15-25.

Kiralan M., Rahimi A., Arslan N., Bayrak A., 2013. Volatiles in an endemic Allium species: *Allium tuncelianum* by headspace solid phase microextraction. J. Essent Oil Bear Pl 16:3, p. 417-420.

Kizil S., Icgil D.Y., Khawar K.M., 2014. Improved *In vitro* regeneration and propagation of Tunceli garlic (*Allium tuncelianum* L.) from sectioned garlic cloves, leaves and root explants. J. Hortic. Sci. Biotech. 89:4 p. 408-414.

Kizil S., Khawar K.M., 2015. Effect of planting depths on some agronomic characteristics of *Allium tuncelianum*. Published in Scientific Papers. Series B, Horticulture, Vol. LIX, p. 229-232.

Miko S., Ahmed M.K., Amans E.B., Falaki A.M., Ilyas N., 2000. Effects of levels of nitrogen, phosphorus and irrigation interval, on the performance and quality of garlic (*Allium sativum* L.). J. Agric. Environ. 1(2), p. 260-264.

Pigliucci M., Murren C.J., 2003.Genetic assimilation and a possible evolutionary paradox: can macroevolution sometimes be so fast as to pass us by? Evolution, 57(7): p. 1455-1464.

Rekowska E., Skupien K., 2007. Influence of flat covers and sowing density on yield and chemical composition of garlic cultivated for bundle-harvest. In: Kosson R., Szwejda J, Gorecka K. (eds). Vegetable Crops Research Bulletin Vol. 66/2007, Research Institute of Vegetable Crops. Skierniewice, Poland, p. 17-24.

Richards A., 2000. Selectable traits to increase crop photosynthesis and yield of grain crops. J. Exp. Bot. 51 (suppl 1): p. 447-458.

Richter S., Kipfer T., Wohlgemuth T., Guerrero C.C., Ghazoul J., Moser B., 2012. Phenotypic plasticity facilitates resistance to climate change in a highly variable environment. Oecologia, 169 (1): p. 269-279.

Salama A.M., Hicks J.R., Nock J.F., 1990. Sugar and organic acid changes in stored onion bulbs treated with maleic hydrazide. Hort. Science 25: p. 1625-1628.

Scheiner S.M., 1998. The genetics of phenotypic plasticity. VII. Evolution in a spatially-structured environment. Journal of Evolutionary Biology, 11(3): p. 303-320.

Stearn W.T., 1992. How many species of Allium are known? Kew Mag. 9, p. 180-182.

Sultan S.E., 2004. Promising directions in plant phenotypic plasticity. Perspectives in Plant Ecology, Evolution and Systematics, 6(4): p. 227-233.

Tuncelisarimsagi, 2016. http://tuncelisarimsagi.net/ shortcode/accordion.html. Accessed 24 January, 2016.

***TVICOM (Tunceli Valiliği Il Cevreve Orman Mudurlugu), 2005. Tunceli Il Cevre Durum Raporu. http://cdr.cevre.gov.tr/icd_raporlari/tunceli%2005.pdf Accessed 20, December 2016).

15

USE OF HAPTEN-CARRIER COMPLEXES FOR BENZIMIDAZOLE PESTICIDES IMMUNOASSAYS DEVELOPMENT

Veronica TANASA[1,2], Radu I. TANASA[3], Madalina DOLTU[2], Gabriela HRISTEA[4], Narcisa BĂBEANU[1]

[1]University of Agronomic Sciences and Veterinary Medicine of Bucharest,
59 Mărăşti Blvd., District 1, Bucharest, 011464, Romania
[2]Institute of Research and Development for Industrialization and Marketing of Horticultural Products - HORTING, 1A Intrarea Binelui, District 4, Bucharest,042159, Romania
[3]National Institute of Research "Cantacuzino", 103 Splaiul Independentei,
District 5, 050096, Bucharest, Romania
[4]National Institute of Research and Development in Electrical Engineering ICPE-CA,
313 Splaiul Unirii, District 3, 030138, Bucharest, Romania

Corresponding author email: vero.tanasa@yahoo.co.uk

Abstract

In order to develop immunoassays for pesticides detection, this work describes the choice of three different haptens that present structural similarity to benzimidazole molecule, methods for coupling them with two carrier proteins in order to make them immunogenic, a protocol for immunization of laboratory rodents with hapten-carrier complexes, and the evaluation of the specific antibody responses against haptens using an in-house developed immunoassay. Three carbendazim (methyl 2-benzimidazole-carbamate) derivatives bearing different functional reactive groups (-NH2, -SH and -COOH), namely 2-(2-Aminoethyl) benzimidazole (AEB),2-Mercaptobenzimidazole (2MB) and 2-Benzimidazole propionic acid (BPA), were coupled to keyhole limpet haemocyanin (KLH) and bovine serum albumin(BSA), respectively, mixed with immuno-adjuvants, and injected four times into Balb/C mice and Wistar rats for induction of specific immune responses.All three chemicals elicited a specific but weak antibodies response upon immunization with hapten-KLH complexes followed by serological testing by indirect ELISA with hapten-BSA complexes, and showed detectabledifferences in antibody titers with regard to number of inoculations, hapten structure and animal species. Whereas the AEB-KLH complex was the strongest, the 2MB-KLH complex was the weakest immunogenic in mice. However, the best animal responders allow the application of technologies for getting monoclonal antibodies against benzimidazoles, which can then be used for immunoassays development.

Key words: hapten-carrier, benzimidazole, immunogenicity, antibodies, pesticides.

INTRODUCTION

Pesticides play a major role in improving agricultural production through control of pest populations such as insects, weeds, and plant diseases. Unfortunately, the toxicological properties of pesticides provide a potential risks to humans, to the environment, and to non-target organisms that might be inadvertently exposed to such chemicals as well. In particular, pesticides pose risks to agricultural workers involved in mixing, loading, and application of pesticides, as well as to those who perform works in agricultural settings where pesticides have been applied (Winter, 2012). Despite their merits, pesticides are considered to be among of the most dangerous environmental contaminants because of their ability to accumulate and their long-term effects on living organisms. The presence of pesticides in the environment is particularly hazardous, and exposure to these pesticides leads to several health problems that range from asthma attacks, skin rashes, severe eye irritation and chronic disorders to neurological diseases (Aragay et al., 2012; Schrenk, 2012). In the European Union the use of pesticides is strictly regulated and all EU Member States apply the same evaluation procedures and authorization criteria, in order to place a plant protection product on the market. In this respect the European Union legislation has established a maximum residue level (MRL) for food and feed of plant and animal origin [Commission Regulations (EC) 396/2005, amended by Commission Regulations (EC) No. 149/2008] which is updated as necessary (Keikotlhaile and Spanoghe, 2011). The identification and quantification of pesticides are generally based on gas chromatography - mass spectrometry (GC-MS), liquid chromatography - mass spectrometry (LC-MS), or high performance liquid chromatography-mass spectrometry (HPLC-MS) (Nunes and Barcelo, 1999). These methods permit precise and accurate detection, and quantification of trace levels of hundreds of these chemicals. However, these conventional methods for pesticides monitoring require multiple steps for sample preparation and analysis, often including derivatization, highly trained personnel, expensive specialized equipment, and are time consuming (Schrenk, 2012). Immunoassays are based on the use of anti-pesticide antibodies (Ab) as the specific sensing element and that can provide concentration-dependent signals. Such assays appear to be appropriate for identification of a single pesticide or, in some cases, small groups of similar pesticides in food, feed and environmental matrices, as they are rapid, specific, sensitive and included in cost-effective analytical devices. Furthermore, they can be used and interpret in the field by operators with minimal training and, generally, do not require sophisticated equipment to be accomplished. Therefore, developing such immunoassays has gained popularity during the recent years (Fan and He, 2011; Liu et al., 2013).

On the other hand pesticides, organic compounds of molecular mass less than 1,000, are usually non-immunogenic, and hence do not elicit an immune response unless coupled with some macromolecules such as proteins. Therefore, it is necessary to modify these chemicals - known as **haptens** - by coupling them with macromolecules - known as **carriers** - in order to make a stable carrier-hapten complex. The carrier-hapten complexes can then be used to generate antibodies against pesticides (Dankwardt, 2000; Raman et al., 2002). Also, because of the small size of these pesticide molecules, a suitable immunoassay technique must be employed for their detection.

In this paper we report the preparation of hapten-carrier immunogenic complexes derived from benzimidazole pesticides and their use for stimulation of antibody responses in laboratory animals.

MATERIALS AND METHODS

a) Haptens selection strategy

We considered as a target structure the pesticide carbendazim (methyl 2-benzimidazole-carbamate), a well known and extensively fungicide used in agriculture and horticulture, and which has not longer been approved for use in the European Union starting with 2015.

In order to select some similar chemical structures that would be able to be used as immunogens, we used the public SuperHapten database (http://bioinformatics. charite.de/superhapten/) that offers details of over 7,250 possible immunogenic haptensand a percentage hierarchy of their 2-D similarity compared with the target structure (Günther et al., 2007).

We also accessed the HaptenDB (http://www.imtech.res.in/cgibin/haptendb/index.html), a bioinformatic database that includes similar information of over 1,080 haptens, including pesticides (Singh et al., 2006).

On the other hand, there have been recently reported results on the induction of antibody responses against carbendazim and similar compounds using commercial chemical structures having - NH_2, -COOH, - SH, -OH as functional groups for conjugation with protein carriers (Moran et al., 2002; Gough et al., 2011; Zikos et al., 2015).

Since customized synthesis of similar compounds to a pesticide target is quite expensive, and after collating all the information provided by the previously-described strategy, we have chosen to use the following three commercially available carbendazim derivatives:

1.2-(2-Aminoethyl) benzimidazole (AEB)

(Sigma-Aldrich, 98% purity), which has - NH_2 as a functional reactive group;

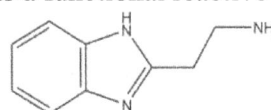

2.2-Mercaptobenzimidazole (2MB) (Sigma-Aldrich, 98% purity), which has - SH as a functional reactive group;

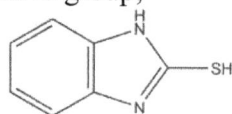

3.2-Benzimidazole propionic acid (BPA) (Sigma-Aldrich, 97% purity), which has - COOH as a functional reactive group.

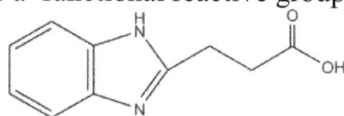

b) Haptens conjugation

Each hapten was conjugated to carrier protein keyhole limpet haemocyanin (KLH; mcKLH - a product of Thermo Scientific) and bovine serum albumin (BSA; the product of Serva Feinbiochemica GmbH &Co, or the Imject® BSA which is a product of Thermo Scientific), following previously established conjugation chemistry steps (Singh et al., 2004; Hermanson, 2013).

The AEB-KLH and AEB-BSA complexes were prepared through glutaraldehyde-mediated link chemistry, using a previously-described protocol (Hermanson, 2013).

The 2MB-KLH and 2MB-BSA complexes were prepared using commercial carriers (mcKLHand BSA Imject®, Thermo Scientific) with sulfo-SMCC (succinimidyl 4- [N-maleimidomethyl] cyclohexane-1-Carboxylate) (Pierce, Thermo Scientific) as a heterobifunctional linker, as directed by the manufacturer.

The BPA-KLH and BPA-BSA complexes were obtained using commercial kits (EDC Imject® Carrier Protein Kits Thermo Scientific), through EDC (1-ethyl-3- [3-dimethylaminopropyl] carbodiimide hydrochloride) - mediated chemistry, as directed by the manufacturer.

We used a molar ratio of 1:40 and 1:900 (carrier:hapten) for preparation of BSA-hapten complexes and KLH-hapten complexes, respectively.

Hapten-carrier complex formation was evaluated by UV-VIS spectrophotometry (Abad et al., 1999), recording the spectra in the regions of the maximum absorbance of the unconjugated and conjugated protein (λ max = 275-280 nm) and hapten, respectively, or by using 2,4,6-trinitrobenzene 1-sulfonic acid (TNBS) reagent (Sashidhar et al., 1994).

c) Animals and immunizations

Immunizations were carried out using female Balb/C mice (4 animals/group) and Wistar rats (2 animals/group) of 6-8 weks of age. The animals were reared in clean standard environment, withfood and water supplyad libitum. The experimental protocol with animals was performed in accordance with relevant institutional and national guidelines and regulations, and was approved by the Ethics Committee of The National Research Institute „Cantacuzino" (Application CE/ 36/04.02.2015).

For induction of antibody responses against haptens, the mice were inoculated with KLH-hapten complexes only, via subcutaneous (s.c) route first, and then with three booster immunisations via intraperitoneally (i.p.) route, every two weeks apart, with a combination of hapten-carrier complex (30-100 µg protein) adsorbed on immuno-adjuvants - [Al(PO$_4$)$_3$] plus Gerbu adjuvant MM (GERBU Biotechnik GmbH, Heidelberg, Germany) - ina 0.05-0.2 ml final volume/animal. There were four groups of mice used for immunization, of which 3 groups were inoculated with each KLH-hapten complex (KLH-AEB, KLH-2MB, KLH-BPA) and another one was inoculated with a mixture of all three complexes. The rats (one group) were immunized with a mixture of all KLH-hapten complexes + adjuvants, using volumes of 0.2-0.5 ml/animal. Other negative control groups of mice and rats were mock immunizedwith KLH + adjuvants only.

All the animals were bled from the tail veins before the immunization schedule first, and then one week apart from the second (day 14) and the forth injection (day 42). The serum was separated from blood by centrifugation and used for evaluation of the antibody response against haptens by ELISA (see below).

d) Hapten antibody response evaluation

Because KLH and BSA do not induce a detectable cross-reactive immuno-response, the BSA-hapten complexes were used as antigens for in vitro evaluation of antibody responses against haptens, by indirect ELISA. MaxiSorp

ELISA plates (Nunc, Roskilde, Denmark) were coated overnight at 4°C with the corresponding BSA-hapten complex (5 ug/ml), in carbonate-bicarbonate buffer (pH-9.6). After blocking with 1% caseinate in phosphate-buffered saline (PBS) and washing with PBS-Tween 20 (PBST, 4 times), twofold serial dilutions of the sera (in PBS), starting from 1/10, were incubated for 1 h at 37°C. After washing (4 times), the plates were incubated for 1 hour with either anti-mouse-IgG or anti-rat-IgG peroxidase conjugated secondary antibodies (SouthernBiotech, Birmingham, AL, U.S.A.), diluted (1/1000) in PBS. After incubation (1 hour at 37°C) and washing (4 times), the color reaction was developed with SigmaFast OPD (Sigma-Aldrich) according to the manufacturers instructions, for 15-30 min at 37°C, and absorbance was measured with a plate reader (Infinite F200, Tecan Austria GmbH) at $\lambda = 450$ nm.

RESULTS AND DISCUSSIONS

a) Haptens selection and conjugation

We found the hapten bioinformatics databases very useful for rapid orientation and down-narrowing of the screening, in order to find suitable hapten candidates. Particularly, SuperHapten Database offers 2-D/3-D structural details of possible immunogenic haptens, their scientific and commercial information, physicochemical properties, and a percentage hierarchy of their 2-D similarity compared with the target structure (Günther et al., 2007). These information are very important when choosing a suitable chemical structure able to be coupled with carriers and then to induce a suitable antibody response that can be further exploited for the development of reliable immunoassays.

In this way, in our experiments, we selected a total of five preliminary candidates that have similar structure to that of carbendazim and, therefore, possibly to be used as immunogens (Table 1). However, we decided to select the three final haptens (AEB, 2MB and BPA) after further taking into account the previously published results on this topic, and of the relevant trade information provided by the well known life science chemical substances manufacturers, also.

By scanning the absorbance of proteins, haptens and conjugates we found some subtle deviations from the unconjugated proteins, especially in the case of BSA-AEB complex formation (Figure 1) but obvious changes in absorbance spectra were not obtained with the some other conjugates, in agreement with another report (Gough et al, 2011).However, we found evidence that conjugation had taken place using TNBS reagent that strongly reacted with the ε-amino groups of L-lysine present in free carrier proteins, but less after hapten-protein cross-linking (data not shown).

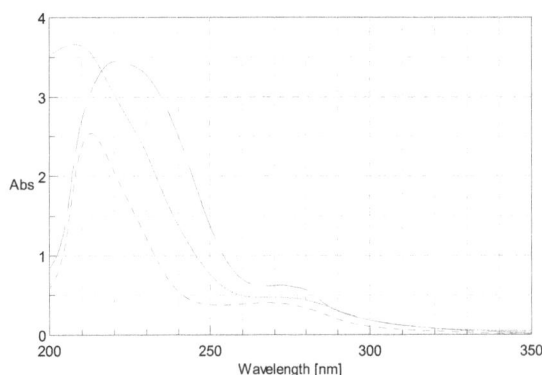

Figure 1. Evidence of a hapten-protein conjugation through spectrophotometry. Overlapped UV-VIS spectra demostrate a shifting from the spectrum of the BSA protein alone (in red), in comparison to BSA-AEB complex, either before (in blue) and after dialysis (in green)

By assuming that the molar absorptivity of haptens was the same for free and conjugated forms (Abad et al., 1999), apparent molar ratio was estimated as ~10, in the case of BSA-hapten complexes. We did not estimate this ratio in the case of KLH-complexes but, we assumed that because we used the same protocols, and the KLH is a very large protein (MW 4.5×10^5 to 1.3×10^7) with over 4,600 functional groups available for conjugation/mole in comparison to BSA (MW 67,000) that has over 100 such functional groups (Hermanson, 2013), it was enough hapten bound to carrier to induce an immune response.

b) Antibody responses against haptens

By ELISA, we detected antibodies that reacted with the corresponding hapten, albeit of low intensity, in all groups of mice, except the negative control group (Figures 2, 3, 4 and 5).

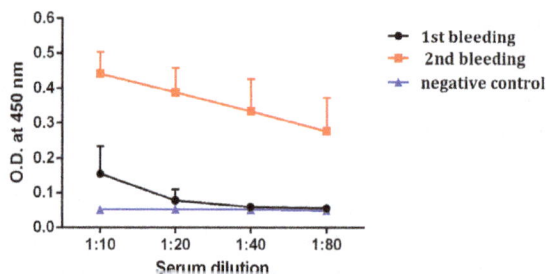

Figure 2. Antibody responses in Balb/C mice immunized against2-(2-Aminoethyl) benzimidazole (AEB). 1st bleeding was done after two inoculations and the 2nd bleeding was done after four inoculations. The results are presented as the mean optical density (O.D.) by ELISA with standard deviation bars of n = 4 mice/group

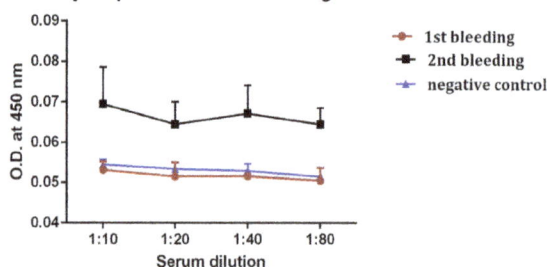

Figure 3. Antibody responses in Balb/C mice immunized against2-Mercaptobenzimidazole (2MB). 1st bleeding was done after two inoculations and the 2nd bleeding was done after four inoculations. The results are presented as the mean optical density (O.D.) by ELISA with standard deviation bars of n = 4 mice/group

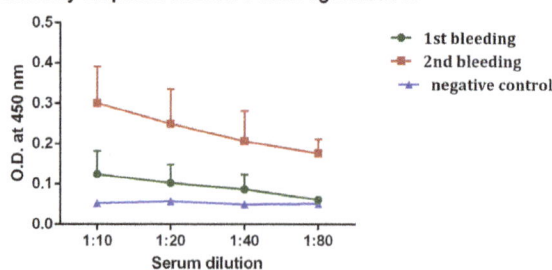

Figure 4. Antibody responses in Balb/C mice immunized against2-Benzimidazole propionic acid (BPA). 1st bleeding was done after two inoculations and the 2nd bleeding was done after four inoculations. The results are presented as the mean optical density (O.D.) by ELISA with standard deviation bars of n = 4 mice/group

Furthermore, we clearly found an increased response in antibodies, by indirect ELISA, after the 4th inoculation relative to the 2nd inoculation that is relevant for the immune maturation process that took place inside the body after repeated antigenic stimulation.

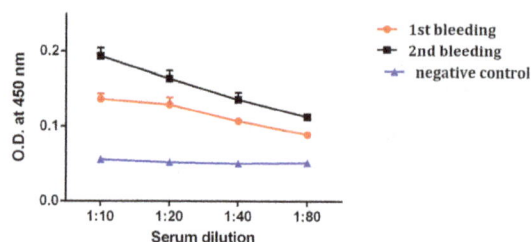

Figure 5. Antibody responses in Balb/C mice immunized against a mixture of AEB+2MB+BPA haptens. 1st bleeding was done after two inoculations and the 2nd bleeding was done after four inoculations. The results are presented as the mean optical density (O.D.) by ELISA with standard deviation bars of n = 4 mice/group

On the other hand, there were differences in the antibody response against each hapten, with better results when used AEB-carrier and a very low response against 2MB-carrier complex, though potent adjuvants for B-cell stimulation and differentiation were employed. Even when used a mixture of all there haptens we obtained the same poor results (Figure 5), that is the general characteristic of the immune responses against haptens.

The rats elicited a better antibody responses against a mixture of all haptens (Figure 6) in comparison to mice, the most probably due to differences in immunoreactivity between these animal species, a well known phenomenon.

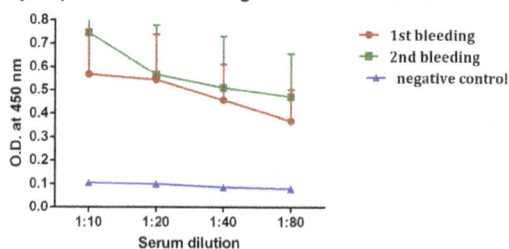

Figure 6. Antibody responses in Wistar rats immunized against a mixture of AEB+2MB+BPA haptens. 1st bleeding was done after two inoculations and the 2nd bleeding was done after four inoculations. The results are presented as the mean optical density (O.D.) by ELISA with standard deviation bars of n = 2 rats/group

All immunogenic complexes were well tolerated by the Balb/C mice, except KLH-AEB that induced a moderate, nodular dermatitis at the s.c. inoculation sites, but was letal within 24-48 hours post-innoculation when administered via i.p. route into an animal. Therefore, we followed the immunization protocol with KLH-AEB complex *via* s.c. route only, in both animal species.

How can one explain the differences in immunoreactivity to haptens within the same species? It has been shown that small molecules very often show low immunogenicity that is mainly due to the rapid breakdown of the molecule *in vivo* or clearance via the renal pathway (Moran et al., 2002).

Therefore, both the hapten selection and the choice of carriers have a qualitatively and quantitatively influence on the immune responses, including the secretion of antibodies. Because of these reasons, some rules have been established that likely would lead to make an immunogenic hapten close to the ideal (Goodrow and Hammock, 1998; Tong et al., 2007; Song et al., 2010; Goel, 2013). With this regard, the hapten should (i) have the structure, conformation and physicochemical properties as close to perfection as compared to the target chemical structure(s); (ii) have in its structure aromatic rings/hetero-aromatic rings/branched radicals, and at least one reactive functional group (-NH$_2$, -COOH, -OH, -SH) for attachment by covalent bonds to the carrier; (iii) keep the original conformation after coupling to a carrier and, if coupled to a carrier

molecule *via* a linking spacer, the latter must be immunologically unresponsive.

Therefore, the very low antibody response against 2MB can partially be explained by the simpler structure of this chemical compound and a less degree of similarity to carbendazim by comparison to AEB and BPA (see also the Table 1). Another possibility is that the linking reaction efficiency was much lower for 2MB in comparison to AEB and BPA, respectively. It is also possible that the hapten chemical structures were not properly exposed for recognition by the immune system cells. As a consequence, there were less B-cell epitopes available for processing that ultimately led to a low antibodies response. On the other hand, even in the case of poor antibody responses against haptens, it is possible to isolate monoclonal antibodies with the desired specificity by employing high-throughput strategies for fusion, screening and cloning (Chiarella and Fazio, 2008).

Currently, we have ongoing experiments for getting monoclonal antibodies recognizing these haptens, and which can then be used in immunoassays development for detection of benzimidazole pesticide residues in food and feed.

Table 1. Fivehapten candidates from top 30 structures most 2D-similar to carbendazim and possible to be used as immunogens based on the information provided by the Super Hapten database (http://bioinformatics.charite.de/superhapten/)

Name	Structure	ID	2D-Similarity
methyl 2-benzimidazolecarbamate (carbendazim)		2426	100.00
2-succinamidobenzimidazole		2425	79.89
2-aminobenzimidazole		2309	65.75

2-amino-5-(propylthio)benzimidazole		2302	61.74
2-mercaptobenzimidazole		10107	61.00
4`-hydroxyfenbendazole		2306	60.36

CONCLUSIONS

In order to prepare immunogenic haptens for developing antibodies against benzimidazole pesticides, a working algorithm involving checking of public bioinformatics databases was applied for selection of similar chemical structures to carbendazim (methyl 2-benzimidazole-carbamate) and bearing different reactive groups available for conjugation to protein carriers.

Three commercial chemical compounds, namely2-(2-Aminoethyl) benzimidazole (AEB), 2-Mercaptobenzimidazole (2MB) and 2-Benzimidazole propionic acid (PBA),were coupled to carrier protein keyhole limpet haemocyanin (KLH), mixed with immuno-adjuvants, and injected four times into Balb/C mice and Wistar rats, respectively.

All haptens induced a weak but specific antibody response, as evidenced by an in-house developed indirect ELISA with haptens coupled to bovine serum albumin (BSA) as coating reagent, and with detectable differences with regard to number of inoculations, chemical structure and animal species.

The AEB molecule induced the strongest and the 2MB molecule induced the weakest antibody responses in mice.

Our works showed evidence that further inoculations were necessary in order to properly stimulate the immune responses for generation of monoclonal antibodies against benzimidazoles.

ACKNOWLEDGEMENTS

This work was supported by the Romanian Ministry of Education and Scientific Research - Executive Agency for Higher Education, Research, Development and Innovation Funding (UEFISCDI), under the National R&D&I Plan II - Partnering Program, Grant PN II-PT-PCCA-2013-4-0128, Contract no. 147/2014 to G.H., R.I.T. and M.D. Some expenditures for dissemination of results were supported by the Doctoral School in Engineering and Plant and Animal Resources Management of the University of Agronomical Sciences and Veterinary Medicine in Bucharest, for V.T. and N.B.
All the authors declare no conflict of interest.

REFERENCES

Abad A., Moreno M.J., Montoya A., 1999. Development of monoclonal antibody-based immunoassays to the N-methylcarbamate pesticide carbofuran. J. Agric. Food Chem. 47: p. 2475-2485.

Aragay G., Pino F., Merkoçi A., 2012. Nanomaterials for sensing and destroying pesticides. Chem. Rev. 112: p. 5317-5338.

Chiarella P., Fazio V.M., 2008. Mouse monoclonal antibodies in biological research: strategies for high-throughput production. Biotechnol. Lett. 30: p. 1303-1310.

Dankwardt A., 2000. Immunochemical assays in pesticide analysis. In: Encyclopedia of Analytical Chemistry (RA Meyers, Ed.). John Wiley & Sons Ltd, Chichester, p. 1-27.

Fan M., He J., 2011. Pesticides immunoassay. In: Pesticides - Strategies for Pesticides Analysis, (M Stoytcheva, Ed.), In Tech, Rijeka, p. 293-314.

Goel P., 2013. Immunodiagnosis of pesticides: a review. Afr. J. Biotechnol. 12: p. 7158-7167.

Goodrow M.H., Hammock B.D., 1998. Hapten design for compound-selective antibodies: ELISAS for environmentally deleterious small molecules. Analytica Chimica Acta, 376: p. 83-91.

Gough K.C., Jarvis S., Maddison B.C., 2011. Development of competitive immunoassays to hydroxyl containing fungicide metabolites. J. Environ. Sci. Health B 46: p. 581-589.

Günther S., Hempel D., Dunkel M., Rother K., Preissner R., 2007. SuperHapten: a comprehensive database for small immunogenic compounds. Nucleic Acids Res 35(suppl. 1): p. D906-D910.

Hermanson G.T., 2013. Bioconjugate Techniques. 3rd ed. Elsevier Inc. Academic Press, Amsterdam-Boston-Heidelberg-London, 1146 p.

Keikotlhaile B.M., Spanoghe P., 2011. Pesticide residues in fruits and vegetables. In: Pesticides - Formulations, Effects, Fate (M. Stoytcheva, Ed.), In Tech., Rijeka, p. 243-252.

Liu S., Zheng Z., Li X., 2013. Advances in pesticide biosensors: current status, challenges, and future perspectives. Anal. Bioanal. Chemistry 405: p. 63-90.

Moran E., O'Keeffe M., O'Connor R., Larkin A.M., Murphy P., Clynes M., 2002. Methods for generation of monoclonal antibodies to the very small drug hapten, 5-benzimidazolecarboxylic acid. J. Immunol Methods 271: p. 65-75.

Nunes G.S., Barcelo D., 1999. Analysis of carbamate insecticides in foodstuffs using chromatography and immunoassay techniques. Trends Anal. Chem. 18: p. 99-106.

Raman S.C., Raje M., Varshney G.C., 2002. Immunosensors for pesticide analysis: antibody production and sensor development. Crit Rev Biotechnol. 22: p. 15-32.

Sashidhar R.B., Capoor A.K., Ramana D., 1994. Quantitation of epsilon-amino group using amino acids as reference standards by trinitrobenzene sulfonic acid. A simple spectrophotometric method for the estimation of hapten to carrier protein ratio. J. Immunol. Methods 167: p. 121-127.

Schrenk, 2012. Chemical contaminants and residues in food. Woodhead Publishing Limited, Oxford, 577 p.

Singh K., Kaur J., Varshney G.C., Raje M., Suri C.R., 2004. Synthesis andcharacterization of hapten-protein conjugates for antibody production againstsmall molecules. Bioconjugate Chem. 15, p. 168-173.

Singh M.K., Srivastava S., Raghava G.P., Varshney G.C., 2006. HaptenDB: a comprehensive database of haptens, carrier proteins and anti-hapten antibodies. Bioinformatics 22: p. 253-255.

Song J., Wang R.M., Wang Y.Q., Tang Y.R., Deng A.P., 2010. Hapten design, modification and preparation of artificial antigens. Chin. J. Anal. Chem. 38: p. 1211-1218.

Tong D., Hesheng Y., Wang J., 2007. Recent advances in the synthesis of artificial antigen and its application in the detection of pesticide residue. Am J. Agri & Bio Sci. 2 (2): p. 88-93.

Zikos C., Evangelou A., Karachaliou C.E., Gourma G., Blouchos P., Moschopoulou G., Yialouris C., Griffiths J., Johnson G., Petrou P., Kakabakos S., Kintzios S., Livaniou E., 2015. Commercially available chemicals as immunizing haptens for the development of a polyclonal antibody recognizing carbendazim and other benzimidazole-type fungicides. Chemosphere 119 (Suppl): p. S16-S20.

Winter C.K., 2012. Pesticide residues in food. In: Chemical contaminants and residues in food (D. Schrenk, Ed.). Woodhead Publishing Limited, Oxford, p. 183-200.

***2008. Commission Regulation (EC) No. 149/2008 of 29 January 2008 amending Regulation (EC) No 396/2005 of the European Parliament and of the Council by establishing Annexes II, III and IV setting maximum residue levels for products covered by Annex I thereto.

16

THE INFLUENCE OF FERTILIZING SCHEMES AND THERMAL REGIME ON WHEAT YIELDS IN THE NORTH-WEST REGION OF ROMANIA

Andra PORUŢIU[1], Iulia MUREŞAN[1], Felix ARION[1], Tudor SĂLĂGEAN[1], Teodor RUSU[1], Raluca FĂRCAŞ[2]

[1]University of Agricultural Sciences and Veterinary Medicine, 3-5 Mănăştur Street, 400372, Cluj-Napoca, Romania, Email: andra.porutiu@usamvcluj.ro

[2]Technical University, 128-130 21 Decembrie 1989 Avenue, 400604 Cluj-Napoca, Romania, Email: raluca.farcas@mtc.utcluj.ro

Corresponding author email: andra.porutiu@usamvcluj.ro

Abstract

The current research is based on the production results obtained on wheat crops (Dumbrava Variety), cultivated following corn and following soy, in long term experiments conducted on an argyle chernozem as a representative soil for the north-western region of Romania, especially Cluj County. The production data are obtained from these experiments, which hold objectives that target both the effect of differentiated fertilizations on wheat yields and the quantity of the productions obtained.The goal of this research is to exhibit the differentiated fertilization systems involved in obtaining high productions for wheat (grown following soy, respectively following corn) in the reference area. In this study it was tracked the effect of the nitrogen-phosphorous interaction and the effect of the thermal regime during three experimental years (2011, 2012, 2013) in achieving the wheat productions.

Key words: fertilization systems, nitrogen-phosphorous interaction, thermal regime, wheat crops, wheat yield.

INTRODUCTION

Agricultural performance as a requirement of contemporary society requires assimilation into production of all elements of technical progress - scientific and economic - advanced in all its sides that determine quantitative and qualitative productivity, higher yields, economic efficiency obtained under optimum systems involved in their implementation, food safety and consumer protection (Cheţan et al., 2016).

Promoting during the last decades of sustainable agriculture and sustainable concepts entails the application of the principles that lead to productive agricultural technologies, technically and economically efficient solutions with effective protection of the environment and consumers that ensure not only productivity but also real optimization of production, social and environmental components and causes a new quality of life (Kurtinecz and Rusu, 2007; Ciontu et al., 2012, Marin et al., 2016).

Optimizing agrochemical soil - plant system is meeting the essential objective in a higher degree and the crops requirements to soil reaction and representation of elements and nutrients in specific concentrations and ratios between them (Borlan and Hera, 1984; Rusu et al., 2009; Mărghitaş et al., 2011).

In this study was tracked the effect of the nitrogen-phosphorous interaction in achieving wheat and corn productions. The research presents the stated results as annual (partial) values and it will continue with them as being reference values for further experimental years (as stages in long term experiments) and with approaches that will substantiate the suggested solutions. The production data are obtained from such experiments, framed in the "long term experiments system" from Fundulea network, which hold objectives that target both the effect and efficiency of differentiated fertilizations on productions and also the impact of fertilizers on the soil fertility evolution, on the quality of the productions obtained (Poruţiu et al., 2013).

In the context of the optimization of soil-plant system, an important scientific and practical role is played by the agrochemical optimization alternatives that harmonize the fertilizing components of the soil with the demands of the vegetal species that can exploit better the production capacity of the soil and genotypes cultivated in order to obtain high vegetal productions that are consumable in large quantities, having superior quality indices, in terms of maintaining an equilibrium in the environment and determining food safety and security (Rusu et al., 2005).

MATERIALS AND METHODS

Experimental approaches were performed under Agricultural Research and Development Station from Turdaconditions, using the experimental protocol of long term experiences, located in the agricultural year 1966/1967, for wheat-corn-soy rotation (Haş, 2006).

Dumbrava wheat variety has the following biological, agronomical and productive characteristics: plants height - 85-95 cm, exhibit a white, 9-11 cm long ear. The grain is medium-sized, oval and red. One thousand grain weight (MMB) is quite high, within 45-50 g, the hectoliter mass (MH) of 75-80 kg/hl (Turda, SCDA).

The field experience which underpins the achievement of objectives is a bi-factorial structure that tracks the effect of the NP interaction on wheat and corn according to the following experimental schemes: A factor - phosphorus doses (kg P_2O_5/ha): 0, 40, 80,120, 160 with annual application to wheat; B factor - nitrogen doses (kg N/ha): 0, 50, 100, 150, 200 with annual application to wheat after corn andnitrogen doses (kg N/ha) 0, 40, 80, 120, 160 with annual application to wheat following soy. According to soil mapping, pedological and agrochemical study and from the soil quality monitoring results, this soil fits the argic chernozem type, in the pedological class of cernisoils (SRTS, 2012).

The fertilizers used in the experiments are represented by a complex fertilizer 20:20:0 which is a solid, granulated nitrophosphate, which holds when applied, the effect of the interaction of the two elements from its composition (N:P), here in balanced concentrations and reports (1:1:0) and nitrate of ammonium which is a simple nitrogen mineral fertilizer that holds the active substance in nitrate and ammonia state.

RESULTS AND DISCUSSIONS

Wheat crops respond positively to the NP levels applied to the soil in the experience, the production effects are at the level of 3-6 t grains per surface unit with production differences (increases) that are very distinctly significant for all nitrogen-phosphorous combinations applied (Table 1).

Table 1. Summary Indicators of Fertilizer Applied to Wheat Crops (Dumbrava Variety)

Year	Crop	Maximum production obtained (kg/ha)	NP Dose	Significance of factors influence [x]
2011	Wheat grown following corn	5533.33	N150P80	NP - v. d. s.; N - v. d. s.; P - n. s.
	Wheat grown following soy	5400.00	N120P160	NP - f. d. s.; N - v. d. s.; P - v. d. s.
2012	Wheat grown following corn	6696.13	N150P120	NP - v. d. s.; N - v. d. s.; P - v. d. s.
	Wheat grown following soy	5755.80	N120P120	NP - v. d. s.; N - v. d. s.; P - v. d. s.
2013	Wheat grown following corn	6945.33	N150P120	NP - v. d. s.; N - v. d. s.; P - d. s.
	Wheat grown following soy	6564.00	N160P160	NP - v. d. s.; N - v. d. s.; P - s.
Mean	Wheat grown following corn	6391.60	N150P106	
	Wheat grown following soy	5906.60	N133P146	

[x] v.d.s. - very distinctly significant; d.s. - distinctly significant; s. - significant; n.s. - insignificant

Table 2. Report on Yield and Maximum Increases to the Content of Active Substance/Hectare (N+P)

Year	Crop	Maximum yield	NP Dose	Dose sum N+P	Production/NP dose	Prod. Dif. (M)/NP dose
2011	Wheat after corn	5533.33	N150P80	230	24	9.6
	Wheat after soy	5400.00	N150P160	310	19	6.7
2012	Wheat after corn	6696.13	N150P120	270	25	14.2
	Wheat after soy	5755.80	N120P120	240	24	8.9
2013	Wheat after corn	6945.33	N150P120	270	26	16.8
	Wheat after soy	6564.00	N160P160	320	21	10.6
Mean	Wheat after corn	6391.60	N150P106	256	25	10.1
	Wheat after soy	5906.60	N133P146	279	21	8.9

Wheat production results in the experimental years 2011 - 2012 - 2013 allow a synthesis of their analysis regarding some production effective approaches through differential fertilizing systems based on the NP complex effect, a high priority and often used technology (Table 2).

Technical results obtained as the mean of the years 2011, 2012, 2013 prove the possibility of obtaining maximum yields of wheat, Dumbrava variety, of 6391.60 kg/ha using N150P106 fertilizer effort (crop after corn) and 5906.60 kg/ha with a fertilizer dose of N133P146 (crop after soy).

Wheat crops which emphasize a technical and productive response very distinctly significant to the effect of NP interaction and mostly very distinctly significant to the influence of N factor action, respond to the effects mentioned by units and production curves according to the polynomial model $y = a + bx - cx^2$ (Figure 1, Figure 2).

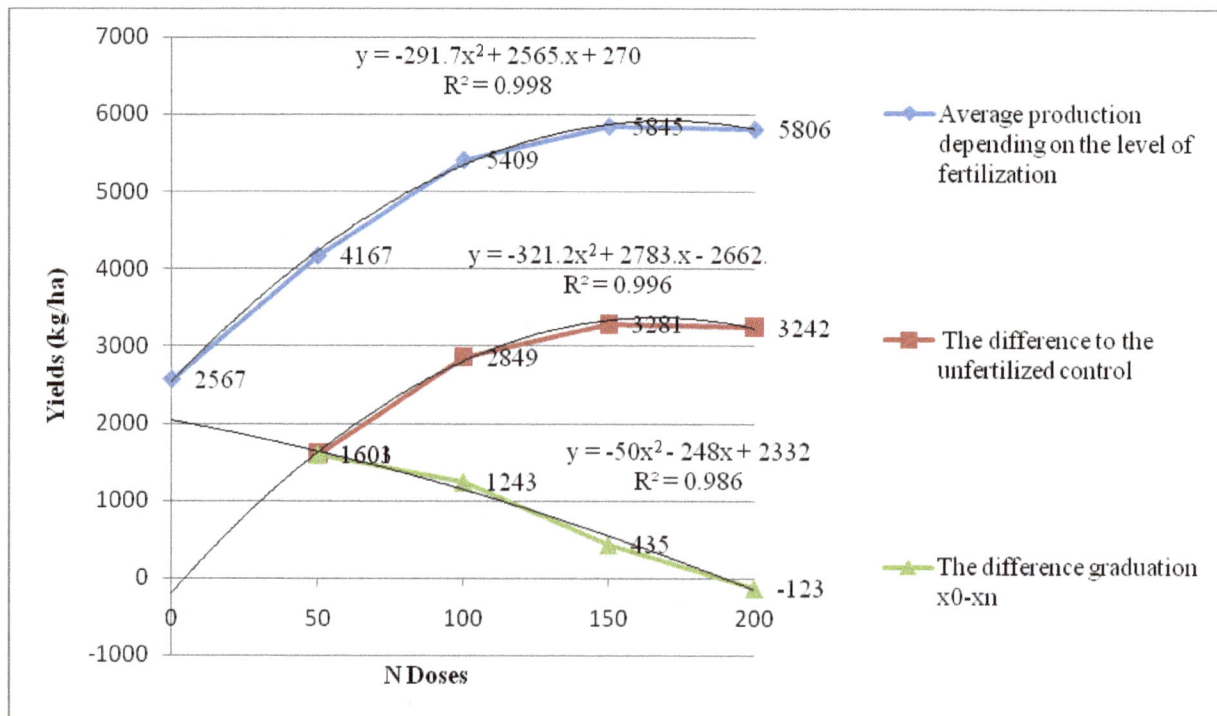

Figure 1. Curves of the Average Yield of Wheat (Dumbrava Variety) Caused by the Variation of N and NP (Previous Plant - Corn)

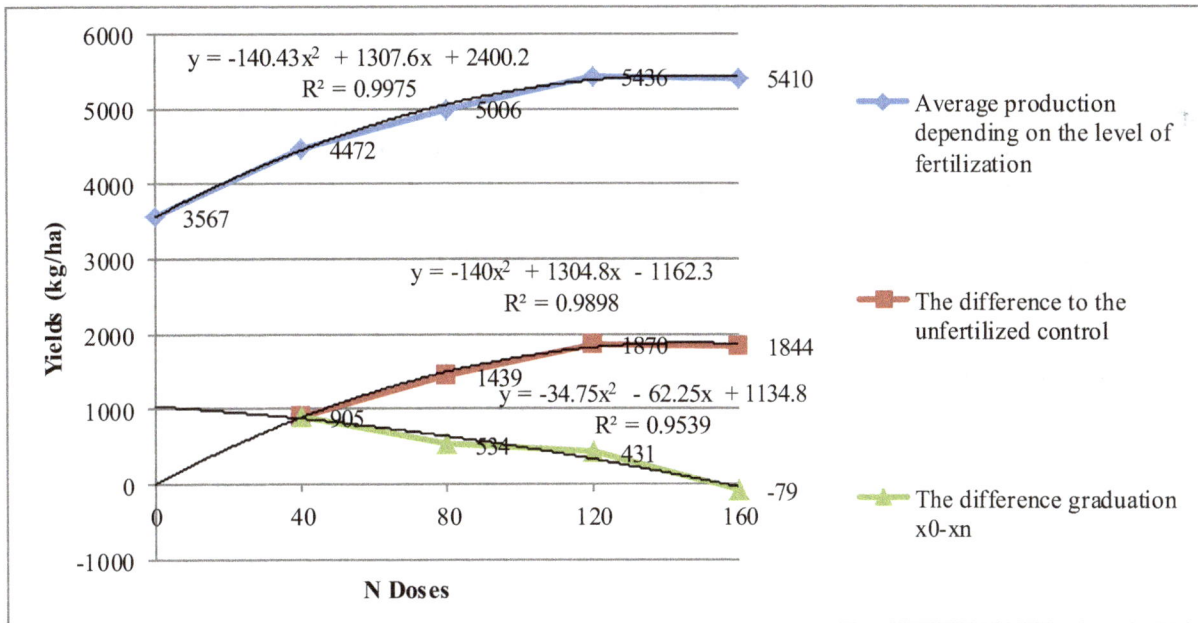

Figure 2. Curves of the Average Yield of Wheat (Dumbrava Variety)
Caused by the Variation of N and NP (Previous Plant - Soy)

The thermal regime during 2011, 2012, 2013 exhibit the characteristics of the Transylvanian Plain, with local specificity (Agricultural Research and Development Station from Turda) and particular specificity of the research period (years) (Figure 3).

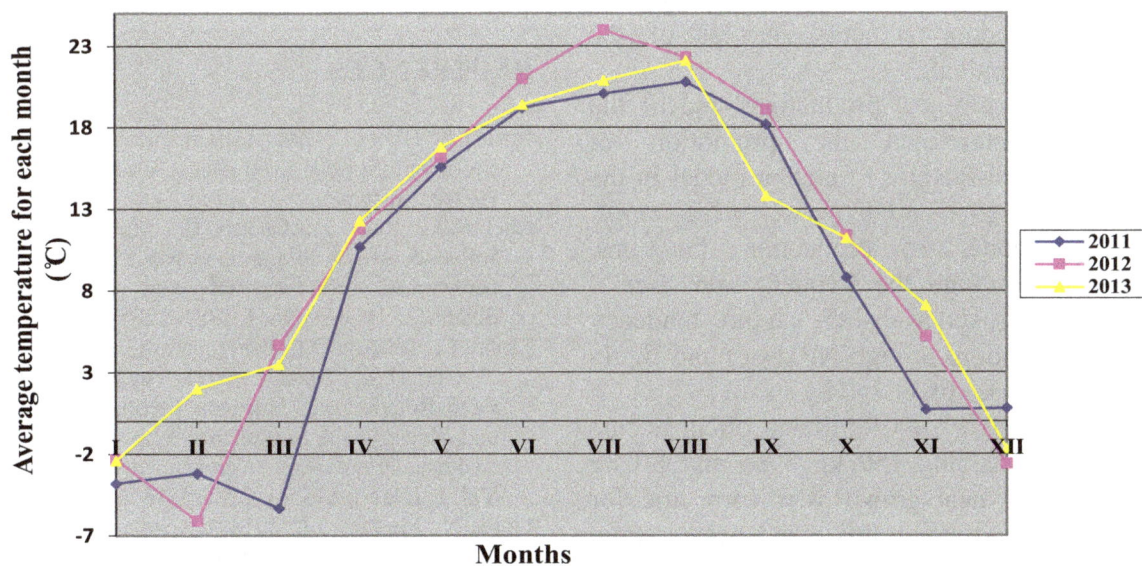

Figure 3. Thermal Regime during 2011-2012-2013

2011 is generally characterized by average monthly temperatures above normal in spring and during those with active vegetative activity (March-September) which makes the average annual temperature to be 9.4°C with 0.5°C above the normal average temperature of the research area. July and August have the highest temperatures. These deficits and disorders jeopardize the current vegetation and the following crops. This year is warm with temperature, surpluses during summer 2012 exhibits a surplus of thermal conditions and therefore is warmer in spring - summer - autumn, with July and August the warmest. The average annual temperature is 10.4°C with 1.4°C over the annual average of this indicator.

In terms of heat 2012 it is warmer than the previous year. 2013 exhibits the characteristics of previous years, with surplus heat in spring - summer when the vegetation is in full swing. The average annual temperature is 10.4°C with 1.4°C than the annual average, meaning that this year continues the trend of general and local heating.

Regarding the experimental years 2011 - 2012 - 2013 heating can be stated as a general feature, especially in the summer months which are the warmest. The consequences of climate anomalies related to excess heat especially reflected in yields obtained, to which were found large variations from year to year and between the experimenys variants.

CONCLUSIONS

Statistically it was proven to be essential and very distinctly significant the effect of NP interaction for wheat crops grown after corn and after soy, followed by the individual action of nitrogen and less of the phosphorus.

Wheat grown after corn has a higher apparent response and a more constant one to NP interaction, then to N, the previous plant here proves to induce a better harness of the fertilization applied;

The level of physical productions and of the increases determined the promotion of polynomial (bi-factorial) function model in the following form: $y = a \pm bx_1 \pm cx_1^2 \pm bx_2 \pm cx_2^2$. In this model of production functions, expressing through the frequent term $(-cx_1^2)$ and $(-cx_2^2)$, proved as real the capping tendency of the productions at high NP doses and firstly at nitrogen doses (over 150 kg a.s./ha).

Capping trends of the fertilizing effect occurs on doses exceeding 150 kg N/ha and 80 kg P_2O_5/ha for wheat grown after corn, and for wheat grown after soy this effect occurs when exceeding the following doses: N-120 kg/ha, P_2O_5-40-80 kg/ha.

It turns out that the effect of phosphorus contributes to a better use of high and very high doses of N. Therefore, the NP interaction remains the primarily effect, followed by the effect of nitrogen, as essential and then the effect of phosphorus.

In the set of the mentioned alternatives of fertilization with the mentioned doses, grain yields can be obtained of 5-6-6.5 t/ha for wheat crops on argic chernozem at Agricultural Research and Development Station from Turda, specific to the ecological conditions of the Transylvanian Plain.

For wheat grown after corn, in 2011, the values of technical optimum doses are: DOT_N - 153.75 kg/ha, DOT_P - 127.4 kg/ha and in 2013, DOT_N - 188 kg/ha, DOT_P - 126.25 kg/ha.

For wheat grown after soy, in 2011, the values of technical optimum doses are: DOT_N - 159.61 kg/ha, DOT_P - 50 kg/ha and in 2013, DOT_N - 301.13 kg/ha, DOT_P - 363.75 kg/ha.

When using these fertilizing schemes, excess heat is also beneficial for both wheat cultivated following corn and for wheat cultivated following soy. The maximum yields obtained are higher in 2012 and 2013, when the average temperature for each month exceeded the normal than in 2011, when the temperatures recorded were closer to the normal ones.

ACKNOWLEDGEMENTS

This research was carried out with the support of the Agricultural Research and Development Station from Turda, Cluj County.

REFERENCES

Borlan Z., Hera C., 1984. Agrochemical Optimization of Soil-Plant System. Romanian Academy Publishing House, Bucureşti.

Chetan F., Rusu T., Chetan C., Moraru P.I., 2016. Influence of soil tillage upon weeds, production and economical efficiency of corn crop. AgroLife Scientif. J., Vol. 5, No. 1, ISSN 2285-5718, p. 36-43.

Ciontu C., Săndoiu D.I., Penescu A., Gâdea M., Şchiopu T., Obrişcă M., Dincă L., 2012. The Essential Role of Crop Rotation and Nitrogen Fertilization in Wheat and Maize in the Sustainable Agriculture System of Reddish Preluvosoil. AgroLife Scientific Journal, Vol. 1, Issue 1, ISSN-L 2285-5718, p. 68-77.

Florea N., Munteanu I., 2012. Romanian System of Soil Taxonomy. Sitech Publishing House, Craiova.

Haş I., 2006. Seed Production for Agricultural Plants. Academic Press Publishing House, Cluj-Napoca.

Kurtinecz P., Rusu M., 2007. Certain Possibilities for the Interpretation of Analytical Results from Long-Term Experiments Regarding the Fertilization and Liming of Acid Soils in North-Western Transylvania. First Internaţional Conference Metagro, 14-16 june 2007, Cluj-Napoca; Proceedings.

Lixandru G., 2006. Ingrated Fertilizing Systems in Agriculture. Pim Publishing House, Iaşi.

Marin D.I., Mihalache M., Rusu T., Ilie L., Ciontu C., 2016. Tillage Efects on Some Properties of Chromic Luvisol and the Maize Crop Yield. 16t[h] International Multidisciplinary Scientific Geoconference SGEM 2016, Book 3 - Water Resources, Forest, Marine and Ocean Ecosystems, Conference Proceedings, Vol. II, p. 449-454.

Mărghitaş M., Razec I., Mihai M., Toader C., Moldovan L., Roman G., 2011. Manual of Best Practices in the Fertilizing Technology of Agricultural Plants. AcademicPres Publishing House, Cluj-Napoca.

Poruţiu A., Rusu M., Mărghitaş M., Toader C., Moldovan L., Deac V., Cheţan F., 2013. Research Concerning the Agrochemical Optimization of the Fertilization System for Wheat Crops on an Argic Phaeozem Soil in the Transylvanian Plain. Research Journal of Agricultural Science, 45 (1), Timişoara.

Rusu M., Mărghitaş M., Oroian I., Mihăiescu T., Dumitraş A, 2005. Agrochemistry Treaty. Ceres Publishing House, Bucharest.

Rusu M., Mărghitaş M., Toader C., Kurtinecz P., Mihai M., Moldovan L., 2009. Agrochemical Modification in Long-Term Experiments with Amendaments and Fertilizers. Lucr. Conf. XIX SNRSS, Iaşi.

***Turda SCDA (Agricultural Research and Development Station), 2011-2013. Scientific Research Reports. SCDA Archives, Turda.

17

PRODUCTIVITY OF COTTON CULTIVAR DARMI UNDER THE INFLUENCE OF FERTILIZATION AT LONG-TERM FIELD TRIAL

Galia PANAYOTOVA[1], Svetla KOSTADINOVA[2], Neli VALKOVA[3], Lubov PLESKUTA[1]

[1]Trakia University, Faculty of Agriculture, 6000 Stara Zagora, Bulgaria
[2]Agricultural University, Faculty of Agronomy, Mendeleev 12, 4000 Plovdiv, Bulgaria
[3]Field Crops Institute, 6200 Chirpan, G. Dimitrov 2, Bulgaria

Corresponding author email: galia_panayotova@abv.bg

Abstract

The aim of this study was to evaluate the effects of different application rates of nitrogen and phosphorus on growth, yield, earliness and quality of cotton (Gossypium hirsutum L.) cultivar Darmi, grown in Chirpan, Bulgaria. The cotton was grown during 2007-2010 in crop-rotation with durum wheat under non-irrigated conditions. The experimental design was a randomized complete block with four replications. Single and combined nitrogen and phosphorus fertilizers in rates 0; 40; 80; 120 and 160 kg.ha⁻¹ were tested. Values were established for the September and total seed-cotton yield, earliness, lint percentage, number of bolls per plant, boll size, plant height, 1000 seeds weight, fibre length. Under the influence of N fertilization the total yield increased with 10.0- 23.1% compared to the control and under phosphorous fertilization - with 1.4- 6.6% (P_{120}). Productivity increased most under combined fertilization $N_{120-160}$ P_{40-80} - 27.7-36.3% more than the unfertilised with a very good share of the September yield. The increase of nitrogen rates decreased earliness with 4.2% (N_{120}) to the check. Average for the 20-year period a high effective yield was formed under moderate N rates combined with low to moderate P levels, whereat the cotton yield increased with 17.8-24.4%. The fertilization significantly increased the height of the plant, boll number (55.6% over control) and boll weight (21.8% more). There was a tendency for decrease in lint percentage with the increase of fertilization rate. No significant changes were found in terms of fiber length and it ranged within 23.76-25.05 mm.

Key words: cotton, fertilization, nitrogen, phosphorus, yield.

INTRODUCTION

The cotton productivity and quality varies to a wide range depending on weather, cultivar, soil fertility, agrotechnology. Without adequate amount at each growth stage, the maximum yield potential of cotton can not be achieved (Gushevilov and Karev, 2000; Karthikeyan and Jayakumar, 2001; Zhao and Oosterhuis, 2000). Fertilizers are essential component of modern cotton production that affects plant growth, fruiting and yield (Sawan et al., 2006). The application of such agrotechnical methods that would guarantee high levels and stability of the economic parameters under different conditions is of importance for the production (Coker et al., 2009).

Without adequate amount of nutrient elements at each growth stage, the maximum potential of cotton cannot be achieved (Kirchmann & Thorvaldsson, 2000). Recommendations for cotton fertilization range from relatively low to very high rates (Clawson et al., 2008; McConell et al., 1993). The nitrogen level is one of the determinants of cotton productivity (Ali at al., 2003; Christidis, 1985; Geric et al., 1998). N-deficient cotton plants are likely to have suppressed vegetative and reproductive growth, prematurely senesce (Stewart et al., 2010) and low yields (Radin and Mauney, 1986). In contrast, excessive N can have negative impacts on yield and can result in economic loss. At the early boll filling stage, excessive N may inhibit fruit production due to the promotion of vegetative growth (Gerik et al., 1998).

According to Clawson et al. (2006), Munir et al. (2015) and other authors nitrogen increased lint yield, plant height, main stem nodes, and both whole-plant and subset individual boll weight, but lint percentage was not affected by nitrogen. Pettigrew and Adamczyk (2006) reported that N treatments had no effect on lint yield or any dry matter partitioning components. Nitrogen influenced both vegetative and reproductive growth (Surya et al., 2010) as its deficiency decreased yield by accelerating premature leaf senescence (Fageria and Baligar, 2005) and early cut-out (Read et al., 2006), while, N in excess can delay crop maturity and promote boll shedding, diseases and insect damages (Howard et al., 2001; Oosterhuis, 2001). Diagnosing and correction of nitrogen deficiency is not difficult while excess of N is more difficult to detect and rectify. The cotton cultivars evolved in different agroclimatic regions behave differentially to application of mineral fertilizers (Prasad and Siddique, 2004).

According to Girma et al. (2007) the nitrogen, phosphorus and potassium fertilizer use in cotton production remains important, as N has a decisive influence, while phosphorus has less effect.

Cahill et al. (2008), Sawan et al. (2008), Mitchell (2000) and other authors reported that P fertilizer generally was not effective, and significant differences were not observed for cotton yield. Saleem et al. (2010) reported that phosphorus levels (30, 60 and 90 kg.ha^{-1}) significantly affected almost all the characters related to earliness and yield. According to Cahill et al. (2008) phosphorus deficiency violates the nitrogen nutrition.

Several factors including soil type affect cotton response to phosphorus. The critical level of P is a function of actual concentration of the labile pool that in turn determines the available P during the growth of cotton (Crozier et al., 2004). Bronson et al. (2003) founded that several variables including early P accumulation, biomass, and lint yields positively responded to P fertilization in calcareous soils. The nitrogen uptake is reduced in plants with phosphorus deficiency (Breitenbeck and Boquet, 1993). The phosphorus requirements of cotton are considered very low because of its deep root system and indeterminate growth habit (Malik et al., 1996). According to Gill et al. (2000) and Cope (1984) there are cases where cotton response to phosphorus has been positive and economical. Application of NPK nutrients had some effect on lint yield, although most of the response was attributed to N (all cultivars) and to some extent P (Girma et al., 2007). The use of only starter-N is more cost effective than using both N and P starter fertilizer (Cahill et al., 2008).

P significantly enhanced crop growth, N and K uptake, total chlorophyll concentration and dry matter yield of cotton plant (Sawan et al., 2008). Deshpande and Lakhdive (1994) reported that P application increased P uptake and content in leaf, stem and reproductive part like seed. P is essential for cell division and has a stimulating effect on a number of flower buds and bolls per plant. Kaynak (1995) reported that positive correlation exists between seed cotton weight per boll and seed cotton yield per plant.

Dorahy et al. (2008) reported that phosphorus fertilizer application only increased P concentration in the plants during leaf expansion, but had no effect on biomass production, P uptake at later growth stages sampled, or lint yield. Acording to Leffler (1986) and Bassett et al. (1970) the phosphorus fertilization increased dry matter production and nutrient uptake.

Bronson et al. (2003) reported that phosphorus fertilizer did not affect lint yields at Lamesa.

Reiter and Kreig (2000) established some positive and notable P effects on lint fiber quality factors although both lint yield and lint quality were driven more by moisture availability than phosphorus.

According to some authors the cotton varieties manifest specific nutrient requirements (Karamanidis et al., 2004; Fritschi et al., 2003; Clement-Bailey and Gwathmey, 2008; Javaid et al., 2001; McConnell et al., 2003), while according to others (Kostadinova & Panayotova, 2003; Mullins and Burmester, 1990; Panayotova et al., 2007) the differences in the level of mineral nutrition of genotypes with close origin are insignificant.

Weather conditions and fertilizers exert great influence on the cotton yield. Cotton yield under different conditions is a desirable characteristic because Bulgaria is located on the northern cotton-cultivating boundary. In Bulgaria there are suitable soil and climatic conditions, tradition, experience and advanced research for cotton growing. The foreign cultivars in Bulgaria realise late maturity and fail to manifest their yield and quality potential. According to Christidis (1985) the NPK content in the cotton correlated significantly with the environmental conditions.

The nutrient requirements of the new varieties are often questioned by producers.

Optimizing fertilization for cotton cultivars is one possible way of tailoring production practices to achieve optimal economic returns.

The aim of this study was to evaluate the effects of different application rates of nitrogen and phosphorus on growth, yield, earliness and quality of cotton (*Gossypium hirsutum* L.) cultivar Darmi, grown in the region of Central-South Bulgaria.

MATERIALS AND METHODS

The experiment was carried out on the field of the Field Crops Institute, Chirpan, situated in a major cotton-growing region of Bulgaria. Data from a long-term experiment initiated in 1967 was used. The cotton (*Gossypium hirsutum* L.) cultivar Darmi was grown in double crop-rotation with durum wheat under non-irrigated conditions. The experimental design was a randomized complete block with four replications. Individual plots consisted of six 8.33 m rows spaced 0.60 m apart with a net plot size of 50 m^2. Twenty-five treatments containing different rates of N-P were evaluated. Single and combination of N and P_2O_5 fertilizers in rates of 0; 40; 80; 120 and 160 kg.ha^{-1} were tested. The source of the N was ammonium nitrate, of P_2O_5 - triple superphosphate.

The applied agrotechnical practices were complied with the technology established for the region. Cotton seeds were sown within 20-30 April. The plant population reached as much as 160,000 plants.ha^{-1}, approximately. Weeds were controlled by preplant and preemergence herbicides, interrow cultivation and hand chipping. Defoliants were not applied. There were two harvests made by hand from four middle rows (20 m^2). At maturity the seedcotton yield from each plot was weighed and ginned on a rollergin.

The cultivar Darmi was established in Bulgaria in 2007 with improved quality of the fiber. It was created by crossing the breeding line № 268 (with genplasm of species G. barbadense L.) x C-9070 (Uzbek variety). The cultivar staking high fruiting bolls.

There were determined the September yield and total seed-cotton yield (kg. ha^{-1}); index of earliness as correlation between two harvests; lint percentage (%); the structural elements of yield: number of bolls per plant by accurate count; boll size (g), which was determined as seed cotton weight/ number of bolls; plant height (cm); 1000 seeds weight (g); fibre length measured handily by "butterfly" method (mm). Analysis of variance (ANOVA) was performed to evaluate differences and interaction among the nitrogen and phosphorus fertilization and years.

The studied years were with different meteorological conditions during the vegetation period (May-October) (Table 1). With regard to temperature and rainfall supply 2007 was warm and moderately wet; 2008 was warm and moderately dry; 2009 - moderately warm and dry, without precipitation in the august and insufficient rainfall during flowering and boll formation; 2010 - moderately warm and wet, with sufficient rainfall during critical stage of the cotton development (Table 1).

The soil type at the Institute region was Pellic Vertisols (FAO), defined by the sandy-clay composition, with high humidity capacity and small water-permeability. The soil in the field was with neutral soil reaction in the 0-60 cm soil layer, medium supplied with organic matter, moderately N provided, with low content of mobile P_2O_5 and well supplied with available K_2O (Table 2).

RESULTS AND DISCUSSIONS

The results for cotton yield showed good effectiveness of the applied fertilization despite unfavorable weather conditions during the cotton vegetation. Additionally yield was realized as a result of direct fertilization and of soil fertility created by long-term fertilization.

Table 1. Meteorological data recorded at the region of Chirpan, Bulgaria during the vegetative period of cotton, 2007-2010

| Year | Months | | | | | | Σ IV-IX | Σ VI-VIII | Σ V-IX |
	IV	V	VI	VII	VIII	IX			
	Temperature sum, Σ t $^\circ$C								
1928/2010	343	519	622	720	711	561	3476	2053	3133
2007	351	579	693	825	753	527	3728	2271	3377
2008	386	522	636	717	792	555	3608	2145	3222
2009	357	569	648	751	725	571	3621	2124	3264
2010	364	554	625	706	798	582	3629	2129	3265
	Rainfall, Σ W mm								
1928/2010	45	63	65	52	41	34	300	158	255
2007	19	53	39	0	62	128	301	101	282
2008	66	36	95	36	3	91	327	134	261
2009	17	16	14	89	35	58	229	138	212
2010	63	27	82	114	22	48	356	218	293
	Hydrothermal coefficient (by Selyaninov)								
1928/2010	1.31	1.21	1.05	0.72	0.58	0.61	0.86	0.77	0.81
2007	0.54	0.92	0.56	0	0.82	2.43	0.81	0.45	0.84
2008	1.71	0.69	1.49	0.50	0.04	1.64	0.91	0.63	0.81
2009	0.48	0.28	0.22	1.19	0.48	1.02	0.63	0.65	0.67
2010	1.73	0.48	1.31	1.62	0.28	0.83	0.98	1.,02	0,90

Table 2. Content of humus, mineral nitrogen and mobile phosphorus and potassium in the Pellic Vertisols, Chirpan

| Treatment | Depth, cm | N_{min}, kg.ha^{-1} | | | P_2O_5, mg/100 g | K_2O, mg/100 g | Humus, % |
		$N-NH_4$	$N-NO_3$	Total			
$N_0P_0K_0$	0–30	20	14	34	2.4	17.4	2.28
	30–60	13	10	23	1.6	16.5	2.24
P_{160}	0–30	17	21	38	19.3	18.2	2.32
	30–60	21	15	36	18.7	16.9	2.28
N_{80}	0–30	44	30	74	3.2	16.6	2.39
	30–60	60	37	97	2.4	16.5	2.33
N_{160}	0–30	64	43	107	3.7	16.7	2.58
	30–60	57	23	80	3.3	16.2	2.42
$N_{120}P_{120}$	0–30	63	39	102	16.7	28.0	2.56
	30–60	61	26	87	15.0	22.5	2.47

The total seed-cotton yield was significantly influenced by the environmental conditions of the year, type of fertilizer and fertilization level (Table 3). The uncontrolled year conditions had greatest share in the total variation of the factors - 87.0%. The fertilization level led to significant differences - 8.31%. The influence of nitrogen was 6.92% and of phosphorus - 0.85% of total variation. No significant differences in the total seed-cotton yield were occurred as a function of the interaction between nitrogen and phosphorus rate. The variance of factors for the September yield was similar.

The total seed-cotton yield without fertilization was 1.52 t.ha^{-1} average for 4-year period (Table 4). Under the influence of alone N fertilization the yield increased by 10.0 (N_{40}) to 23.1% (N_{160}) compared to the control, and at phosphorous fertilization - by 1.4 (P_{160}) to 6.6% (P_{120}). Productivity increased the utmost at combined fertilization $N_{120-160}P_{40-80}$ - by 27.7-36.3% in more than unfertilised.

The total seed-cotton yield without fertilization was 1.52 t.ha^{-1} average for 4-years period (Table 4).

Table 3. Analysis of variance for total seed cotton yield, 2007-2010.

Source of variation	df.	Sum of squares, SQ	Sum of squares SQ, %	Mean squares MS	F
Total	99	312553.8	100.00	-	-
Years	3	271927.0	87.00	90642***	444.8
Fertilization	24	25953.2	8.31	1081***	5.3
N	4	21620.0	6.92	5405***	26.5
P	4	2647.2	0.85	661 *	3.2
N x P	16	1686.0	0.54	105 n.s.	0.5
Error	72	14673.5	4.69	203	

F-ratio amonhg the variables

Table 4. Effect of NP fertilization on the total seed-cotton yield, average for 2007-2010, t.ha^{-1}

Fertilization	2007	2008	2009	2010	Average		
					t.ha^{-1}	%	Agronomic efficiency, kg cotton
$N_0P_0K_0$	1.40	1.53	1.04	2.09	1.52	100.0	-
N_{40}	1.48	1.62	1.09	2.48	1.67	110.0	3.80
N_{80}	1.67	1.71	1.17	2.49	1.76	116.0	3.02
N_{120}	1.81	1.71	1.28	2.64	1.86	122.5	2.84
N_{160}	1.75	1.68	1.28	2.80	1.87	123.1	2.19
P_{40}	1.48	1.61	1.05	2.10	1.56	102.9	1.10
$N_{40}P_{40}$	1.61	1.73	1.07	2.88	1.82	120.2	-
$N_{80}P_{40}$	1.84	1.84	1.16	3.01	1.96	129.3	-
$N_{120}P_{40}$	1.94	1.90	1.21	3.22	2.07	136.3	-
$N_{160}P_{40}$	1.79	1.89	1.22	3.22	2.03	134.0	-
P_{80}	1.58	1.52	1.10	2.18	1.59	105.1	0.96
$N_{40}P_{80}$	1.58	1.70	1.14	2.65	1.77	116.5	-
$N_{80}P_{80}$	1.66	1.71	1.22	2.75	1.83	120.8	-
$N_{120}P_{80}$	1.82	1.76	1.26	2.91	1.94	127.7	-
$N_{160}P_{80}$	1.94	1.79	1.34	2.95	2.00	132.1	-
P_{120}	1.62	1.58	1.12	2.13	1.61	106.6	0.81
$N_{40}P_{120}$	1.74	1.80	1.14	2.71	1.85	121.8	-
$N_{80}P_{120}$	1.83	1.88	1.33	2.78	1.96	128.9	-
$N_{120}P_{120}$	1.96	1.83	1.52	2.71	2.00	132.0	-
$N_{160}P_{120}$	1.43	1.77	1.45	2.74	1.85	121.8	-
P_{160}	1.48	1.48	1.09	2.10	1.54	101.4	0.14
$N_{40}P_{160}$	1.64	1.80	1.13	2.53	1.77	116.9	-
$N_{80}P_{160}$	1.70	1.81	1.40	2.69	1.90	125.2	-
$N_{120}P_{160}$	1.92	1.85	1.34	2.98	2.02	133.3	-
$N_{160}P_{160}$	1.78	1.75	1.38	2.75	1.91	126.2	-
Average	1.70	1.73	1.22	2.66	1.83	-	-
GD 5 %	0.352	0.459	0.263	0.566	0.201	13.26	-
GD 1 %	0.466	0.603	0.347	0.751	0.267	17.61	-
GD 0.1%	0.604	0.761	0.475	0.973	0.346	22.83	-

Under the influence of alone N fertilization the yield increased by 10.0 (N_{40}) to 23.1% (N_{160}) compared to the control, and at phosphorous fertilization - by 1.4 (P_{160}) to 6.6% (P_{120}). Productivity increased the utmost at combined fertilization $N_{120-160}P_{40-80}$ - by 27.7-36.3% in more than unfertilised. The yield was significantly lower in 2009 - an average 1.22 t.ha^{-1} under the influence of unfavorable weather conditions for cotton, while in other years the average yield was from 1.70 t.ha^{-1} (2007) to 2.66 t.ha^{-1} (2010). The results show that at NP fertilization combined with favorable weather conditions Bulgarian cotton varieties realize their potential for high productivity. The combine fertilization proved a much better influence on the cotton yield with very good share of the September yield (Table 5). The increase of nitrogen rates decreased earliness index by 4.2% (N_{120}) to the check. Dong et al. (2012) also showed that increasing N rate reduced earliness. Setatou and Simonis (1996) found that N fertilization caused a delay in the maturity of cotton plants, ranging from 0.2 to 2.5 days in comparison to the control in most of the experiments. Yang et al. (2011) reported that N ratio promoted an earlier squaring and flowering but delayed the opening stage, so prolonged the boll setting and filling period.

The efficiency of fertilization and the effect of 1 kg fertilizer were significantly higher for nitrogen (2.19-3.80 kg seed-cotton) as compared to phosphorus (0.14-0.96 kg) and the effect decreased at higher nutrient levels.

The cotton productivity increased as a result of long-term fertilization and soil fertility.

The average cotton yield for the last 20 years without fertilization was 1.56 t.ha^{-1} (Table 6).

Average for the period a high, economically effective and stable yield was formed under moderate N rates combined with low to moderate P levels, whereat the cotton yield increases with 17.8-24.4% to the unfertilized.

The high rates effect in most of the years was close to the one received by moderate fertilization caused by the extended vegetative period of cotton and failure of most of the formed bolls to ripe, especially in years with more rainfall and lower temperatures.

Table 5. September seed-cotton yield (t.ha^{-1}) at nitrogen-phosphorus fertilization, average for 2007-2010

Treatment	Year				Average		
	2007	2008	2009	2010	kg.ha^{-1}	%	Earliness, %
$N_0P_0K_0$	0.88	0.88	0.92	1.57	1.06	100.00	70.0
N_{40}	0.98	0.97	0.96	1.86	1.19	112.23	71.4
N_{80}	1.06	0.97	1.06	1.54	1.16	109.09	65.8
N_{120}	1.18	0.94	1.12	1.72	1.24	116.88	66.8
N_{160}	1.14	0.91	1.10	1.91	1.27	119.29	67.8
P_{40}	0.98	1.07	0.98	1.56	1.15	108.43	73.7
$N_{40}P_{40}$	1.06	1.02	0.99	1.96	1.26	118.45	69.0
$N_{80}P_{40}$	1.16	1.02	1.05	2.07	1.32	124.79	67.5
$N_{120}P_{40}$	1.18	1.06	1.10	2.14	1.37	129.04	66.2
$N_{160}P_{40}$	1.08	1.16	1.13	2.09	1.37	128.64	66.0
P_{80}	1.04	0.94	1.00	1.76	1.19	111.71	74.4
$N_{40}P_{80}$	1.03	1.06	1.04	1.83	1.24	116.95	70.3
$N_{80}P_{80}$	1.08	1.01	1.12	1.96	1.29	121.87	70.6
$N_{120}P_{80}$	1.19	1.05	1.16	1.98	1.34	126.57	69.4
$N_{160}P_{80}$	1.28	1.10	1.22	1.92	1.38	130.26	69.0
P_{120}	1.07	1.08	0.99	1.56	1.18	110.69	72.8
$N_{40}P_{120}$	1.11	1.24	1.03	1.85	1.31	123.09	70.7
$N_{80}P_{120}$	1.15	1.32	1.22	2.13	1.45	136.88	74.3
$N_{120}P_{120}$	1.20	1.19	1.37	1.94	1.42	134.22	71.1
$N_{160}P_{120}$	1.00	1.08	1.35	2.04	1.37	128.71	73.9
P_{1600}	0.96	0.93	0.97	1.60	1.11	105.02	72.4
$N_{40}P_{160}$	1.10	1.23	1.00	1.83	1.29	121.42	72.7
$N_{80}P_{160}$	1.12	1.22	1.26	1.86	1.36	128.42	71.8
$N_{120}P_{160}$	1.25	1.18	1.22	2.11	1.44	135.85	71.3
$N_{160}P_{160}$	1.18	1.10	1.24	1.88	1.35	127.35	70.6
Average	**1.10**	**1.07**	**1.10**	**1.87**	**1.28**	-	**70.38**
GD 5 %	0.247	0.306	0.348	0.632	0.348	32.83	
GD 1 %	0.327	0.464	0.489	0.838	0.489	46.13	
GD 0.1%	0.424	0.586	0.612	1.085	0.612	57.74	

Table 6. Effect of long-term fertilization on total seed-cotton yield for last 20 years (1991-2010)

Treatment	t.ha^{-1}	%	Agronomical efficiency, kg	Variation
$N_0P_0K_0$	1.56	100.0	-	0.76 – 2.09
N_{80}	1.75	112.0	2.34	0.89 – 2.86
N_{120}	1.88	120.4	2.67	0.91 – 2.79
N_{160}	1.90	121.8	2.12	0.79 – 2.63
P_{80}	1.68	107.7	1.50	0.78 – 2.26
P_{120}	1.71	109.4	1.25	0.79 – 2.44
P_{160}	1.66	106.6	0.62	0.81 – 2.49
$N_{80}P_{80}$	1.91	122.2	-	0.99 – 3.16
$N_{120}P_{80}$	1.84	117.8	-	0.89 – 2.84
$N_{120}P_{120}$	1.94	124.4	-	1.01 –3.10
$N_{160}P_{80}$	1.81	116.0	-	0.92-3.08
$N_{160}P_{120}$	1.92	123.2	-	0.95 – 2.91
$N_{160}P_{160}$	1.92	122.8	-	0.99 – 3.05
GD 5%	0.167	10.70		
GD 1%	0.189	12.12		
GD 0.1%	0.196	12.56		

The fertilization efficiency was strongly dependent on the weather conditions, which was signified by the significant yield variance by years. The nitrogen fertilization exerted a decisive influence on the cotton productivity, signified by the significantly higher values for its effect (2.12-2.67 kg) compared to phosphorus (0.62-1.50 kg).

This showed that an alone phosphorus fertilization is not an effective agrotechnical measure.

Average for the studied four years the weight of 1 boll was 3.77 g, without fertilization - 3.40 g, and reached a maximum value of 4.14 g at $N_{80}P_{120}$ (Table 7).

At higher fertilizer rate the number of bolls per 1 plant increased and at fertilization with $N_{120}P_{120}$ the number of mature bolls per plant was 4.56, by 55.6% above unfertilized (2.93). The highest weight of 1 boll and number of bolls per 1 plant were formed in 2010 - an average of 4.25 g (in the range 3.76-4.66 g) and 5.43 (from 4.02 to 6.32) respectively.

Rashidi et al. (2011) reported that 200 kg.ha^{-1} N application rate resulted significant increased in the boll number (19.8), and boll weight (6.26 g) compared to low rates.

Table 7. Structural elements of cotton yield, height in maturity, fiber length and output at fertilization, average for 2007-2010

Treatment	Weight of 1 boll		Bolls per 1 plant		Height at maturity		Fiber length		Lint percentage	
	g	%	number	%	cm	%	mm	%	%	% то check
$N_0P_0K_0$	3.40	100.0	2.93	100.0	44.6	100.0	24.29	100.0	38.1	100.0
N_4	3.46	101.8	3.58	122.2	50.2	112.6	24.36	100.3	37.9	99.5
N_8	3.58	105.3	3.67	125.3	53.4	119.7	24.38	100.4	37.8	99.2
N_{12}	3.62	106.5	4.40	150.2	60.5	135.6	24.64	101.4	37.4	98.2
N_{16}	3.76	110.6	4.30	146.8	60.9	136.6	24.70	101.7	37.3	97.9
P_4	3.46	101.8	3.47	118.4	47.7	106.9	24.63	101.4	38.0	99.7
N_4P_4	3.55	104.4	4.15	141.6	52.3	117.3	24.30	100.0	38.0	99.7
N_8P_4	3.81	112.1	4.21	143.7	56.1	125.8	24.83	102.2	37.9	99.5
$N_{12}P_4$	3.76	110.6	4.26	145.4	60.5	135.6	25.05	103.1	37.7	98.9
$N_{16}P_4$	3.69	108.5	4.15	141.6	62.6	140.4	24.63	101.4	37.6	98.7
P_8	3.64	107.1	3.69	125.9	47.3	106.0	23.89	98.4	38.0	99.7
N_4P_8	3.86	113.5	3.94	134.5	51.0	114.4	24.55	101.1	37.9	99.5
N_8P_8	3.82	112.4	4.19	143.0	56.9	127.6	24.83	102.2	37.7	98.9
$N_{12}P_8$	3.97	116.8	4.38	149.5	61.6	138.1	24.50	100.9	37.7	98.9
$N_{16}P_8$	3.96	116.5	4.05	138.2	64.1	143.7	24.36	100.3	37.5	98.4
P_{12}	3.64	107.1	3.16	107.8	48.1	107.8	24.00	98.8	37.9	99.5
N_4P_{12}	3.80	111.8	4.24	144.7	51.4	115.2	24.41	100.5	37.7	98.9
N_8P_{12}	4.14	121.8	4.17	142.3	56.7	127.1	24.08	99.1	37.6	98.7
$N_{12}P_{12}$	4.02	118.2	4.56	155.6	63.7	142.8	24.20	99.6	37.3	97.9
$N_{16}P_{12}$	4.00	117.6	4.29	146.4	63.7	142.8	24.14	99.4	37.1	97.4
P_{16}	3.59	105.6	3.81	130.0	47.6	106.7	24.61	101.3	37.9	99.5
N_4P_{16}	4.11	120.9	4.02	137.2	53.5	120.0	24.38	100.4	37.9	99.5
N_8P_{16}	3.88	114.1	4.47	152.6	56.2	126.0	23.96	98.6	37.6	98.7
$N_{12}P_{16}$	3.88	114.1	4.55	155.3	62.4	139.9	23.76	97.8	37.3	97.9
$N_{16}P_{16}$	3.86	113.5	4.29	146.4	64.3	144.2	24.01	98.8	37.3	97.9
Средно	3.77	-	4.04	-	55.89	-	24.38	-	37.68	-

Seilsepour and Rashidi (2011) and Rashidi and Gholami (2011) also reported that the seed-cotton yield, the fiber yield, weight of 1 boll, the number of bolls per 1 plant, the weight of seeds in 1 boll and N concentrations in the leaves significant increase at nitrogen fertilization. Several research findings reported that the yield advantages due to optimal N application were attributed to larger bolls at a greater number of fruiting sites (Boquet and Breitenbeck, 2000; McConnell et al., 1998; Moore, 1999).

Plant height is a genetically controlled factor but nutritional disorder may also influence the height of plant (Ahmed et al., 2009). A field study showed that difference in plant height was due to the cultivars and number of monopodial branches per plant decreased while number of sympodial branches per plant increased with increasing P levels (Copur, 2006).

In our study alone and combined nitrogen fertilization significantly increased the height of cotton plants in maturity (from 44.6 cm without fertilization to 64.3 cm at $N_{160}P_{160}$). In 2010 under the influence of fertilization, combined with rainfall throughout the cotton vegetation, the growth rate of plants was the most intense and the height at the end of the vegetation from 53.8 cm without fertilization reached 91.5 cm at $N_{120}P_{120}$. Hallikeri et al. (2010) showed that cotton height was significantly affected by application of N levels, as taller plants were observed with N up to $120\ kg \cdot ha^{-1}$.

Under the influence of various types of fertilizers and applied rates no significant changes were found in terms of the fiber length and it was in a range 23.76-25.05 mm. There was a tendency for a decrease of lint percentage with increasing the rate of fertilization, connected to the higher weight of 1000 seeds (108-113 g) (Figure 1). The conditions over the years proved to be a strong influence on the fiber length and lint percentage in comparisson with fertilization. Rashidi and Gholami (2011) and Sawan et al. (2006) also found that the effect of nitrogen on fiber properties were small and inconsistent. Fritschi et al. (2003) found that lint yield was increased linearly each year with N fertility levels, attaining a maximum yield at the 224 kg N ha^{-1} rate and they found that with increased N, gin turnout was decreased.

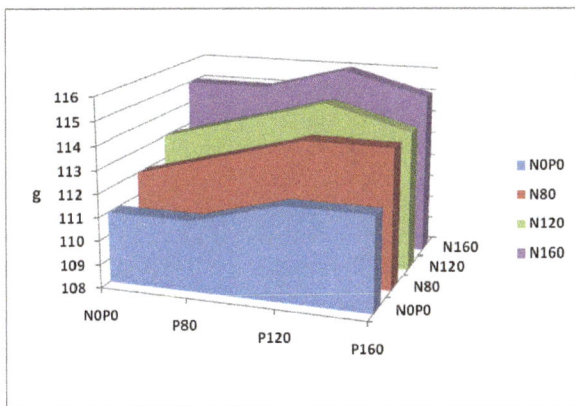

Figure 1. 1000 seeds weight at N-P fertilization, 2007-2010

CONCLUSIONS

The total seed-cotton yield without fertilization was 1.52 t.ha^{-1} average for 4-year period. Under the influence of N fertilization the yield increased by 10.0 (N$_{40}$) to 23.1% (N$_{160}$) compared to the control and at phosphorous fertilization - by 1.4 (P$_{160}$) to 6.6% (P$_{120}$). Productivity increased the utmost at combined fertilization N$_{120-160}$P$_{40-80}$ - by 27.7-36.3% in more than unfertilised with very good share of the September yield.

The increase of nitrogen rates decreased earliness by 4.2% (N$_{120}$) to the check.

The agronomic efficiency was significantly higher for nitrogen (2.19-3.80 kg seed-cotton) as compared to phosphorus (0.14-0.96 kg) and the effect decreased at higher nutrient levels.

The average cotton yield for the last 20 years without fertilization was 1.56 t.ha^{-1}. Average for the long period a high effective yield was formed under moderate N rates combined with low to moderate P levels, whereat the cotton yield increases with 17.8-24.4% to the unfertilized.

The fertilization significantly increased the height of the cotton plant (44.2% over unfertilized), boll number (55.6% over control) and boll weight (with 21.8 % more).

There was a tendency for a decrease of lint percentage with increasing the rate of fertilization.

No significant changes were found in terms of the fiber length and it was in a range 23.76-25.05 mm.

REFERENCES

Ahmed H., Ali R., Zamir S. and Mahmood N., 2009, Growth, yield and quality performance of cotton cultivar BH-160 (*G. hirsutum* L.) as influenced by different plant spacing. JAPS, no. 19, p. 189-192.

Bang M., Milory S. and Roberts G., 2006. Managing for crop maturity. Aust. Cotton grower. 27 (7), p. 53-56.

Bassett D., Anderson W. and Werkoven C., 1970. Dry matter production and nutrient uptake in irrigated cotton (*Gossypium hirsutum*). Agron. J., 62, p. 299-303.

Breitenbeck G. and Boquet D., 1993. Effect of N fertilization on nutrient uptake by cotton. Proceedings of the Beltwide Cotton Conferences, Jan. 10-14 1993, New Orleans, LA., p. 1298-1300.

Bronson K.F., Keeling J.W., Booker J.D., Chua T.T., Wheeler T.A., Boman R.K. and Lascano R.J., 2003. Influence of Landscape position, soil series and phosphorus fertilizer on cotton lint yield. Agron. J., 95, p. 949-957.

Boquet D.J. and Breitenbeck G.A., 2000. Nitrogen rate effect on partitioning of nitrogen and dry matter by cotton. Crop Sci, 40, p. 1685-1693.

Cahill Sh., Johnson A., Osmond D.L, Hardy D.H., 2008. Response of Corn and Cotton to Starter Phosphorus on Soils Testing Very High in Phosphorus. Agronomy Journal, 100 (3), p. 537-542.

Christidis N, 1985. The cotton. Ed. Athens, p. 312-322 (Gr).

Clawson E.L., Cothren J.T. and Blouin D.C., 2006. Nitrogen fertilization and yield of cotton in ultra-narrow and conventional row spacings. Agron. Journal, 98, p. 72-79.

Clawson E., Cothren J.T., Blouin D.C. and Satterwhite J.L., 2008. Timing of Maturity in Ultra-Narrow and Conventional Row Cotton as Affected by Nitrogen Fertilizer Rate. Agronomy Journal, Vol. 100, Issue 2, p. 421–431.

Clement-Bailey J. and Gwathmey C., 2008. Earliness, partitioning and yield responses to potassium in cotton cultivars. Paper 1302, World Cotton Res. Conf.-4, 10-14 Sep 2007, Lubbock TX. ©2008 Omnipress [CD-ROM].

Coker D., Oosterhuis D. and Brown R., 2009. Cotton Yield Response to Soil- and Foliar-Applied Potassium as Influenced by Irrigation. The Journal of Cotton Science, 13, p. 1-10.

Cope J., 1984. Relationships among rates of N, P and K, soil test values, leaf analysis and yield of cotton at 6 locations. Commun. Soil Sci. Plant Anal., 15, p. 253-276.

Copur O., 2006. Determination of yield and yield components of some cotton cultivars in semi arid conditions. Pak. J. Biol. Sci., 9, p. 2572-2578.

Crozier C., Walls B., Hardy D. and Barnes J., 2004. Response of Cotton to P and K Soil Fertility Gradients in North Carolina. J. Cotton. Sci., 8, p. 130-141.

Dong H., Li W., Eneji A., Zhang D., 2012. Nitrogen rate and plant density effects on yield and late-season leaf senescence of cotton raised on a saline field. Field Crops Research, 126, p. 137-144.

Deshpande R. and Lakhdive B., 1994. Effect of plant growth substances and P levels on yield and P uptake by cotton. PKV Res. J., 18, p. 118-121.

Dorahy Ch., Rochester I., Blair Gr., Till A., 2008. Phosphorus Use-Efficiency by Cotton Grown in an Alkaline Soil as Determined Using ^{32}Phosphorus and ^{33}Phosphorus Radio-Isotopes. J. of Plant Nutrition, Vol. 31, Issue 11, p. 1877-1888.

Fageria N.K. and Baligar V.C., 2005. Enhancing nitrogen use efficiency in crop plants. Adv. Agron. 88, p. 97-185.

Fritschi F.B., Roberts B.A., Travis R.L., Rains D.W. and Hutmacher R.B., 2003. Response of irrigated Acala and Pima cotton to nitrogen fertilization: Growth, dry matter partitioning, and yield. Agron. J., 95, p. 133-146.

Javaid R., Ranjha A., Waheed T. and Ahmed I., 2001. Genotypic Differences Among Cotton Genotypes for Phosphorus Use Efficiency and Stress Factor. Intern. J. of Agriculture & Biology, 03-2, p. 186-187.

Gerik T.J., Jackson B.S., Stocckle C.O. and Rosenthal W.D., 1994. Plant nitrogen status and boll load of cotton. Agron. Journal, 86, p. 514-518.

Gerik T., Oosterhuis D. and Torbert H., 1998. Managing cotton nitrogen supply. Advances in Agronomy, 64, p. 115 -147.

Gill K., Sherazi S., Iqbal J., Ramzan M., Shaheen M. and Ali Z., 2000. Soil Fertility Investigations on Farmers Fields in Punjab. Soil Fertility Research Institute, Department of Agriculture, Govt. of Punjab, Lahore, Pakistan, p. 133-135.

Girma K., Teal R., Freeman K., Boman R., Raun W., 2007. Cotton Lint Yield and Quality as Affected by Cultivar and Long-Term Applications of N, P, and K Fertilizers. 15th annual 2017 NUE Conference, Baton Rouge, LA, p. 12-19.

Gushevilov J. and Karev K., 2000. Sustainable fertilizing influence on the yields and quality of cotton grown on leached cinnamonic forest soil. Plant Sci., 37, p. 150-156.

Hallikeri S.S., Halemani H.I., Patil V.C., Palled Y.B., Patil B.C. and Katageri I.S., 2010. Effect of nitrogen levels, split application of nitrogen and detopping on seed cotton yield and fibre quality in Bt-cotton. Karnataka J. Agric. Sci., 23 (3), p. 418-422.

Howard D.D., Gwathmey C.O., Essington M.E., Roberts R.K., and Mullen M.D., 2001. Nitrogen fertilization of no-till cotton on loess- derived soils. Agron. J., 93, p. 157-163.

Karamanidis G., Nechev Hr. and Stoilova A., 2004. Effect of fertilization and plant density on Bulgarian and Greece cotton cultivars growth in Orestiada district - Greece. Proc. Conference USB – St. Zagora, June 3-4 2004, p. 67-70.

Karthikeyan P. and Jayakumar R., 2001. Nitrogen and chlormequat chloride on cotton cultivar, p. 806-807. In: W.J. Horst et al. (ed.) Plant nutrition: food security and sustainability of agro-ecosystems through basic and applied research. Intl. Plant Nutr. Colloq., 14th, Hannover, Germany. 27 July-3 Aug.

2001. Kluwer Academic Publishers, Dordrecht, Netherlands.

Kaynak M., 1995. A research on the correlation and path coefficient analysis of the yield, yield components and fiber characteristics in cotton (G. hirsutum L.). J. Faculty Agric. Harran University, 1, p. 20-31.

Kirchmann H. and Thorvaldsson G., 2000. Challenging targets for future agriculture, European Journal of Agronomy, Vol. 12, Issues 3-4, p. 145-161.

Kostadinova S. and Panayotova G., 2003. Agronomic Efficiency of Nitrogen Fertilization at Bulgarian Cotton Cultivars. Ecology and Future, vol. I, 1, p. 43-46.

Leffler H., 1986. Mineral compartmentation within the boll. Chapter 21, p. 301-309. In J.R. Mauney and J.McD. Stewart (eds.), Cotton Physiology. The Cotton Foundation. Memphis, TN.

Malik M., Chaudhry F. and Makhdum M., 1996. Investigation on phosphorus availability and seed cotton yield in silt loam soils, J. An. Plant Science, 6 (12), p. 21-23.

McConnell J.S., Baker W., Miller D., Frizzell BS. and Varvil J.J., 1993. Nitrogen Fertilization of Cotton Cultivars of Differing Maturity. Agronomy Journal, Vol. 85, No. 6, p. 1151-1156.

McConnell J.S., Baker W.H., and Kirst, R.C.Jr., 1998. Yield and petiole nitrate concentrations of cotton treated with soil-applied and foliar-applied nitrogen [Online]. J. Cotton Sci., 2, p. 143-152.

McConnell J., Meyers B. and Mozaffari M., 2003. Varietal Responses of Cotton to Nitrogen Fertilization. In: Wayne E. Sabbe, Arkansas Soil Fertility Studies, p. 32-33.

Mitchell C.C., 2000. Cotton response to P in Alabama's long-term experiments. In: Proceeding Beltwide Cotton Conference, Vol. 2, National Cotton Council of America, Memphis, TN, p. 1420-1425.

Moore S.H., 1999. Nitrogen effect on position of harvestable bolls in cotton. J. Plant Nutr., 22, p. 901-909.

Munir M., Tahir M., Saleem M. and Yaseen M., 2015. Growth, yield and earliness response of cotton to row spacing and nitrogen management. The Journal of Animal & Plant Sciences, 25(3), p. 729-738.

Oosterhuis D., 2001. Physiology and nutrition of high yielding cotton in the USA. Informações agronômicas, 95, p. 18-24.

Panayotova G., Stoilova A. and Valkova N., 2007. Response of New Cotton Varieties to Fertilizing Levels at Stationary Trial Condition. Field Crops Studies, vol. IV - 2, p. 327-332.

Pettigrew W.T. and Adamczyk J.J., 2006. Nitrogen Fertility and Planting Date Effects on Lint Yield and Cry1Ac (Bt) Endotoxin Production. Agron. J., 98, p. 691-697.

Prasad M. and Siddique M.R., 2004. Effect of nitrogen and mepiquat chloride on yield and quality of upland cotton (Gossypium hirsutum). Indian J. Agric. Sci., 74, p. 560-562.

Rashidi M., Gholami M., 2011. Nitrogen and boron effects on yield and quality of cotton (Gossypium hirsutum L.). International Res J. Agri. Sci. and Soil Sci, 1, p. 118-125.

Read J.J., Reddy K.R. and Jenkins J.N., 2006. Yield and fiber quality of Upland cotton as influenced by nitrogen and potassium nutrition. Europ. J. Agron., 24, p. 282-290.

Reiter J.S., and Kreig D.R., 2000. Texas research shows fertigation is a viable option to save cotton growers both time and money on fertilizer inputs. Fluid Journal, Issue 29, 8 (2), p. 20-22.

Rutto E., Arnall B., May J., Butchee K. and Raun W., 2013. Ability of cotton (*Gossypium hirsutum* L.) to recover from early season nitrogen deficiency. J. Cot. Sci., 17, p. 70-79.

Saleem M., Shakeel A., Bilal M., Shahid M. and Anjum S., 2010. Effect of different phosphorus levels on earliness and yield of cotton cultivars. Soil & Environ., 29 (2), p. 128-135.

Sawan Z., Mahmoud M. and El-Guibali A., 2006. Response of Yield, Yield Components, and Fiber Properties of Egyptian Cotton (*Gossypium barbadense* L.) to Nitrogen Fertilization and Foliar-applied Potassium and Mepiquat Chloride. The J. of Cotton Science, Vol. 10, Issue 3, p. 224-234.

Sawan Z., Mahmoud M. and El-Guibali A., 2008. Influence of K fertilization and foliar application of zinc and P on growth, yield and fiber properties of Egyptian cotton. J. Plant Ecol., no. 1, p. 259-270.

Seilsepour M. and Rashidi M., 2011. Effect of different application rates of nitrogen on yield and quality of cotton (*Gossypium hirsutum*). American-Eurasian J. Agric. & Environ. Sci., 10 (3), p. 366-370.

Setatou H.B. and Simonis A.D., 1996. Effect of Time and Rate of Nitrogen Application on Cotton. Fertilizer Research, 43, p. 49-53.

Stewart J.McD., 1986. Integrated Events in the Flower and Fruit. In: J.R. Mauney and J. Mcd. Stewart (Eds.). Cotton Physiology. Cotton Foundation, Memphis, Tenn., p. 261-297.

Stewart J., Oosterhuis D., Heitholt J. and Mauney J. Editors. 2010. Cotton Physiology, Springer Publishing Co., New York.

Surya K., Bi Y. and Rothstein S., 2010. Understanding plant response to nitrogen limitation for the improvement of crop nitrogen use efficiency. J. Exp. Bot., 62(4), p. 1499-1509.

Yang G., Tang H., Nie Y., Zhang X., 2011. Responses of cotton growth, yield, and biomass to nitrogen split application ratio. European Journal of Agronomy, Vol. 35, Issue 3, p. 164-170.

Zhao D. and Oosterhuis D., 2000. Nitrogen application effect on leaf photosynthesis, nonstructural carbohydrate concentrations and yield of field-grown cotton. Spec. Rep. 198. Arkansas Agric. Exp. Stn., Fayetteville, AR.

18

RESEARCH REGARDING THE INFLUENCE OF NITROGEN AND PHOSPHORUS FERTILIZATION ON THE YIELD OF GRAIN SORGHUM HYBRIDS

Cristina Andreea OPREA, Doru Ioan MARIN, Ciprian BOLOHAN, Aurelian PENESCU

University of Agronomic Sciences and Veterinary Medicine of Bucharest, 59 Marasti Blvd, District 1, Bucharest, Romania

Corresponding author email: oprea.andreeacristina@yahoo.com

Abstract

The fertilization is an essential technological factor in improving yield. The purpose of this research was to evaluate the effect of fertilization and the answer of grain sorghum hybrids in terms of yield and its components. Research conducted over two agricultural years (2013/2014-2014/2015) was located in the South-Eastern Romania area and was based on a split plot design. The paper presents the results regarding the following tested factors: Factor A- grain sorghum hybrids: Alize, Aquilon, Arack; Factor B - fertilization levels: N_0P_0, $N_{90}P_0$, $N_{90}P_{60}$, $N_{120}P_0$, $N_{120}P_{60}$. Regarding fertilization, both the use of nitrogen (N) and the fertilization with complex fertilizers (nitrogen and phosphorus) generated the increase of yield and its components with values statistically assured for the three researched hybrids. The average yield was 7.83 t/ha, varying from 5.61 t/ha (Arack hybrid unfertilized) to 9.22 t/ha (Aquilon hybrid fertilized $N_{120}P_{60}$). Hectolitre mass was highly significant influenced by the use of fertilizers, the highest value (80.74 kg/hl) being recorded by the hybrid Arack fertilized $N_{120}P_{60}$. The maximum value of a thousand seeds weight was recorded by the hybrid Aquilon fertilized $N_{120}P_{60}$ (24.15 g) and the lowest value was recorded by the hybrid Alize unfertilized (19.03 g) with variations between +0.95 g (Arack hybrid fertilized $N_{90}P_0$) and +3.56 g (Aquilon hybrid fertilized $N_{120}P_{60}$).

Key words: Sorghum, fertilization, grain yield, yield components.

INTRODUCTION

Sorghum bicolor (L.) is well known as a grain with a high capacity of exploiting natural resources (Varvel, 2000; Almondares, 2008; Borghia, 2013). Nevertheless grain sorghum production and yield are highly influenced by the interaction of the plants with environmental factors (Showemimo, 2007) and technological factors. Hydric stress, poor soils and a poor fertilization management are the main limiting factors of the yield (Aleminew, 2015). Water stress at panicle differentiation and during flowering determines the yield reduction, following the decrease in number of seeds by 45% (Tolka, 2013). Lack of nitrogen fertilization determines a yield reduction by nearly 19% (Smith, 1990 cited by Tucker, 2009). Thus, agricultural practices regarding fertilization management are essential in terms of increasing yield. Nitrogen (N) is the main determinant nutrient of yield (Baozhen, 2014; Akdeniz, 2006; Saber-Rezaii, 2009). Long term research (40years) conducted in Western

Kansas highlighted the positive influence of fertilization with different doses of N (up to 225 kg/ha) and P (up to 20 kg/ha) on the grain yield of sorghum. On average, during the 40 years, the grain yield increased by 31%. The highest average yield was recorded for the application of $N_{180}P_{20}$ (Stewart, 2005). Along with nitrogen (N), phosphorus (P) is important for the development of sorghum plants (Khalili, 2008; Sahrawat, 1999), P fertilization with a dose of 40 kg/ha once every two years satisfying the plant requirements under semi-arid agro-climatic conditions (Sahrawat, 2000).Phosphorus fertilization in irrigated cropping conditions has a positive influence on the yield and its elements (No. of seed/panicle, a thousand seeds weight) (Afshar, 2014). In Romania, research results underline the fact that sorghum reacts to fertilization rates of N_{50-80}, P_{30-60}, K_{40} kg/ha active substance (a.s.) (Sin, 2005 cited by Nica, 2011)and to fertilization doses of N_{60-120}, P_{60}, K_{60} kg/ha a.s. (Oprea, 2015).Research conducted on a sandy soil (in Craiova area, at Tâmbureşti) reveals the obtaining of better yields under the fertilization with $N_{160}P_{80}K_{80}$ (Matei, 2011).Other authors (Munteanu L.M. and Tabără V., 2011, 2012) also show the fertilization influence on the yield of grain sorghum cultivated on a brown soil, the maximum yield being obtained for the fertilization with $N_{240}P_{80}K_{80}$, respectively 8,214 kg/ha in 2010 and 6,770 kg/ha in 2011.

The purpose of this research was to determine the influence of applying different doses of nitrogen and phosphorus on the grain yield of some sorghum hybrids (*Sorghum bicolor* (L.) Moench, in the field conditions of South-Eastern Romania.

MATERIALS AND METHODS

The research was conducted over two agricultural years (2013/2014 and 2014/2015) in the South-Eastern area of Romania, in Ilfov County. The experiments were placed at the Experimental Field of Moara Domnească Didactic Farm (44°30' North latitude, 26°13' East longitude, 90 m altitude) on a chromic luvos oil with a moderately acidic reaction.

The experimental design was developed in order to study the effect of different fertilization doses of nitrogen and phosphorus over the grain yield of sorghum hybrids. The experiments were based on the split plot method and the tested factors were the following:

- Factor A -grain sorghum hybrids: a_1 - Alize, a_2 - Aquilon, a_3 - Arack;
- Factor B -fertilization treatment with the following graduations: b_1 - N_0P_0, b_2 - $N_{90}P_0$, b_3 - $N_{120}P_0$, b_4 - $N_{90}P_{60}$, b_5- $N_{120}P_{60}$.

In the two agricultural years of research the sorghum hybrids were sown at the distance of 70 cm between rows, on the 2nd of May 2014 and on the 24th of April 2015. The fertilization treatments were applied when preparing the field for sowing and during the vegetative stages, at 36 days after sowing (2013/2014) and 41 days after sowing (2014/2015). Regarding the maintenance of the crop in both years Dual Gold 1.5 l/ha (S-metolachlor 960 g/l) herbicide (pre-emergence) and Dicopur Top 1 l/ha (2.4-D dimethylamine salt 600 g/l) herbicide (after emergence) were used to control weeds, two mechanical hoes were also applied. Pests control (*Tanymecus dilaticollis)* was carried out using Calypso (150 ml/ha).

Normal climatic conditions of the research area, presented in Table 1, are characterized by rainfalls with values of 301 mm and average temperatures of 19.5°C. In the agricultural year 2013/2014, during the plants growing season, rainfalls recorded a value similar to the normal standard of the area (291 mm) but with different distribution. In terms of temperature, the average value during the growing season was 2° C higher compared to the multiannual values. In the agricultural year 2014/2015 the average temperature during sorghum vegetation period recorded values with 2.8° C higher than the normal temperature of the season and 0.8° C higher compared to the same interval in the previous year. Regarding rainfalls, these recorded in the growing season values by approx. 49% lower compared to the normal precipitations values characteristic to the same period (144 mm).

Table 1. Climatic conditions in the agricultural years 2013/2014 and 2014/2015 at Moara Domnească, Ilfov

Month	Temperature (°C)			Rainfall (mm)		
	Year 2013-2014	Year 2014-2015	Normal	Year 2013-2014	Year 2014-2015	Normal
October	14.0	11.8	11.0	81.7	64.2	35.8
November	8.3	6.0	5.3	17.6	49.1	40.6
December	-0.2	1.4	0.4	1.2	84.6	36.7
January	-0.5	-1.5	-3.0	33.2	33.4	30
February	1.2	2.0	-0.9	7.6	21.4	32.1
March	8.9	6.2	4.4	37.3	65	31.6
April	13.4	11.7	11.2	116.0	2	48.1
May	19.3	18.6	16.5	88.0	33.6	67.7
June	19.9	21.0	20.2	113.0	56.8	86.3
July	22.8	25.3	22.1	38.0	5.2	63.1
August	24.1	24.3	21.1	26.2	48.4	50.5
September	21.3	-	17.5	25.8	-	33.6
Avg./Sum	12.7	11.5	10.5	585.6	463.7	556.1
Avg./Sum* May–Sept	21.5	22.3	19.5	291.0	144.0	301.2

*May-15 Sept (2014); 24 Apr-1 Sept. (2015)

RESULTS AND DISCUSSIONS

Hybrid influence on grain sorghum yield and yield components, Moara Domnească

Grain yield of sorghum hybrids researched in the field conditions of South - Eastern Romania, in the agricultural years 2013/2014 - 2014/2015recorded an average value of nearly

7.8 t/ha (Table 2). Under hybrid's influence grain sorghum yield recorded variations from - 3.8% to +4.4%, but without being statistically assured. Aquilon hybrid recorded positive yield increases (between 2.3% and 4.6%) compared to Control. Alize and Arack hybrids recorded mainly negative yield variations compared to Control, the average yield being by 2.4% and 0.8% lower. Arack hybrid unfertilized had the minimum yield (5.61 t/ha) and Aquilon hybrid

fertilized $N_{120}P_{60}$ had the maximum yield of 9.22 t/ha. Compared to Alize, Aquilon hybrid recorded yield increases statistically assured for the fertilization with complex fertilizers ($N_{90}P_{60}$ and $N_{120}P_{60}$) while compared to Arack the yield had a significant variation for the fertilization treatment $N_{90}P_{60}$. Yield differences between the hybrids Alize and Arack varied depending of the fertilization treatment from -0.23 t/ha to +0.42 t/ha and were not statistically assured.

Table 2. Hybrid influence on sorghum grain yield (t/ha)
(Moara Domnească, Average 2014-2015)

Variants	Alize (a₁)		Aquilon (a₂)		Arack (a₃)		Average Hybrids			Diff. between hybrids		
	GY (t/ha)	%	GY (t/ha)	%	GY (t/ha)	%	GY (t/ha)	%	Signf.	a₂-a₁ (t/ha)	a₃-a₁ (t/ha)	a₃-a₂ (t/ha)
N₀ P₀ (b₁)	5.83 ns	100.5	5.98 ns	103.0	5.61 ns	96.6	5.81	100	Ct	0.15 ns	-0.23 ns	-0.37 ns
N₉₀P₀ (b₂)	7.68 ns	98.1	8.09 ns	103.4	7.72 ns	98.6	7.83	100	Ct	0.41 ns	0.04 ns	-0.37 ns
N₁₂₀ P₀ (b₃)	8.02 ns	97.2	8.44 ns	102.3	8.30 ns	100.6	8.26	100	Ct	0.42 ns	0.28 ns	-0.14 ns
N₉₀ P₆₀ (b₄)	8.04 ns	96.8	8.69 ns	104.6	8.18 ns	98.5	8.30	100	Ct	0.65*	0.14 ns	-0.50 o
N₁₂₀ P₆₀ (b₅)	8.65 ns	96.3	9.22 ns	102.7	9.06 ns	101.0	8.98	100	Ct	0.57*	0.42 ns	-0.16 ns
Avg.	7.64 ns	97.6	8.08 ns	103.2	7.77 ns	99.2	7.83	100	Ct	0.44 ns	0.13 ns	-0.31 ns

LSD 5% = 0.50 t/ha; LSD 1% = 0.70 t/ha; LSD 0.1% =0.97 t/ha
GY = Grain Yield at STAS humidity (14%); Ct = Control; ns = no significance

Analyzing the hybrids' influence on a thousand seeds weight (TSW, g), data presented in Table 3 show that Aquilon hybrid recorded the highest value of TSW (24.15 g) for the fertilization treatment $N_{120}P_{60}$; the minimum value was recorded by Alize hybrid un fertilized (19.03 g). Compared to Control the three hybrids recorded variations of the TSW between -0.85 g (Alize fertilized with $N_{90}P_{60}$) and +1.19 g (Aquilon fertilized with $N_{90}P_0$). Increases of a thousand seeds weight statis- tically assured compared to Control were recorded by Aquilon hybrid for 4 fertilization

treatments ($N_{90}P_0$, $N_{120}P_0$, $N_{90}P_{60}$, $N_{120}P_{60}$). Also, Aquilon present statistically assured increases of TSW compared to Alize for all the fertilization variants, with values between 1.6 g and 2.01 g. Differences between the hybrids Arack and Aquilon ranged from -0.86 g and - 1.62 g, being distinctly significant when fertilizers were used ($N_{90}P_0$, $N_{120}P_0$, $N_{90}P_{60}$, $N_{120}P_{60}$). A thousand seeds weight for the hybrid Arack varied negligible in relation to Alize hybrid, the differences were between +0.20 g. and +0.69 g.

Table 3. Hybrid influence on grain sorghum hybrids'a thousand seeds weight (TSW, g)
(Moara Domnească, Average 2014-2015)

Variants	Alize (a₁)		Aquilon (a₂)		Arack (a₃)		Average Hybrids			Diff. between hybrids		
	TSW (g)	%	TSW (g)	%	TSW (g)	%	TSW (g)	%	Signf.	a₂-a₁ (g)	a₃-a₁ (g)	a₃-a₂ (g)
N₀ P₀ (b₁)	19.03ns	96.2	20.59 ns	104.1	19.72 ns	99.7	19.78	100	Ct	1.55**	0.69 ns	-0.86 ns
N₉₀P₀ (b₂)	20.26 ns	96.2	22.25*	105.6	20.67 ns	98.1	21.06	100	Ct	1.98***	0.40 ns	-1.58oo
N₁₂₀ P₀ (b₃)	21.65 ns	97.4	23.34*	105.0	21.72 ns	97.7	22.24	100	Ct	1.69***	0.06 ns	-1.62oo
N₉₀ P₆₀ (b₄)	20.69 ns	96.1	22.70*	105.4	21.21 ns	98.5	21.53	100	Ct	2.01***	0.52 ns	-1.49oo
N₁₂₀ P₆₀ (b₅)	22.54 ns	97.4	24.15*	104.3	22.74 ns	98.3	23.14	100	Ct	1.60**	0.20 ns	-1.40oo
Avg.	20.84 ns	96.7	22.60*	104.9	21.21 ns	98.4	21.55	100	Ct	1.77***	0.38 ns	-1.39oo

LSD 5% =0.89 gr; LSD 1% =1.22 gr; LSD 0.1% =1.65 gr
TSW = a thousand seeds weight; Ct = Control; ns = no significance

The hectolitre mass (HLM) of the grain sorghum hybrids recorded an average value of 78.48 kg/hl (Table 4). Under hybrid's influence the indicator recorded insignificant variations, nevertheless the maximum value of HLM was recorded by Arack fertilized $N_{120}P_{60}$ (80.74 kg/hl), while Alize unfertilized had the minimum value (73.65 kg/hl). Compared to Control the influence of the three sorghum hybrids on the hectolitre mass is not significant, the indicator recording variation's values from -0.94 kg/hl (Alize unfertilized) to 0.86 kg/hl (Arack fertilized $N_{120}P_{60}$). The hybrids' hectolitre mass differences varied between -0.17 kg/hl and +1.60 kg/hl, being mainly insignificant. A comparison between hybrids' HLM reveals differences statistically assured between Aquilon and Alize, Arack and Alize for the unfertilized variant and also between Arack and Aquilon, when fertilization treatments with $N_{120}P_0$ and $N_{120}P_{60}$ were applied.

Table 4. Hybrid influence on grain sorghum hybrids' hectolitre mass (HLM kg/hl)
(Moara Domnească, Average 2014-2015)

Variants	Alize (a_1)		Aquilon (a_2)		Arack (a_3)		Average Hybrids			Diff. between hybrids		
	HLM (Kg/hl)	%	HLM (Kg/hl)	%	HLM (Kg/hl)	%	HLM (Kg/hl)	%	Signf.	a_2-a_1 (Kg/hl)	a_3-a_1 (Kg/hl)	a_3-a_2 (Kg/hl)
$N_0 P_0$ (b_1)	73.65ns	98.7	74.87 ns	100.4	75.25 ns	100.9	74.59	100	Ct	1.21*	1.60**	0.38 ns
$N_{90}P_0$ (b_2)	78.64 ns	99.6	79.00 ns	100.1	79.22 ns	100.3	78.95	100	Ct	0.36ns	0.58 ns	0.22 ns
$N_{120} P_0$ (b_3)	79.55 ns	99.6	79.47 ns	99.5	80.47 ns	100.8	79.83	100	Ct	-0.08 ns	0.92 ns	1.00 *
$N_{90} P_{60}$ (b_4)	78.93 ns	99.7	79.19 ns	100.0	79.35 ns	100.2	79.16	100	Ct	0.26 ns	0.42 ns	0.16 ns
$N_{120} P_{60}$ (b_5)	79.95 ns	100.1	79.77 ns	98.8	80.74 ns	101.1	79.88	100	Ct	-0.17ns	0.79 ns	0.97*
Avg.	78.14 ns	99.6	78.29 ns	99.8	79.01 ns	100.7	78.48	100	Ct	0.32 ns	0.86 ns	0.55 ns

LSD 5% =0.88 gr; LSD 1% =1.22 gr; LSD 0.1% =1.77 gr
HLM = hectolitre mass; Ct = Control; ns = no significance

Fertilization influence on grain sorghum yield components, Moara Domnească

Nitrogen and phosphorus fertilization (Table 5) generated an average yield growth from 34.8% to 54.6%, statistically assured compared to Control (that was not fertilized). The positive influence of fertilization on the grain yield was significant for all the hybrids. Nitrogen fertilization with doses of 90 kg/ha and 120 kg/ha (a.s.) generated significant positive yield increases from 1.85 t/ha (Alize) to 2.70 t/ha (Arack) while $N_{90}P_{60}$ and $N_{120}P_{60}$ fertilization treatments generated an yield growth statistically assured with values between 2.21 t/ha (for the hybrid Alize fertilized $N_{90}P_{60}$) and 3.45 t/ha (for the hybrid Arack fertilized $N_{120}P_{60}$).

Table 5. Fertilization influence on sorghum grain yield (GY t/ha)
(Moara Domnească, Average 2014-2015)

Variants	Alize				Aquilon				Arack				Average hybrids			
	GY (t/ha)	%	Diff. (t/ha)	Signf	GY (t/ha)	%	Diff. (t/ha)	Signf	GY (t/ha)	%	Diff. (t/ha)	Signf	GY (t/ha)	%	Diff. (t/ha)	Signf
$N_0 P_0$ (Ct)	5.83	100.0	Ct	-	5.98	100.0	Ct	-	5.61	100.0	Ct	-	5.81	100.0	Ct	-
$N_{90}P_0$	7.68	131.6	1.85	***	8.09	135.3	2.11	***	7.72	137.6	2.11	***	7.83	134.8	2.02	***
$N_{120}P_0$	8.02	137.5	2.19	***	8.44	141.2	2.46	***	8.30	148.1	2.70	***	8.26	142.2	2.45	***
$N_{90}P_{60}$	8.04	137.8	2.21	***	8.69	145.3	2.71	***	8.18	145.9	2.58	***	8.30	143.0	2.50	***
$N_{120}P_{60}$	8.65	148.2	2.81	***	9.22	154.2	3.24	***	9.06	161.6	3.45	***	8.98	154.6	3.17	***
Avg.	7.64	131.0	1.81	***	8.08	135.2	2.10	***	7.77	138.7	2.17	***	7.83	134.9	2.03	***

LSD 5% = 0.61 t/ha; LSD 1% = 0.81 t/ha; LSD 0.1% = 1.06 t/ha
GY = Grain Yield at STAS humidity (14%); Ct = Control

Considering the influence of nitrogen and phosphorus fertilization treatments on sorghum hybrids' a thousand seeds weight (Table 6), on average this varied between 19.78 g for the unfertilized variant and 23.14 g for the application of $N_{120}P_{60}$. The fertilization treatment $N_{90}P_0$ generated a significant growth of the indicator compared to Control, the application of $N_{90}P_{60}$ determined a distinctly significant increase and the fertilization with

$N_{120}P_0$ and $N_{120}P_{60}$ had a very significant positive influence on TSW.

Nitrogen fertilization with a rate of 90 kg/ha active substance generated an increase of the TSW statistically assured for the hybrid Aquilon (+1.66 g), while for the hybrids Alize and Arackthe indicator variations were not significant. The treatment with complex fertilizers $N_{90}P_{60}$ determined a significant growth of a thousand seeds weight for the hybrids Arack and Alize, with 1.49 g and 1.65 g respectively, compared to the unfertilized variant, and distinctly significant for Aquilon (+2.11 g compared to Control). The use of nitrogen in a dose of 120 kg/ha (a.s.) and the fertilization with $N_{120}P_{60}$ generated TSW's increases statistically assured compared to Control, for all three tested hybrids.

Table 6. Fertilization influence on grain sorghum a thousand seeds weight (TSW, g)
(Moara Domnească, Average 2014-2015)

Variants	Alize				Aquilon				Arack				Average hybrids			
	TSW (g)	%	Diff. (g)	Signf	TSW (g)	%	Diff. (g)	Signf	TSW (g)	%	Diff. (g)	Signf	TSW (g)	%	Diff. (g)	Signf
$N_0 P_0$ (Ct)	19.03	100.0	Ct	-	20.59	100.0	Ct	-	19.72	100.0	Ct	-	19.78	100.0	Ct	-
$N_{90}P_0$	20.26	106.5	1.23	ns	22.25	108.1	1.66	*	20.67	104.8	0.95	ns	21.06	106.5	1.28	*
$N_{120}P_0$	21.65	113.8	2.62	***	23.34	113.4	2.76	***	21.72	110.1	2.00	**	22.24	112.4	2.46	***
$N_{90}P_{60}$	20.69	108.7	1.65	*	22.70	110.3	2.11	**	21.21	107.5	1.49	*	21.53	108.9	1.75	**
$N_{120}P_{60}$	22.54	118.4	3.51	***	24.15	117.3	3.56	***	22.74	115.3	3.02	***	23.14	117.0	3.36	***
Avg.	20.84	109.5	1.80	**	22.60	109.8	2.02	**	21.21	107.6	1.49	*	21.55	108.9	1.77	**

LSD 5% =1.25 gr.; LSD 1%=1.66 gr.; LSD 0.1% = 2.17 gr
TSW = 1,000 seeds weight; Ct = Control; ns = no significance

The use of fertilizers determined increases of the hectolitre mass statistically assured for each of the three researched grain sorghum hybrids (Table 7). The average HLM varied following the application of fertilizers with values between +4.36 kg/hl and +5.29 kg/hl, statistically assured for all the fertilization treatments. Thus, the application of nitrogen in dose of 90 kg/ha and 120 kg/ha (a.s.) determined the growth of the HLM with values ranging from 3.79 kg/hl to 5.90 kg/hl, and the use of complex fertilizers ($N_{90}P_{60}$ and $N_{120}P_{60}$) generated increases of the indicator from 4.10 kg/hl to +6.30 kg/hl, compared to Control.

Table 7. Fertilization influence on grain sorghum hybrids'hectolitre mass (HLM kg/hl)
(Moara Domnească, Average 2014-2015)

Variants	Alize				Aquilon				Arack				Average hybrids			
	HLM Kg/hl	%	Diff. Kg/hl	Signf	HLM Kg/hl	%	Diff. Kg/hl	Signf	HLM Kg/hl	%	Diff. Kg/hl	Signf	HLM Kg/hl	%	Diff. Kg/hl	Signf
$N_0 P_0$ (Ct)	73.65	100.0	Ct	-	74.87	100.0	Ct	-	75.25	100.0	Ct	-	74.59	100.0		-
$N_{90}P_0$	78.64	106.8	4.99	***	79.00	105.5	4.13	***	79.22	105.3	3.97	***	78.95	105.8	4.36	***
$N_{120}P_0$	79.55	108.0	5.90	***	79.47	106.2	4.60	***	80.47	106.9	5.22	***	79.83	107.0	5.24	***
$N_{90}P_{60}$	78.93	107.2	5.28	***	79.19	105.8	4.32	***	79.35	105.5	4.10	***	79.16	106.1	4.57	***
$N_{120}P_{60}$	79.95	108.5	6.30	***	78.94	105.4	4.08	***	80.74	107.3	5.49	***	79.88	107.1	5.29	***
Avg.	78.14	106.1	4.49	***	78.29	104.6	3.43	***	79.01	105.0	3.76	***	78.48	105.2	3.89	***

LSD 5% =1.35 kg/hl; LSD 1%=1.79 kg/hl; LSD 0.1% = 2.35 kg/hl
HM= hectoliter mass; Ct = Control

Correlation between fertilization and grain yield and yield components

The high positive correlation between fertilizer doses (a.s.) applied to the analyzed sorghum hybrids and their grain yield is supported by a coefficient with the value r = 0.9788[***] (very significant positive) (Figure 1). The relationship between grain yield and the quantity of fertilizers active substance applied is described by a regression line with a positive slope, b = 0.0171, meaning that for every additional kg of fertilizer active substance applied per ha an increase of 0.017 t/ha in grain yieldis expected. The y-intercept of 5.9915 shows that in the absence of fertilization (x=0) the expected average grain yield is nearly 5.99 t/ha.

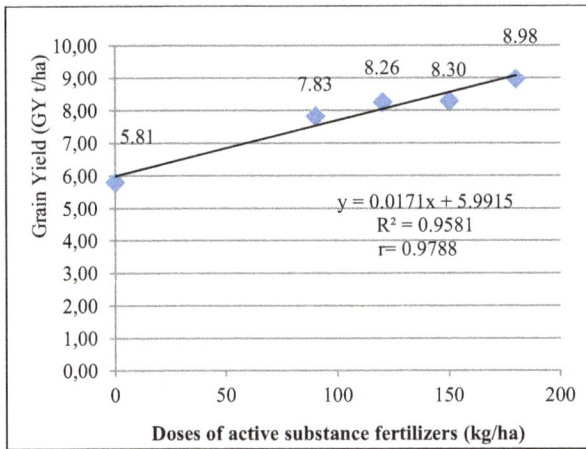

Figure 1. Correlation between the grain yield (GY t/ha) of sorghum hybrids and the doses of active substance fertilizers applied (kg/ha)

The relationship between a thousand seeds weight and the doses of fertilizers (a.s.) is described by a regression line with a positive slope, that allows the association of fertilizers doses (a.s.) to a higher TSW, supported by a high correlation coefficient r = 0.9228[***] (Figure 2). The yield indicator's value is mainly influenced (85%) by the quantity of active substance fertilizers applied, while in the lack of fertilization grain sorghum hybrids record an average TSW of nearly 19.73 g.

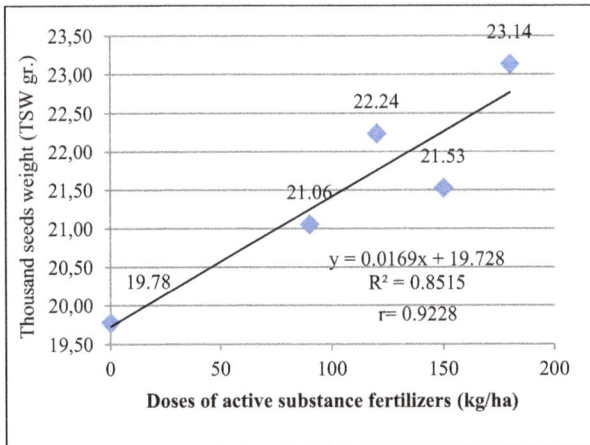

Figure 2. Correlation between a thousand seeds weight (TSW g) of sorghum hybrids and the doses of active substance fertilizers applied (kg/ha)

The influence of the doses of active substance fertilizers on the hectolitre mass of sorghum hybrids (Figure 3) is supported by a correlation coefficient with the value r=0.9113[***].R square (R^2=0.8304) shows that the increase of the hectolitre mass, for the three researched

hybrids, is due in a proportion of 83% to the doses of active substance fertilizers applied. The positive slope of the regression line shows that an increase of the hectoliter mass by 0.0292 kg/hl is expected for every additional kg of fertilizer active substance applied per ha. Much more, in the absence of fertilization the average HLM of grain sorghum hybrids was approx. 75.33 kg/hl.

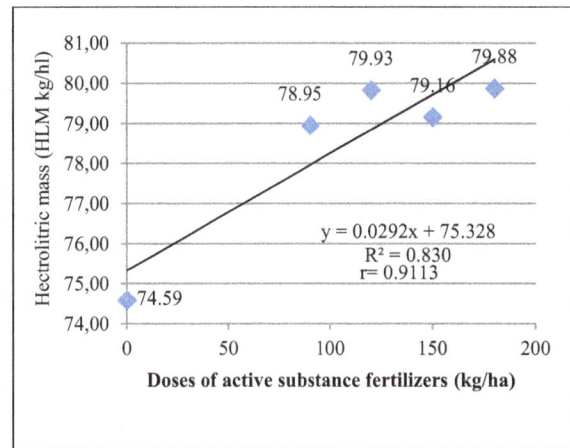

Figure 3.Correlation between hectolitre mass (HLM kg/hl) of sorghum hybrids and the doses of active substance fertilizers applied (kg/ha)

CONCLUSIONS

In the field conditions of South-Eastern Romania, corresponding to the agricultural years 2013/2014 - 2014/2015, research results regarding the influence of fertilization and hybrid on grain sorghum yield and its components conducted to the following conclusions:
- Under the hybrid influence the average grain yield varied between 7.64 t/ha for Alize and 8.08 t/ha for Aquilon.
- Aquilon hybrid had the highest value of a thousand seeds weight (24.15 g), recording significant increases of the indicator compared both to Alize and Arack; the lowest value of the indicator was recorded by the hybrid Alize (19.03 g).
- The average hectolitre mass of the researched hybrids had a value of 78.48 kg/hl, the highest value was recorded by the hybrid Arack (79.01 kg/hl) and the lowest value by thehybrid Alize (78.14 kg/hl).
- Under the hybrid influence the hectolitre mass recorded variations from -0.17 kg/hl to +1.60 kg/hl, with differences statistically assured between the hybrids

Aquilon and Alize, Arackand Alize, Arackand Aquilon.

- Nitrogen fertilization with rates of 90 kg/ha and 120 kg/ha (a.s.) determined yield increases between 1.85 t/ha (Alize) and 2.70 t/ha (Arack).
- Under the influence of different fertilization treatments sorghum hybrids recorded TSW's increases statistically assured with values between 1.66 gand 2.76 gand HLM's increases with values form 3.79 kg/hl and 5.90 kg/hl.
- The application of complex fertilizers likeN$_{90}$P$_{60}$ and N$_{120}$P$_{60}$had a strongly significant positive influence on the grain yield of three researched hybrids, with average increases of 2.5 and 3.17 t/ha. Under the fertilization with N$_{90}$P$_{60}$ the average TSW increased by 1.75 g and HLM recorded a variation of +4.57 kg/hl. N$_{120}$P$_{60}$ fertilization generated TWS's and HLM's growth statistically assured with values of 3.36 g and 5.29 kg/hl respectively.

REFERENCES

Afshar R.K., Jovini M.A., Chaichi M.R., Hashemi M., 2014. Grain Sorghum Response to Arbuscular Mycorrhiza and Phosphorus Fertilizer under Deficit Irrigation.Agronomy Journal 106, p. 1212-1218.

Akdeniz H., Yilmaz I., Bozkurt M.A., Keskin B., 2006. The Effects of Sewage Sludge and Nitrogen Applications on Grain Sorghum Grown (*Sorghum vulgare* L.) in Van-Turkey. Polish Journal of Environmental Studies Vol. 15, No. 1, p. 19-26.

Aleminew A., 2015. Yield Response of Local Long Maturing Sorghum Varieties to Timing of Nitrogen Fertilizer Application in Eastern Amhara Region, Ethiopia. Journal of Biology, Agriculture and Healthcare Vol. 5, No. 3,p. 184-189.

Almodares A., Taheri R., III Min Chung, Fathi M., 2008. The effect of nitrogen and potassium fertilizers on growth parameters and carbohydrate contents of sweet sorghum cultivars. Journal of Environmental Biology 29(6), p. 849-852

BaozhenHao Q.X., 2014. Biomass production, water and nitrogen use efficiency in photoperiod-sensitive sorghum. Biomass and Bioenergy 62, p. 108-116.

Borghia E., CrusciolC.A.C., Nascente A.S., Sousa V.V., Martins P.O., Mateusd G.P, Costae C., 2013. Sorghum grain yield, forage biomass production and revenue as affected by intercropping time. Elsevier, Europ. J. Agronomy 51, p. 130-139.

Khalili A., Akbari N., Chaichi M.R., 2008. Limited Irrigation and Phosphorus Fertilizer Effects on Yield and Yield Components of Grain Sorghum (*Sorghum bicolor* L.var. *Kimia*). American-Eurasian J. Agric. & Environ. Sci., 3 (5): p. 697-702.

Matei Gh., 2011. Research on some technological measures for increasing the yields on grain sorghum cultivated on sandy soils from Tâmbureşti. Analele Universităţii din Craiova, seria Agricultură-Montanologie - Cadastru Vol. XLI 2011/1.

Munteanu L.M., Tabara V., 2011. Influence hybrids and fertilisation on the production of grain sorghum grains (*Sorghum bicolor* var. *Eusorghum*) in the experimental field from Răcăşadia Caraş-Severin. Research Journal of Agricultural Science, 43(4).

Munteanu L.M., Tabără V., 2012. Influence of culture on technology grain sorghum (*Sorghum bicolor* var. *Eusorghum*) in the experimental field from Răcăşadia. Research Journal of Agricultural Science, 44(1).

Nica V., 2011. Studiu comparativ privind productivitatea, calitatea productiei si rentabilitatea culturilor de porumb si sorg in conditiile pedoclimatice din Judetul Ialomita. Theses. A.M.C. - U.S.A.M.V. Bucuresti.

Oprea C.A., Marin D.I., Bolohan C., 2015. Influence of some technological factors on grain sorghum (*Sorghum bicolor* (L.) Moench var. *Eusorghum*) yield grown under the conditions of South-Eastern Romania. AgroLife Scientific Journal - Vol. 4, No. 1,p. 123-130.

Saber-Rezaii M., Amirnia R., Gadimzadeh M., Gorttapeh A.H., (2009). Influence of Nitrogen Foliar Application on Grain Yield and Protein Content of Grain Sorghum. Research Journal of Biological Sciences accessed la www.medwelljournals.com/fulltext/?doi=rjbsci.2009, p. 490-493.

Sahrawat K.L., 1999. Assessing the fertilizer phosphorus requirement of grain sorghum. Commun. Soil Sci. plant anal, 30, p. 1593-1601.

Sahrawat K.L., 2000. Residual phosphorus and management strategy for grain sorghum on a vertisoil. Commun. Soil Sci. plant anal, 30, p. 1593-1601.

Showemimo F.A., 2007. Grain yield response and stability indices in sorghum (*Sorghum bicolor (L.) Moench*). Commun. Biometry Crop Sci. 2 (1), p. 68-73.

Stewart W.M., Dibb D.W., Johnston A.E., Smyth T.J., 2005. The contribution of commercial fertilizer nutrients to food production, Agronomy Journal, Vol. 97, No. 1.

Tolka J.A., Howella T.A., Miller F.R., 2013. Yield component analysis of grain sorghum grown under water stress. Field Crops Research 145, p. 44-51.

Tucker A.N., 2005. Managing nitrogen in grain sorghum to maximize n use efficiency and yield while minimizing producer risk. Theses accessed at http://krex.k-state.edu/dspace/handle/2097/1424.

Varvel G.E., 2000. Crop Rotation and Nitrogen Effects on Normalized Grain Yields in a Long-Term Study. Agron. J. 92, p. 938-941.

19

EVALUATION OF THE MODERN METHODS OF ASSESSING THE DAMAGE SUFFERED BY RAPESEED CROPS AT THE END OF WINTER

Alexandra TRIF, Mihai GÎDEA, Aurelian PENESCU

University of Agronomic Sciences and Veterinary Medicine of Bucharest, 59 Mărăști Blvd, District 1, Bucharest, Romania

Corresponding author email: trif_alexandra@yahoo.com

Abstract

The aim of this paper is to determine the benefit of modern methods of scanning and processing images of a rapeseed crop to determine affected areas at the end of winter. There are where the measurements took place is on the study grounds of "Belciugatele Research Station - Moara Domnească Farm". Three methods were used, all based on the same image provided by a drone as a basemap, and RTK-GPS measurements. For the first method the affected areas were delimited using a RTK receiver, which required a person on the ground to manually provide the required data. The second method uses the same image, but digitizing was done at a later date by using GIS instruments. In the third method the digitizing was also done using software, by classifying the pixels. The result of the study showed that the third method was superior because of higher precision and lower processing time.

Key words: drone, GIS instruments, pixel classification, GPS measuring, affected area, rapeseed crop.

INTRODUCTION

The steady rise in energy consumption as an effort to reduce the use of fossil fuels meant a fast increase in rapeseed crops. In 1994, there were 303 ha of rapeseed crops in Romania. That number grew to 527175 ha in 2010. (http://faostat.fao.org/site/567/default.aspx#anc or). Therefore, specialists have invested in improving technologies that increase production and minimize loss.

One factor that negatively affect rapeseed crops is drought during sowing period, which causes patchy or delayed sprouting. This results in plants not having enough time to accumulate sufficient biomass to survive freezing during the winter. In Romania, the biggest losses occur during spring, as a result of sporadic hot-cold spells (Pirna et al., 2011).

At the end of winter, farmers have tough choice to make: either the crop fares well during the winter and there is potential for profit, or the crop is compromised and maintaining if further would mean financial loss. In the latter case, the crop is abandoned and a new one is started on that area (Verger et al., 2014).

Farmer invest in preparing, sowing and protecting the field until the point a decision can be made. Sowing a compromised crop after the sowing time has passed does not guarantee production at the maximum potential of the field, no matter what the type of crop replaces the rapeseed (Moorthy et al., 2015).

Consequently, a decision must be made quickly, so a fast and accurate way is required to determine the extent of the damage (Verger et al., 2015).

To address this need, we tested three methods of determining the amount of deterioration in rapeseed crops at the end of winter.

MATERIALS AND METHODS

The first method, which is the one used by farmers, was based on RTK measurements. The second method uses GIS instruments, whereas in the third method the classification is done automatically by the computer using pixel classification software (Meyer, 2011).

In all three methods a drone was utilized to capture the orthophoto image (Huang et al., 2013) based on digitization, pixel classification (Ashok et al., 2015) and the superimposing of the zones without vegetation from the RTK measurements (Viña et al., 2011).

In order to evaluate the three methods objectively, research was conducted on the "Belciugatele Research Station - Moara

Domnească Farm", during the 2015-2016 agricultural year.

The experiment had two factors: the A factor was the seeding time (a1-28.08.2015 and a2-15.09.2015). The B factor was the distance between the rows (b1-12.5 cm, b2-25 cm, b3-37.5 cm, b4-50 cm). The determinations using the three methods were done during the vegetation period in the spring, on 01.05.2016.

The orthophoto was first captured with a drone, then geo-referenced. The images were taken with the Phantom 3 Professional drone (Figure 1).

Figure 1. Phantom 3 Professional

Those images represent the orthophoto and the respective basemap which the digitization is based on. Control points were fixed using milestones in the form of an X, for increased visibility from high altitudes (Figure 2).

Figure 2. Control Points + GPS Trimble

After the control points are fixed, the GPS Trimble is set-up for the georeferencing stage. The data processing stage follows those steps: the image from the drone was loaded by using the ArcMap 10.x software; then georeferencing

was done by using the coordinates from the GPS in the field.

The coordinates, which are initially in the WGS84 standard, are converted to Stereo 1970 with the TransDat software acquired from the official ANCPI website. After the control points are added the image was be in the correct position (Figure 3).

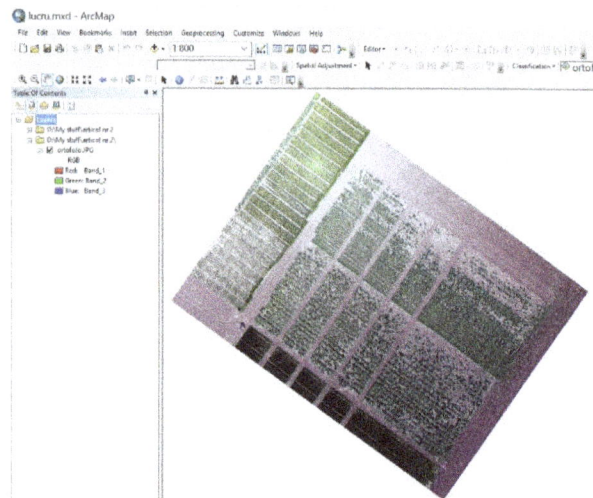

Figure 3. Georeferencing

The first method consists in manually mapping the field using a GTS RTK and the Stop and Go process to delimit each zone without vegetation (Figure 4).

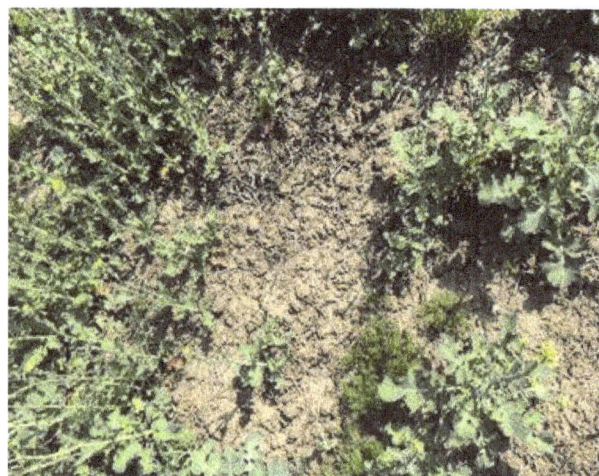

Figure 4. Zone without vegetation

The data is converted in ArcMap (Figure 5) and the surfaces were measured. The data is centralized for each experimental parcel.

The second method is a laboratory method based on manually digitizing zones without vegetation in the orthophoto from the drone.

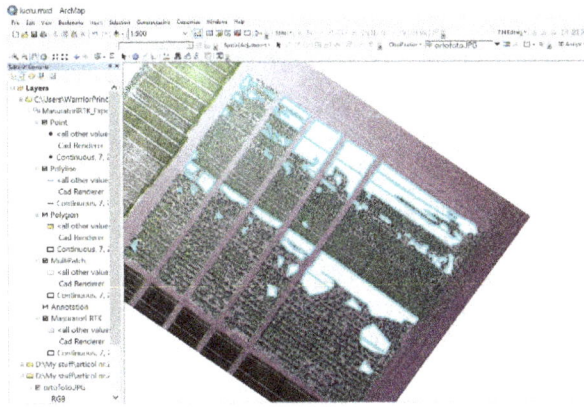

Figure 5. Measures zones applied on the orthophoto

For the second method, two shape file layers are created, one used to digitize the affected areas without vegetation, and the other used to delimit the study areas and determine the total area.

The digitizing is done vertex by vertex, according to the user's skill (Figure 6).

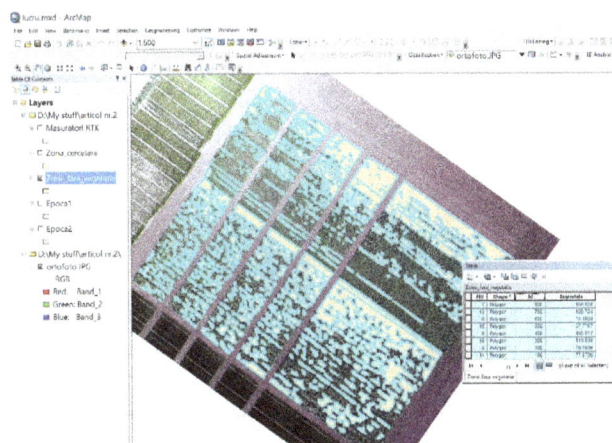

Figure 6. Final edit

Any area can be analysed from the selection. The surface area without any vegetation is determined for a specific lot (Figure 7).

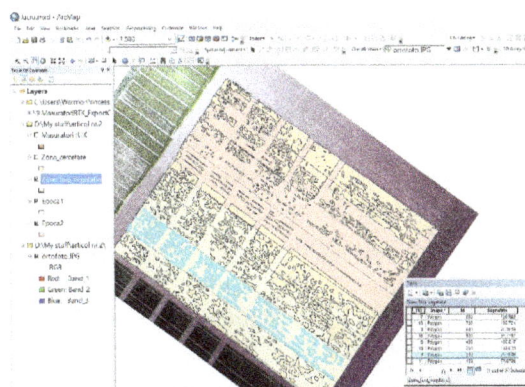

Figure 7. Selection of lot 2 from Epoch A

The third method is a laboratory method based on image processing using software to classify pixels.

In ArcMap, supervised classification results in two groups: Green pixels for zones with vegetation and yellow pixels for the affected areas. The results from seven samples with different parameters are compiled for the most accurate representation.

The resulting image is then cut in eight different images, each correspondent to a lot. A Python script was used inside ArcMap for each image to count the pixels based on colour. The total affected area could be calculated based on the results (Figure 8).

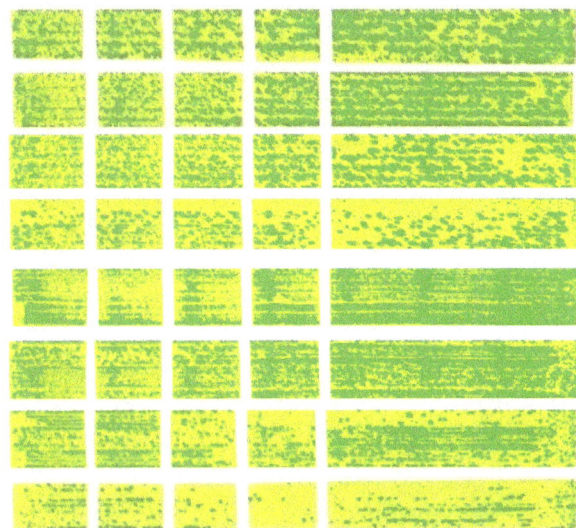

Figure 8. Classifications pixel

In the case of lot 1, the result were two classes of RGB colours: 429691 green pixels ⬛ and 287700 yellow pixels ⬜. Because we know that 324 square meters equals 718286 pixels, it the results was 129.77 square meters of affected area.

RESULTS AND DISCUSSIONS

Looking at the data in Table 1 for the first method we can see the areas affected by climate conditions in the winter varied from 0.6% for the crop sowed during Epoch a1, at 12.5 cm and 52.9% for Epoch a2 at 50 cm distance between rows.

In the case of the manual digitizing in the laboratory the results varied between 11.7% for the crop sowed during Epoch a1 at 12.5 cm and 60.7% for the one in Epoch a2 at 50 cm distance between rows.

For the third method, in which the digitizing was done automatically by the software, the results varied between 33.05% for the crop of Epoch a1 and with 12.5 cm distance between rows and 75.46% for Epoch a2 at 50 cm distance between rows.

The results obtained from the three methods are centralized in table 1.

Table 1. Data centralization for the degree of destruction suffered by the rapeseed crop at winter's end

Epoch	Row Width	GPS RTK Method 1 %	Digitizing Method 2 %	Pixel classification Method 3 %
I	12.5 cm	0.6%	23.8%	35%
	25 cm	1.6%	24.5%	35.12%
	37.5 cm	7.2%	34.2%	49.29%
	50 cm	51.4%	51.2%	67.2%
II	12.5 cm	1.2%	11.7%	33.05%
	25 cm	5.5%	23.5%	34.35%
	37.5 cm	30.0%	39.1%	51.06%
	50 cm	52.9%	60.7%	75.46%

Looking at the results for the same plot, we can observe that the results vary greatly based on the method used. For the first method where GTS-RTK was used, the values obtained were on average 10-20% lower than the other methods.

When the affected area (the area without vegetation) is bigger, the differences between values from the each methods are smaller.

From the agronomical point of view, when a rapeseed crop is destroyed in proportion of up to 30%, the crop has the potential to turn a profit, even if the production is lower. For a degree of destruction over 50%, based on the ability of the plants, the question of abandoning the crop was raised.

In those conditions and with the results obtained, it can be concluded that all three methods provided valuable information to allow a correct decision to be made. Analysing the data obtained from manually mapping with GPS-RTK, we can determine that the plots with 50 cm distance between rows for each Epoch should be abandoned. The same conclusion can be drawn from the second and third method.

Comparing the manual mapping method with the ones using software, we can see much bigger values for the affected area, resulting from the increased ability of the software over the human eye.

Table 2 shows that the time required to determine the affected areas was much lower in the case of automated processing of the image, from 15-20 minutes for the manual mapping to just 1 minute in the case of the pixel classification method.

Table 2. Time needed to determine the affected area of a rapeseed crop

Surface	GPS RTK Method 1	Digitizing Method 2	Pixel classification Method 3
328 mp	15-20 min	5-8 min	1 min

Figure 9 shows that the rapeseed destruction degree was greater in the case of the third method owing to its higher accuracy. Therefore, that method provide the best overview of the affected areas.

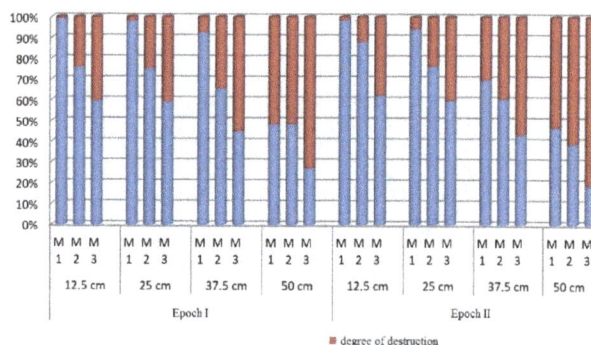

Figure 9. The degree of destruction of rapeseed

CONCLUSIONS

This paper compares three modern methods of determining the affected area of rapeseed crops as a result of cold and frostbite at the end of winter.

The method of manually mapping the field is slow and requires a person to traverse the whole test area and delimit the surface.

Digitizing in the laboratory allows for faster delimitation times of the affected areas with minimal errors and allows for a birds-eye view of the whole crop.

The automated processing of the orthophoto is a fast method with high accuracy.

While the speed and accuracy of the three methods varies, all can be used to make a correct decision on keeping or abandoning a crop.

REFERENCES

Ashok K.D., Prema P., 2015. A study on weed discrimination through wavelet transform, texture feature extraction and classification. International Journal of Computer Science & Information Technology (IJCSIT) Vol. 7, No. 3, p. 41-52.

Huang Y., Thomson S.J., Hoffmann W.C., Lan Y. & Fritz B.K., 2013. Development and prospect of unmanned aerial vehicle technologies for agricultural production management. International Journal of Agricultural and Biological Engineering, 6, p. 1-10.

Meyer G.E., 2011. Machine Vision Identification of Plants, Recent Trends for Enhancing the Diversity and Quality of Soybean Products. Prof. Dora Krezhova (Ed.), InTech, p. 402-414. Available from: http://www.intechopen.com/books/recent-trends-for-enhancing-the-diversity-and-quality-of-soybean-products/machine-vision-identification-of-plants.

Moorthy S., Boigelot B., Mercatoris B.C.N., 2015. Effective segmentation of green vegetation for resource-constrained real-time applications. Wageningen Academic Publishers. Precision agricrop '15, p. 257-266.

Pirna I., Voicu E., Vladut V., 2011. Rapeseed cultivation technology. MLNPSOA Multi-Lingual Network of the Profess. Soc. in Organic Agric., INMA Bucharest. multilingual.bionetsyst.com/images/docs/640067778 1335502327.pdf.

Verger A., Vigneau N., Chéron C., Jean-Marc Gilliotd, Alexis Comar, Baret F., 2014. Green area index from an unmanned aerial system over wheat and rapeseed crops. Article in Remote Sensing of Environment, 152, p. 654-664.

Verger A., Chéron C., Baret F., Vigneau N., 2015. The potential of unmanned aerial vehicles for precision agricrop and phenotyping. Remote Sensing Applications in Agricrop II. 35th Canadian Symposium on Remote Sensing, FR4.10.5, ftp://ftp.legos.obs.mip.fr/pub/tmp3m/IGARSS2014/abstracts/3030.pdf.

Viña A., Anatoly A., Gitelson B., Anthony L., Nguy-Robertson, Yi Peng, 2011. Comparison of different vegetation indices for the remote assessment of green leaf area index of crops. Remote Sensing of Environment 115, p. 3468-3478.

***http://faostat.fao.org/site/567/default.aspx#ancor.

20

MILD ALKALINE PRETREATMENT APPLIED IN THE BIOREFINERY OF SORGHUM BIOMASS FOR ETHANOL AND BIOGAS PRODUCTION

Adrian TRULEA, Teodor VINTILĂ, Nicolae POPA, Georgeta POP

University of Agricultural Sciences and Veterinary Medicine of Banat "Regele Mihai I al României", 119 Calea Aradului, Timişoara, Romania

Corresponding author email: e-mail: tvintila@animalsci-tm.ro

Abstract

Production of bioenergy from lignocellulosic biomass has gained more and more interest over the past years. Due to its high content in carbohydrates, Sorghum bicolor is one of the best suited candidate for bioenergy production. However, the lignocellulosic nature of the sorghum biomass raises difficulties to the access of the microbial enzymes to cellulose and hemicellulose and inhibits processes such as hydrolysis, fermentation and anaerobic digestion of biomass to produce biofuels. In this study, the effects of thermo - chemical pretreatment (mild alkaline pretreatment), applied to improve biodegradability of three different hybrids of sorghum biomass were investigated. Alkaline pretreatment have positive effects on the production of lignocellulosic ethanol, increasing both cellulose and hemicellulose content while breaking the lignocellulosic bonds and reducing lignin content. The achieved methane yields ranged from 320 to 345 l_n CH_4/kg VS for the pretreated biomass, approximately 22% higher than the yields obtained from the untreated biomass. Important increase of methane production has been noticed as well in the case of anaerobic digestion of spent bagasse resulted after ethanol fermentation. Mild alkaline pretreatment is a suitable pretreatment method for conversion of sorghum biomass to ethanol and biogas.

Key words: sorghum, lignocellulose, pretreatment, ethanol, methane.

INTRODUCTION

The dwindling supply and high prices of fossil fuels, energy independence, together with the adverse effects they have on the environment, greenhouse gas emissions and climate change, have pushed the global community towards the development on new, renewable energy sources (Cesaro, 2015; Guo, 2015).

Biomass, as a renewable energy source, is an important substitute for fossil fuels. One of its great advantages is its versatility. Biomass can be used to produce liquid (bioethanol), solid (wood pellets) and gaseous (biogas) biofuels.

Current technologies employed to produce biofuels impede their penetration on the large scale market. To guarantee the economic competitiveness of biofuels, besides constant improvement in conversion technologies, a biorefinery system seems to be an attractive solution (Leitner, 2016).

Different types of raw material are currently being used in the biofuel industry and investigated for their energy potential. Sweet sorghum (*Sorghum bicolor* (L.) Moench) is considered to be one of the most important plants for energy production. Sweet sorghum has a high content of soluble and fermentable sugars, high yields, is drought and flood tolerant and has the ability to grow in a wide range of environmental conditions (Larnaudie, 2016; Maw, 2016).

The aim of this study was to determine the influence and effects of mild alkaline pretreatment on sorghum biomass, when used as feedstock for bioethanol and biogas production. In order to maximize the energy output of the raw material processed, three different conversion pathways were examined:

1. Untreated biomass to biogas via anaerobic digestion.
2. Pretreated biomass to biogas via anaerobic digestion.
3. Pretreated biomass to ethanol via alcoholic fermentation followed by anaerobic digestion of the spent fermentation mass resulted after the alcoholic fermentation.

MATERIALS AND METHODS

Biomass from three different sorghum hybrids, namely *Sorghum bicolor x sudaneze* cv. Jumbo, originally from Australia, *Sorghum bicolor x sudaneze* Sugargraze II, originally from the United States of America and *Sorghum bicolor* var. *saccharatum* cv. F135ST, originally from

Romania, was used as biological raw material. The biomass was air - dried after harvest and then milled to 2 cm theoretical length using a Retsch SM100 mill with 2 cm mesh.

After milling, the biomass was subjected to mild alkaline & steam pretreatment process. The biomass was moistened with 2% NaOH solution (1:6 w/v) and autoclaved for 30 minutes at 121°C (Silverstein, 2007; Zhao, 2008; Chen, 2009). After pretreatment, the biomass was washed with tap water and neutralized with 2% H_2SO_4 solution until pH 5-5.5 ± 0.2. The next step of the process consisted of enzymatic hydrolysis of pretreated biomass catalyzed by the enzymatic complex NS22192, provided by Novozymes®. Pretreated biomass was immersed in citrate buffer pH 5 in concentration of 5% w/v dry matter and autoclaved for 30 minutes at 121°C in order to assure aseptic conditions. Before autoclaving, 2% w/v peptone, 1% w/v yeast extract and 0.5% Tween 80 (Chen 2009), were introduced in the hydrolysis media. All these components are necessary in the fermentation step of the process. The enzymatic complex NS22192 was added in the hydrolysis media in concentration of 15 IU per 1 gram of cellulose, according to the manufacturer's indication. The flasks where incubated at 50°C for 24 hours under constant shaking (150 rpm).

After 24 hours of hydrolysis, the temperature of the incubator was lowered to 30°C and fresh inoculums of *Saccharomyces cerevisiae* CIMT2.21 was aseptically added. The inoculums was obtained from the Collection of Industrial Microorganisms of the USAMVB, Timişoara. The incubation of the flasks continued at 35°C for 24 hours. To each flask two NIR sensors where attached (BlueSens, Germany): one detecting the concentration of ethanol and one sensor detecting the concentration of CO_2. Anaerobic digestion of the biomass was carried out according to the VDI 4630 guidelines (VDI, 2006) in a batch

test. The inoculum used was a combination of four digestates collected from different agriculture biogas plants. Methane concentration in the produced biogas was analysed using a Dräger X-AM 7000 gas analyser. During the experiments dry matter, organic dry mater, cellulose, hemicellulose and acid insoluble lignin content where monitored. Analysis were performed according to NREL standard laboratory procedures (L.A.P., Sluiter, 2005; Sluiter, 2008).

RESULTS AND DISCUSSIONS

The structural characteristics of the untreated and pretreated sorghum biomass are summarized in Table 1. Our data show that in all three cases, mild alkaline/steam pretreatment has a positive effect on the biomass, breaking the lignocellulosic complex, thus liberating cellulose and hemicellulose. Higher concentration in both cellulose and hemicellulose content has been recorded in pretreated biomass. The concentration of cellulose in the pretreated biomass increased from 16.3% to 22.4% and the concentration of hemicellulose increased from 4.3% to 6.2%, depending of the sorghum variety.

The next step in the conversion pathway of sorghum to biofuels is the hydrolysis of lignocelluloses and alcoholic fermentation of the obtained sugars. As shown in figure 1, the fermentation process followed the classic path. In the first hours of incubation the active yeast immediately started the fermentation, after a very short lag phase, producing in three hours the main quantity of ethanol. After six hours the ethanol concentration reached the maximal values, varying only in small margins.

Calculating the ethanol yields, the following yields where obtained in the three sorghum hybrids: 0.33 g/g Jumbo biomass, 0.34 g/g F135ST biomass and 0.38 g/g Sugargraze II biomass, (all reported to dry matter).

Table 1. Structural characterization of sorghum biomass (%)

		Dry matter	Ash	Organic dry matter	Cellulose	Hemicellulose			Lignin (AIL)
						Total	Xylose, galactose and mannose	Arabinose	
Jumbo	untreated	90.67	9.08	90.92	28.76	21.49	19.58	1.91	17.11
	pretreated	18.73	2.09	97.91	51.23	27.72	26.04	1.68	10.33
Sugargraze II	untreated	91.5	8.37	91.63	32.58	22.31	20.54	1.77	18.58
	pretreated	21.42	2.64	97.73	54.38	27.92	25.99	1.93	8.54
F135ST	untreated	90.86	9.69	90.31	32.46	22.92	21.21	1.71	16.95
	pretreated	18.58	5.1	94.9	48.78	27.24	24.63	2.61	10.24

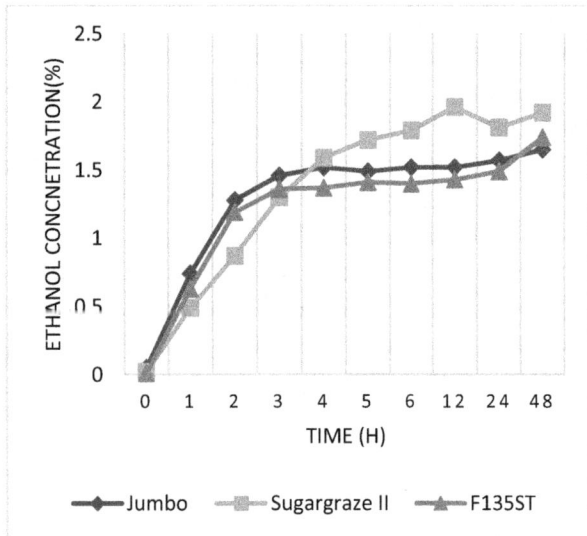

Figure 1. Evolution of ethanol concentration
during fermentation

The final step of the process and the cornerstone of this study was the anaerobic digestion of sorghum biomass in order to observe the effects of mild alkaline/steam pretreatment on the production of biogas.

In figure 2 the quantities of biogas and methane produced during the anaerobic digestion for each sorghum variety as well as the differences between the three conversion pathways are displayed.

The obtained results show important differences in biogas and methane yields obtained in the case of anaerobic digestion of untreated and pretreated biomass. Regarding the methane yield, the pretreatment increases

productions with an average of 61 l_N/kg organic dry matter, which is translated as 21% increase of methane yields.

In the batches performing anaerobic digestion of the spent fermentation mass resulted after alcoholic fermentation of the pretreated sorghum biomass higher yields were obtained. In all three sorghum varieties, the methane yields were on an average 49% higher than in the case of the untreated samples and 22% higher than in the case of the of the pretreated samples.

Although parts of the carbohydrates present in the biomass were degraded and converted into ethanol during the fermentation process, the highest yields of biogas and methane were recorded in the samples containing spent fermented mass. This atypical result can have two reasons at its foundation. First of all, during the alcoholic fermentation of the biomass, yeasts convert the soluble sugar faction into ethanol and other by-products, such as proteins and other substances, that are used as a substrate by the microorganisms employed in the anaerobic digestion process and converted into biogas.

Secondly, the alcoholic fermentation process can be considered as an additional biological pretreatment step that loosens the bonds in the lignocellulosic structure and thus liberating an extra amount of carbohydrates that can be converted in the anaerobic digestion step.

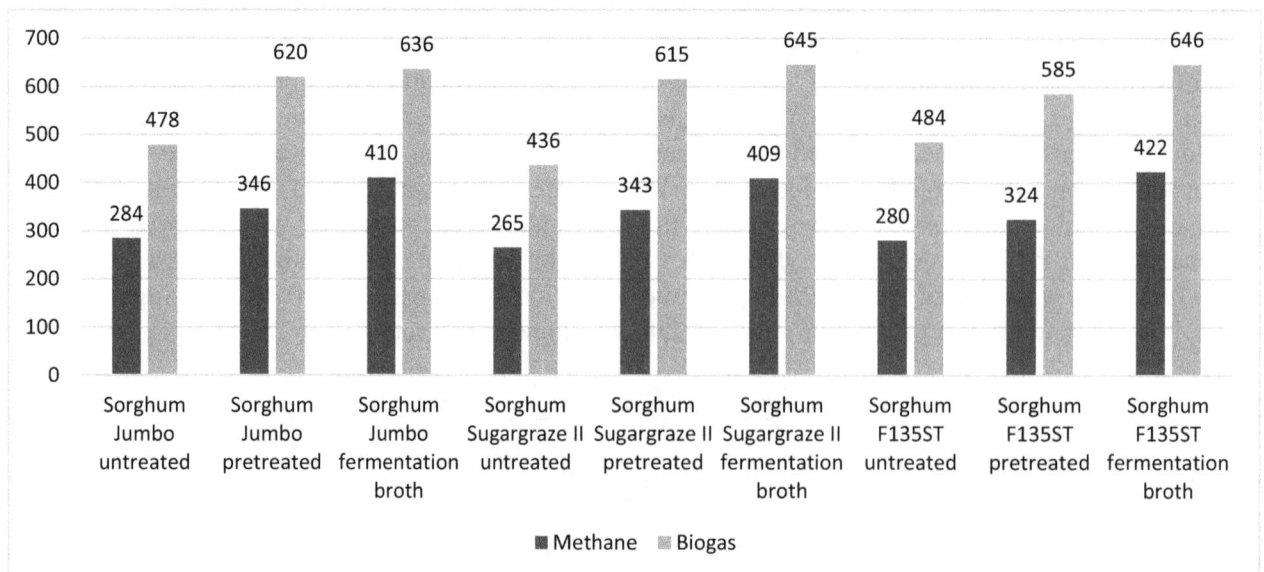

Figure 2. Biogas and methane production during anaerobic digestion

CONCLUSIONS

Mild alkaline/steam pretreatment is a suitable pretreatment method for converting sorghum biomass to ethanol and biogas. The combination of sodium hydroxide solution and steam pretreatment opens up the lignocellulosic bonds, liberating a substantial quantity of cellulose and hemicellulose.

Between the three different conversion pathways studied in this work: (1) Anaerobic digestion of untreated biomass; (2) Anaerobic digestion of pretreated biomass; (3) Anaerobic digestion of bagasse resulted after alcoholic fermentation of pretreated biomass; the highest yields were obtained in the third conversion pathway, by connecting ethanol fermentation to anaerobic digestion. This means that an integrated, cascading, conversion pathway is the most suitable in terms of energy yields from sorghum biomass.

REFERENCES

Chen M., Zhao J., Xia L., 2009. Comparison of four different chemical pretreatments of corn stover for enhancing enzymatic digestibility. Biomass and Bioenergy, 33(10), p. 1381-1385.

Cesaro A., Belgiorno V., 2015. Combined biogas and bioethanol production opportunities for industrial application. Energies, 8, p. 8128-8144.

Guo M., Song W., Buhain J., 2015. Bioenergy and biofuels: History, status and perspective. Renewable and Sustainable Energy Reviews, 42, p. 712-725.

Lanaudie V., Rochon E., Ferrari M.D., Lareo C., 2016. Energy evaluation of fuel bioethanol production from sweet sorghum using very high gravity (VHG) conditions. Renewable Energy, 88, p. 280-287.

Leitner V., Lindorfer J., 2016. Evaluation of technology structure based on energy yield from wheat straw for combined bioethanol and biomethane facility. Renewable Energy, 87, p. 193-202.

Maw M.J.W., Houx J.H., Fritschi F.B., 2016. Sweet sorghum ethanol yield component response to nitrogen fertilization. Industrial Crops and Products, 84, p. 43-49.

Silverstein R.A., Chen Y., Sharma-Shivappa R.R., Boyette M.D., Osborne J.A., 2007. Comparison of chemical pretreatment methods for improving saccharification of cotton stalks. Bioresource Technology, 98(16), p. 3000-3011.

Sluiter A., Hames B., Hyman D., Payne C., Ruiz R., Scarlata C. et al., 2008. Determination of Total Solids in Biomass and Total Dissolved Solids in Liquid Process Samples - Laboratory Analytical Procedure (LAP). National Renewable Energy Laboratory.

Sluiter A., Ruiz R., Scarlata C., Sluiter J., Templeton D., 2005. Determination of Ash in Biomass - Laboratory Analytical Procedure (LAP). National Renewable Energy Laboratory.

Sluiter A., Hames B., Ruiz R., Scarlata C., Sluiter J., Templeton D. et al., 2008. Determination of Structural Carbohydrates and Lignin in Biomass - Laboratory Analytical Procedures (LAP). National Renewable Energy Laboratory.

Zhao X., Zhang L., Liu D., 2008. Comparative study on chemical pretreatment methods for improving enzymatic digestibility of crofton weed stem. Bioresource Technology, 99(9), p. 3729-3736.

***VDI, 2006. Fermentation of organic materials. Characterization of the substrate, sampling, collection of material data, fermentation tests. VDI - Gesselschaft Energie und Umwelt.

21

INFLUENCE OF REARING TECHNOLOGY ON BODY WEIGHT OF YOUNG BROILER BREEDERS

Minodora TUDORACHE[1], Ioan CUSTURA[1], Ilie VAN[1], Andrei MARMANDIU[1], Paul ANTON[2]

[1]University of Agronomic Sciences and Veterinary Medicine of Bucharest, 59 Mărăşti Blvd, District 1, Bucharest, Romania
[2]Aviagen Romania

Corresponding author email: minodoratudorache@yahoo.com

Abstract

This paper describes a sequence from a massive experiment analyzing quality of semen material and breeding efficiency in roosters from hybrid ROSS 308. Study was performed to observe influence of some environmental factors (light intensity and poultry density) and of litter type on body weight in males during rising period (0-18 weeks). Several males were used (14 500 cap) in three experimental procedures (A - with analyze parameters sub-standard and litter made of chopped straws B - with analyze parameters above standard and litter made of rice hulls and C - with analyze parameters at the level recommended by the manufacturer of biological material and litter made of wood shavings). In group A body weight values were mostly under standard growth curve and differences were highly significant statistically. In group B body weight values were above standard growth curve (differences highly significant) and in group C A body weight values were close to standard (differences not significant statistically). Comparison of average body weights of individuals from the three groups has revealed that differences between groups are highly significant statistically during whole raising period except weeks 15-16. Therefore it is advisable the usage of analyzed parameters at values slightly above standard for the possibility to obtain poultries with higher body weight more able to resist to transfer stress de and a smaller mortality during next period to create the ground for good breading results of future adults.

Key words: litter, rosters, density, light intensity, body weight.

INTRODUCTION

Body weight is one of the most important factors for optimum breeding results as birds are being essentially broilers whose weight gain potential must be kept under control to be able to accomplish their breeding potential (Blokhuis and de Wit, 1992). There is a negative relationship between weight gain and breeding performances. For this reason we should not allow birds to gain weight up to their genetic potential otherwise breeding would became uneconomical. For this reason feeding programs are designed and adapted to restrict quantitative and qualitative feed inteke to limit body growth and weight gain (Fairchild and Czarick, 2009). So far the only way to controla birds body weight is feed restriction and so this method is a source of major management unless it is replaces by adding of feed intake inhibiting agents or additives with no nutritive value in feed.

MATERIALS AND METHODS

Researches have been performed during two years on chicks of ROSS 308 hybrid for studying influence of some environmental factors (light intensity, bird density) on body weight gain of young broiler breeders (hens) (Watkins S., 2013). Studied parameter has been analyzed in three different experimental circumstances (three trial series):
- experiment procedure A in which some environmental factors (light intensity and poultry density) are being at sub-standard values and litter is being made of chopped straws;
- experiment procedure B in which parameters are being raised over standard limits and litter made from rice hulls is being used;
- experiment procedure C with parameters at standard values and litter made from wood shavings.

Works were performed inside three farms with one farm for each experiment procedure: Avicola Călăraşi, S.C. Agrafood S.A. and Avicola Focşani.

Experiment procedure A was performed based on results from 4100 ROSS 308 male commercial hybrids during rising period (0-18 weeks).

Environmental parameters considered were:

- litter: chopped straws;
- sub-standard light intensity: 7 lux at 1-6 weeks, 20 lux at 6-9 weeks, 7 lux at 10-20 weeks, and 30 lux over 20 de weeks;
- sub-standard poultry density: 3 males/m^2;

Experiment procedure B was performed based on results from 6000 ROSS 308 male commercial hybrids during rising period (0-18 weeks). Environmental parameters considered were:

- litter: rice hulls;
- over standard light intensity: 30 lux at 1-6 weeks, 60 lux at 6-9 weeks, 30 lux at 10-20 weeks, 70 lux over 20 de weeks;
- over standard poultry density: 5 males/m^2;

Experiment procedure C was performed based on results from 4400 d ROSS 308 male commercial hybrids during rising period (0-18 weeks). Environmental parameters considered were:

- litter: wood shavings;
- standard light intensity: 15 lux at 1-6 weeks, 40 lux at 6-9 weeks, 15 lux at 10-20 weeks, 40 lux over 20 weeks;
- standard poultry density: 4 males/m^2;

Poultry were raised in uniform conditions inside the three units (for the three experiment procedures) on litter bed and in up-to-date houses and with feed and water delivered according to technical book of the hybrid. Poultry used in the three experiment procedures were fed the same way for results to be compatible (Aviagen, 2005).

Liveweight was the parameter observed during rising period (0-18 weeks).

Classical statistical methods were used for phonotypical identification of groups as following (Sandu, 1995):

- *Student* test to compare evenness of two samples averages;
- *Fisher* test was used for several samples after a variance analyze. Calculated value *F* was obtained by referring square averages between samples to samples square average;
- χ^2 test was used to verify evenness of an empirical distribution (of observed frequency O_j) with a theoretical distribution (of frequency T_j).

RESULTS AND DISCUSSIONS

To emphasize the possible influence of environmental factors and litter type on live weight we are showing average values of analyzed parameter for the three experimental procedures and statistical significance of differences observed between average figures. Observations and records were performed weekly during whole raising period (0-18 weeks).

Values obtained for live weight from individuals in experiment procedure A during raising period are presented in Table 1 and graph from Figure 1.

Table 1. Average values for live weight in the growth period, for first experience series

Week	$\bar{X} \pm s_{\bar{X}}$ (g)	s	c.v.%	Standard
1	87 ± 0.14	8.7	10.00	150
2	180 ± 0.67	43.2	24.00	310
3	400 ± 1.17	75.2	18.80	505
4	700 ± 1.61	102.9	14.70	720
5	740 ± 1.10	70.3	9.50	900
6	920 ± 1.34	85.56	9.30	1075
7	1100 ± 1.48	94.6	8.60	1230
8	1280 ± 1.70	108.8	8.50	1375
9	1440 ± 1.80	115.2	8.00	1510
10	1550 ± 1.94	124	8.00	1640
11	1680 ± 2.07	132.72	7.90	1770
12	1920 ± 2.43	155.52	8.10	1900
13	2080 ± 2.05	131.04	6.30	2030
14	2160 ± 2.43	155.52	7.20	2160
15	2320 ± 2.90	185.6	8.00	2290
16	2460 ± 3.11	199.26	8.10	2430
17	2640 ± 3.46	221.76	8.40	2575
18	2780 ± 3.08	197.38	7.10	2725
Differences significance	$\chi^2 = 192.18^{**}$ $\chi^2_{17;0,05} = 27.59$; $\chi^2_{17;0,01} = 33.41$			

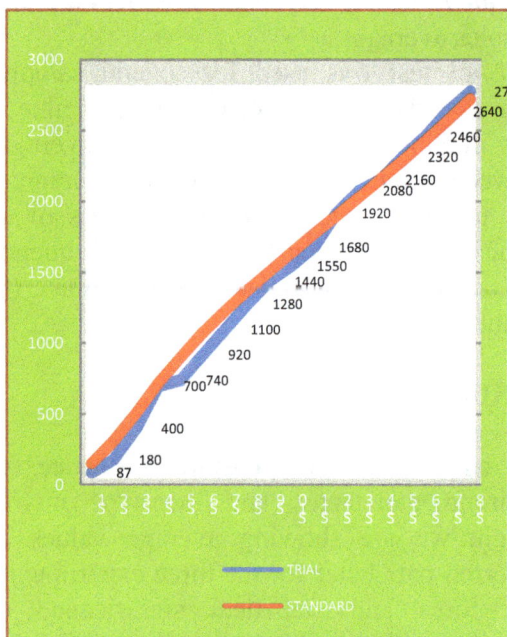

Figure 1. Average values for live weight in the growth period, for first experience series

Table 2. Average values for live weight in the growth period, for second experience series

Week	$\bar{X} \pm s_{\bar{X}}$ (g)	s	c.v.%	Standard
1	128 ± 0.15	11.52	9.00	150
2	267 ± 0.39	30.17	11.30	310
3	418 ± 0.73	56.85	13.60	505
4	622 ± 1.20	93.30	15.00	720
5	877 ± 1.09	84.19	9.60	900
6	999 ± 1.17	90.91	9.10	1075
7	1231 ± 1.38	107.10	8.70	1230
8	1409 ± 1.60	123.99	8.80	1375
9	1578 ± 1.87	145.18	9.20	1510
10	1753 ± 1.95	150.76	8.60	1640
11	1958 ± 2.10	162.51	8.30	1770
12	2021 ± 2.32	179.87	8.90	1900
13	2122 ± 2.58	199.47	9.40	2030
14	2195 ± 2.72	210.72	9.60	2160
15	2328 ± 2.70	209.52	9.00	2290
16	2461 ± 2.83	219.03	8.90	2430
17	2697 ± 2.96	229.25	8.50	2575
18	2818 ± 2.95	228.26	8.10	2725
Differences significance	$\chi^2 = 97.56^{**}$ $\chi^2_{17;0.05} = 27.59$; $\chi^2_{17;0.01} = 33.41$			

It is noticed that average live weight falls between normal limits of the specie with a small variability during the 18 weeks which suggest the presence of uniform conditions of feeding and management.

It is noticed that in ROSS 308 hybrids males from experiment procedure A (Avicola Călăraşi) liveweight values are mostly under standard hybrid's growth curve. These noticed differences between averages of analyzed parameter during the 18 weeks and hybrid's growth curve were tested for statistical significance and value of χ^2 test (192.18) pointing to some highly significant statistical differences between the two allowances. These registered differences might be also explained by litter type and value of environmental parameter values which are sub-standard in this experiment procedure.

Liveweight values from birds from experiment procedure B during growth periode are being shown in Table 2 and graph from Figure 2.

Analyze of results is revealing that average live weight in poultry from experiment procedure B is being inside normal species limits with just a small variability during the 18 weeks revealing also uniform feeding and management conditions.

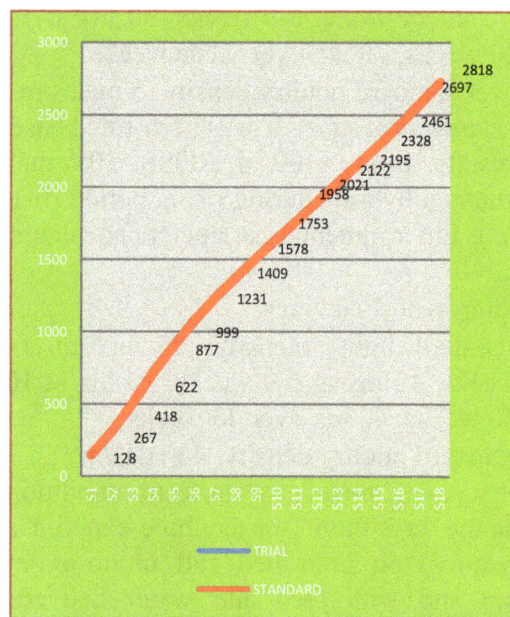

Figure 2. Average values for live weight in the growth period, for second experience series

So in ROSS 308 hybrids males from experiment procedure B (Agrafood) average live weight values are mostly above standard hybrid's growth curve which is favorable for biological and economical efficiency of farm. These differences noticed between average values of character from the 18 weeks analyzed and hybrid's standard values were tested for statistical significance and value of χ^2 test (97.56) pointing to some highly significant statistical differences between values actually recorded and those recommended de hybrid's standard. As poultry from the three experiment procedures received same feeding conditions and greater live weight registered in roosters from experiment procedure B might be assigned to litter type and to environmental parameters above standard.

Values obtained for live weight from experiment procedure C from growth period are presented in Table 3 and graph from Figure 3.

Table 3. Average values for live weight in the growth period, for third experience series

Week	$\overline{X} \pm s_{\overline{X}}$ (g)	s	c.v.%	Standard
1	144 ± 0.17	11.52	8.00	150
2	308 ± 0.46	30.8	10.00	310
3	514 ± 0.85	56.54	11.00	505
4	732 ± 1.26	83.448	11.40	720
5	912 ± 1.07	71.136	7.80	900
6	1094 ± 1.32	87.52	8.00	1075
7	1250 ± 1.55	102.5	8.20	1230
8	1396 ± 1.66	110.28	7.90	1375
9	1530 ± 1.78	117.81	7.70	1510
10	1662 ± 2.00	132.96	8.00	1640
11	1794 ± 2.30	152.49	8.50	1770
12	1926 ± 2.41	159.86	8.30	1900
13	2060 ± 2.42	160.68	7.80	2030
14	2196 ± 2.45	162.5	7.40	2160
15	2330 ± 2.49	165.43	7.10	2290
16	2470 ± 2.83	187.72	7.60	2430
17	2616 ± 3.16	209.28	8.00	2575
18	2766 ± 3.42	226.81	8.20	2725
Differences significance	$\chi^2 = 6.67^{NS}$ $\chi^2_{17;0,05} = 27.59; \chi^2_{17;0,01} = 33.41$			

It is noticed that average live weight values in individuals from experiment procedure C is being inside normal species limits with just a small variability during the 18 weeks revealing also uniform feeding and management conditions. However in individuals from experiment procedure C variability coefficients have lowest values for analyzed character most probably due to taking care of ROSS 308 hybrid's standard.

For ROSS 308 males from experiment procedure C (Avicola Focşani) average live weight values are entirely hybrid's standard growth curve.

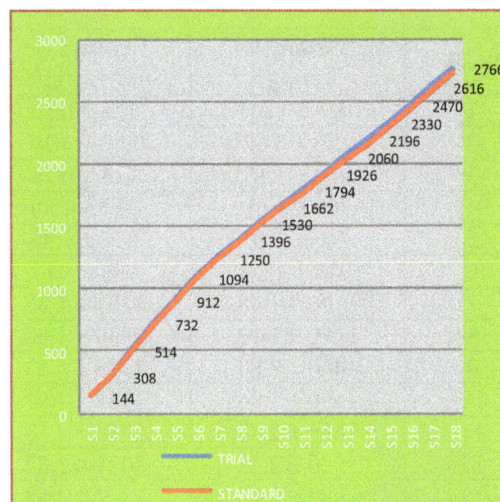

Figure 3. Average values for live weight in the growth period, for third experience series

This statement is corroborated by statistically testing din the observed differences between average values of the character during the 18 analyzed weeks and standard values and test χ^2 (6.67) pointing to differences which might be due to chance or individual variation and with no statistical significance. As birds from all the three experiment procedures received the same feeding conditions being and staying on growth curve might be attributable to litter type and abiding to environmental parameters standard. Next in line would be pointing to and evaluating differences between average live weight values in ROSS 308 males in all the three experiment procedures and the path of these differences and testing their statistical significance.

Table 4 and graph from Figure 4 are showing noticed differences between average values registered in all the three experiment procedures for the analyzed character.

Calculated values of Student test are higher than the presumed values which are revealing the existence of some differences with a high degree of statistical significance between the average values of the analyzed character (live weight), for all combinations, with the exceptions o weeks 15-16, when the growth curves are having a tendency to overlap themselves in all the 3 experiment procedures (as it can be also noticed in graph from Figure 4).

Table 4. Differences between experimental series for live weight

Week	Group A $\bar{x} \pm s_{\bar{x}}$ (g)	Group B $\bar{x} \pm s_{\bar{x}}$ (g)	Group C $\bar{x} \pm s_{\bar{x}}$ (g)	Observed differences		
				A-B (g)	A-C (g)	B-C (g)
1	87 ± 0.14	128 ± 0.15	144 ± 0.17	-41	-57	-16
2	180 ± 0.67	267 ± 0.39	308 ± 0.46	-87	-128	-41
3	400 ± 1.17	418 ± 0.73	514 ± 0.85	-18	-114	-96
4	700 ± 1.61	622 ± 1.20	732 ± 1.26	78	-32	-110
5	740 ± 1.10	877 ± 1.09	912 ± 1.07	-137	-172	-35
6	920 ± 1.34	999 ± 1.17	1094 ± 1.32	-79	-174	-95
7	1100 ± 1.48	1231 ± 1.38	1250 ± 1.55	-131	-150	-19
8	1280 ± 1.70	1409 ± 1.60	1396 ± 1.66	-129	-116	13
9	1440 ± 1.80	1578 ± 1.87	1530 ± 1.78	-138	-90	48
10	1550 ± 1.94	1753 ± 1.95	1662 ± 2.00	-203	-112	91
11	1680 ± 2.07	1958 ± 2.10	1794 ± 2.30	-278	-114	164
12	1920 ± 2.43	2021 ± 2.32	1926 ± 2.41	-101	-6	95
13	2080 ± 2.05	2122 ± 2.58	2060 ± 2.42	-42	20	62
14	2160 ± 2.43	2195 ± 2.72	2196 ± 2.45	-35	-36	-1
15	2320 ± 2.90	2328 ± 2.70	2330 ± 2.49	-8	-10	-2
16	2460 ± 3.11	2461 ± 2.83	2470 ± 2.83	-1	-10	-9
17	2640 ± 3.46	2697 ± 2.96	2616 ± 3.16	-57	24	81
18	2780 ± 3.08	2818 ± 2.95	2766 ± 3.42	-38	14	52

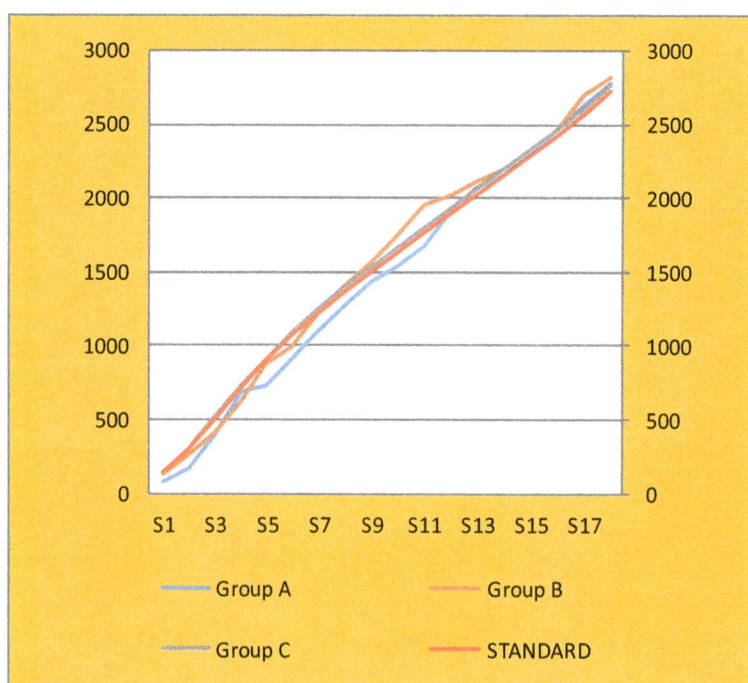

Figure 4. Differences between experimental series for live weight

As in experiment procedure C in which environmental parameters values are maintained at the standard level and litter type used is wood shavings values obtained for live weight are having the tendency to get close and even to overlap the growth curve of ROSS 308 hybrid results obtained look as if they are pleading in favour of experiment procedure B. Inside this procedure usage of rice hulls as litter and of some values of environmental parameters (density and light intensity) above standard seems to have a positive influence on live weight and males are having weekly average values over standard growth curve.

As a consequence it seems to be recommendable usage of some environmental factors with values higher than standard and a litter made from rice hulls with the aim to obtain a higher live weight or at least to ensure a position above the growth curve at the moment of transfer to production house and as consequence to reduce in this way the effects of transport stress and of course to reduce mortality linked by this moment for a more efficient breeding activity of future parent flocks.

CONCLUSIONS

Considering the live weight might come to following conclusions:

1. In ROSS 308 hybrids males cluster from grouping A average live weight values are mostly under standard hybrid's growth curve with differences highly significant statistical.

2. In ROSS 308 hybrids males cluster from grouping B average live weight values are mostly above standard hybrid's growth curve with differences highly significant statistical.

3. In the case of cluster from procedure C average live weight values are in entirely on the standard hybrid's growth curve.

4. It is noticed the existence of some differences with a high degree of statistical significance between the average values of live weight), for all combinations, with the exceptions o weeks 15-16, when the growth curves are having a tendency to overlap themselves in all the three experiment procedures.

5. Usage of rice hulls as litter and of some environmental factors (density and light intensity) with values higher than standard seems to have a positive influence on live weight and males are having weekly average values which are being above standard growth curve.

REFERENCES

Blokhuis H.J., de Wit W., 1992. Housing systems for layers as welfare determining factor. Proc. XIX World's Poultry Congres, 20-24 September, Amsterdam, p. 315-319.xc.

Fairchild B., Czarick M., 2009. Broiler Tip. Getting chicks off to a good start, College of Agricultural and Environmental Sciences / Athens, Georgia.

Sandu Gh., 1995. Modele experimentale în zootehnie. Ed. Coral Sanivet.

Watkins S., 2013. Lighting Programs in Broiler Breeders and Dark Out. Center of Excellence for Poultry Science University of Arkansas System Division of Agriculture Fayetteville, Arkansas.

***Aviagen, 2005. Environmental Management in the broiler breeder laying house.

22

GRAIN LEGUMES - MAIN SOURCE OF VEGETAL PROTEINS FOR EUROPEAN CONSUMPTION

Gheorghe Valentin ROMAN[1], Lenuta Iuliana EPURE[1], Maria TOADER[1], Antonio-Romeo LOMBARDI[2]

[1]University of Agronomic Sciences and Veterinary Medicine of Bucharest, 59 Mărăşti Blvd, District 1, 011464, Bucharest, Romania,
[2]SC Lombardi Agro SRL, Macin, Tulcea County

Corresponding author email: romangv@yahoo.com

Abstract

Grain legumes (pulses) are currently grown on 1.8% of the EU's arable land (compared to 4.7% in 1961) and as a result, Europe is dependent on protein imports totalling annually about 20 mil. tons of soybean meals and 12 mil. tons of soybean grains. Only about 2.5% of soybean meal consumed is produced in the EU. This imbalance between production and consumption creates economic and trade problems; in addition, imported soybean is mostly GM, which is not accepted by many European consumers. Since 2013, the Common Agricultural Policy included as a priority increasing the production of vegetal proteins by subsidies to grain legumes crops (including soybean non-GM), forage crops (alfalfa, clover) and oil crops (rapeseed, sunflower). In this framework encompasses the initiative Donau Soja (Danube Soya), which promotes the cultivation of conventional soybean in the Danube region and the development of yields processing and valorisation network. Romania falls well into these trends by traditions in cultivating soybean (over 500 thou ha before 1990), the favourability of natural conditions, the existence of biological material adapted to the specific natural conditions (varieties developed in ARDS Turda and NARDI Fundulea). Romanian farmers are interested in expanding soybean cultivation for ameliorating soil fertility and as a very good previous crop for winter cereals, but by providing an efficient weeds control, supplying water by irrigation and treating the seeds with bacterial preparations. Also, taking soybean harvest in the food networks involves a very rigorous quality control in terms of protein content and of avoiding contamination by GMOs. It is estimated that it can reach 700 thou ha with soybean, which can produce about 0.51 mil. tons proteins, which would add about 0.40 mil. tons of sunflower proteins (from 800 thou ha) and about 0.25 mil. tons of rapeseed proteins (from 500 thou ha).

Key words: protein supply, Common Agricultural Policy, grain legumes, soybean.

INTRODUCTION

Compared with other major agricultural regions of the world, the EU is characterized by a low level of vegetal protein production.
The dominance of cereals in European agriculture combined with the import of large quantities of soybean grains and meal enable self-sufficiency in livestock products. Increasing the cultivation of protein crops (grain legumes) would be an important contribution to the sustainable development of European agriculture and food systems.
The direct farm, regional and global level environmental benefits of increased grain legumes production, combined with the indirect benefits arising from the better balance of EU agriculture and trade, justify public intervention in this sense (European Parliament, 2013).

In 2013, provisions of new Common Agriculture Policy (CAP) include as a priority the promotion of protein crops to cover a large proportion of the protein consumption from own production.
Grain legumes crops (or pulses), species of *Fabaceae* (*Leguminosae*) family - first pea, field beans and soybean, but broad beans, chick pea, lentils and lupine too -, are now grown on only 1.8% of arable land in the EU, compared with 4.7% in 1961. This decline is the result of diverse economic and policy factors. The direct human consumption of pulses has declined and this had resulted in a reduction in the area of food legumes. Only 11-15% of pea and 9-14% of broad beans are now used for human consumption.
Grain legumes yields are unsteady and technological and breeding efforts are necessary to improvement, as these crops

compete better with cereals. They are less competitive in the fight against weeds, some are susceptible to dropping and shaking of the seeds, water stress, pests and diseases attack.

Legumes crops require small quantities or no nitrogen fertilizer and are effective in the use of soil phosphorus reserves. On the other hand, plant residues rich in nitrogen and left on the soil helps to reduce the need for nitrogen fertilizers of next crops. Grain legumes increase biodiversity, reduce consumption of fossil fuels by reducing the need for fertilizers, produce fewer polluting emissions.

Soybean is currently one of the most important agricultural crop in the world, holding over 100 mil. ha sown annually, the 4th place after wheat, maize and rice. Extent that soybean cultivation has taken in recent decades is due to: chemical composition of the crop, rich in biochemical constituents with high biological value (approximately 20% lipids and proteins around 40%) formed a significant proportion of essential fatty acids and amino acids for human body; very varied possibilities of processing and use of crop (edible fats, animal feed concentrates, protein preparation for food, bio-fuels, other uses as row material in different industries); importance as leguminous plant for land fertility improvement in crop rotations. In this context, decreasing soybean growing in Romania, while in the 80s it had reached over 500 thou ha (the area cultivated was more that over all other European countries together) is an unwanted process with multiple causes and negative consequences.

Our approach comes in the context of decision at EU level on the not-acceptance of GM soybean growing in Europe and new CAP which intend to promote protein crops. In addition, is coming the launch of „Donau Soja" initiative according to a document signed by agriculture ministers from 18 European countries to support traditional soybean varieties growing, including organic soybean crop.

MATERIALS AND METHODS

In the studies carried out they were collected and analyzed statistical data on cultivation of soybean and other crops that produce protein at global, European and national levels, provided by specialized institutions on statistical data collection and processing. There were studied documents in the last decades on European policies regarding soybean and other protein crops.

A particular attention was paid to evaluate the competitiveness of protein crops under economic aspect, using data from different sources and comparing them with data from wheat and maize, crops which are expected to be replaced in some extend by grain legumes crops. To this end, in specific documents data were collected at EU level and in different European countries. Crops value was assessed using average content in starch, lipids and proteins and using the market value of these products, including market prices for grain legumes (pea, field beans, soybean, lupines, alfalfa), sunflower and rapeseed.

In this study were included data obtained in our own research with soybean crop in order to illustrate the level of the competitiveness of this crop in the specific conditions of South Romania.

RESULTS AND DISCUSSIONS

The current situation of grain legumes growing in the world. Most protein crops are components of the grain legumes group, belonging to the *Fabaceae* (*Leguminosae*) family and include dozens of cultivated plant species in the world, especially in Asia, Latin America, Africa; of these 9 species with economic importance stand out (Table 1). Global area with these legumes totals 201,728 thou ha, of which detaches soybean with 111,271 thou ha (55.2% of world total), field beans with 26,827 thou ha (13.3%), peanuts with 25,408 thou ha (12.6%), chick pea with 13,529 thou ha (6.7%) and cow pea with 11,274 thou ha (5.6%).

Of the legumes, soybean dominates by the production of 276,405 thou tons (72.1% of global) and by productivity – 2,485 kg/ha, and is followed by peanuts with 45,160 thou tons (11.8%), field beans with 22,855 tons (6%) and chick pea with 13,103 thou tons (3.4%). Great growers and producers of soybean are found in South and Central America (52,106 thou ha and 146,149 thou tons, mainly Brazil and Argentina), North America (32,523 thou ha and

94,681 thou tons, mainly the USA) and Asia (20,629 thou ha and 27,294 thou tons, mainly China).

Grain legumes crops growing in Europe. Traditional grain legumes for the European continent, with food and feed use, are pea, field beans, lentils, broad bean, soybean and lupines. In Europe, grain legumes are grown on only 5,726 thou ha in 2015 (2.8% of the global area). The most important legumes crops on the European continent are soybean with 3,176 thou ha (5,943 thou tons and 1,871 kg/ha) and pea with 1,723 thou ha (3,021 thou tons and 1,753 kg/ha).

Protein crop minor role in European agriculture reflects the large imbalances in the European agro-food system. The decline in grain legumes production and cereals production increase is due to higher productivity of cereals, by comparison with grain legumes crops grown in Europe. Thus, after 1961, of relatively equal production of protein crops, cereals production has almost doubled from those of grain legumes. Coupled payments that took into account protein crops have slowed down to some extent the decline in protein crops production, but not much changed the interest of businesses agents.

In recent years, protein crops production prices have increased slightly faster than the price of wheat, soybean protein feed have become more expensive and fertilizer prices also increased significantly. As a result, competing legumes position strengthened in terms of economic return (European Parliament, 2013).

As noted, since 2013 promoting protein crops is a priority of the Common Agricultural Policy. The reason for this decision was the EU's dependence on imports of proteic agricultural products and reduce protein crops diversity in European agriculture.

Table 3 contains a summary of the EU balance of feed rich in proteins and shows that the EU is totally dependent on imports of 70%, and for soybean meal dependence is over 97%. This illustrates the risks associated with the growing requirements of vegetal protein products, the sustainability of the European agricultural system is questionable and the evolution of prices on the international market is unpredictable (Schreuder, de Visser, 2014).

The competitiveness of protein crops and European agricultural policies to support these crops. As noted, the EU is heavily dependent on imports of vegetal products rich in proteins. The main reason is that protein crops in the EU is not competitive with other crops commonly grown as productivity, steady production and sometimes growing technology costs. This competition was assessed using data from various protein crops and comparing them with those from wheat and maize, which are expected to be replaced with the growing of protein crops, including grain legumes.

Table 4 presents the situation for wheat and Table 5 the situation for maize. From tables results, in order to become competitive with wheat or maize, the protein crops productivity should increase considerably. Thus, for soybean, the main source of vegetal proteins, incresing of average production should be about 30% (from 2.7 t/ha to 3.4 t/ha) to compete with wheat and about 63% to compete with maize (from 2.7 t/ha to 4.3 t/ha). For other legumes (pea and field beans) increases should be much higher, eg 76-69% in the competition with wheat and 120-112% in the competition with maize.

It notes that, over the past few decades in Europe were initiated a series of measures designed to support protein crops including: price protection, subsidies and direct coupled payments, agriculture-environment schemes; thus: between 1958 and 1992, various schemes have been developed to support prices for soybean, pea, and lupine; in 1989, they were introduced area payments to chick pea, lentils and pea; in 1992, production subsidies were reduced and replaced by area payments, these payments varying according to the type of crop, and soybean received less than other protein crops; in 2012, 17 EU Member States introduced "The protein award", in the protein crops major growing countries; moreover, Lithuania, Poland and Slovenia have used special measures available to the new Member States to support protein crops (European Parliament, 2013).

Table 1. Grain legumes growing area in the world (thou ha)
(Gh.V. Roman, 2015)

Species	North America	South and Central America	Europe	Africa	Asia	Oceania	Total
Pea	1,634	152	1,723	812	1,875	181	6,377
Field beans	616	5,957	260	5,695	14,237	62	26,827
Soybean	32,523	52,106	3,176	1,797	20,629	41	111,272
Lentils	1,095	20	84	178	2,820	146	4,343
Chick pea	158	161	74	483	12,079	574	13,529
Broad bean	0	163	238	570	964	112	2,047
Lupine	0	34	153	14	0	450	651
Peanuts	421	686	11	12,405	11,871	14	25,408
Cow pea	16	16	7	11,075	160	0	11,274
TOTAL	36,463	60,295	5,726	33,029	64,635	1,580	201,728

Table 2. Grain legumes total production in the world (thou tons)
(Gh.V. Roman, 2015)

Species	North America	South and Central America	Europe	Africa	Asia	Oceania	Total
Pea	4,558	188	3,021	720	2,229	263	10,979
Field beans	1,317	5,590	500	4,860	10,635	53	22,855
Soybean	94,681	146,149	5,943	2,246	27,294	92	276,405
Lentils	2,108	12	71	186	2,246	327	4,950
Chick pea	327	270	94	531	11,068	813	13,103
Broad bean	0	192	663	738	1,494	297	3,381
Lupine	0	55	251	21	1	459	787
Peanuts	1,893	1,759	8	11,547	29,951	28	45,160
Cow pea	29	19	24	5,422	193	0	5,687
TOTAL	104,913	154,134	10,575	26,271	85,108	2,306	383,307

Table 3. EU balance of protein-rich feeds in 2012
(R. Schreuder, C. de Visser, 2014)

Material	EU production (Mt)		EU consumption (Mt)	
	Product	Protein	Product	Protein
Soybean/meal	1,189	452	34,134	15,904
Rapeseed and sunflower seed/meal	27,481	5,213	19,721	6,329
Grain legumes (Pulses)	3,045	670	2,800	616
Dried forage	4,056	771	3,900	741
Miscellaneous plant sources	2,877	654	5,859	1,260
Sub-total	38,648	7,760	66,414	24,850
Fish-meal	398	275	599	433
Total	39,046	8,035	67,013	25,283

Table 4. Indication of required yield level increase to match wheat yield based on EU-member states average
(R. Schreuder, C. de Visser, 2014)

Crop	Yield (tons)			Oil production (Mton)	Starch production (Mton)
	Actual	Increased	% increase		
Soybean	2.7	3.4	30%	3.9	0.0
Rapeseed	3.1	3.1	0%	13.8	0.0
Sunflower	2.2	2.9	31%	20.3	0.0
Lupine	1.0	4.2	334%	1.9	0.0
Pea	2.7	4.8	76%	0.0	15.5
Field beans	2.7	4.5	69%	0.0	11.1
Alfalfa	22.9	24.8	8%	0.0	0.0

Table 5. Indication of required yield level increase to match maize based on EU-member states average (R. Schreuder, C. de Visser, 2014)

Crop	Yield (tons)			Oil produc-tion. (Mton)	Starch produc -tion (Mton)
	Actual	Increa-sed	% increase		
Soybean	2.7	4.3	63%	3.9	0.0
Rapeseed	3.1	3.9	25	13.8	0.0
Sunflower	2.2	3.6	64	20.3	0.0
Lupine	1.0	52	443	1.9	0.0
Pea	2.7	6.0	120%	0.0	15.5
Field beans	2.7	5.7	112%	0.0	11.1
Alfalfa	22.9	31.0	36%	0.0	0.0

The evolution of grain legumes in Romania and the current situation. In Romania, the assortment of grain legumes covered, over time, about 10 species, some of economic importance (pea, field beans, soybean, chick pea, lentils), some grown on small areas and of regional significance (for example, broad beans, peanuts, white lupine, yellow lupine, cow pea). The areas sown with grain legumes have been quite volatile over the last century (Figure 1): about 99 thou ha in 1938 (mainly pea and field beans); 167 thou ha in 1950; 194 thou ha in 1963; 672 thou ha in 1987 (390 thou ha of soybean, 170 thou ha of field beans, 100 thou ha of pea, 10 thou ha of chick pea, and 0.7 thou ha of lentils); 250 thou ha in 1992 (166 thou ha of soybean, 58 thou ha of field beans, 22 thou ha of pea); 125 thou ha in 2013 (67 thou ha of soybean, 21 thou ha of field beans, 30 thou ha of pea).

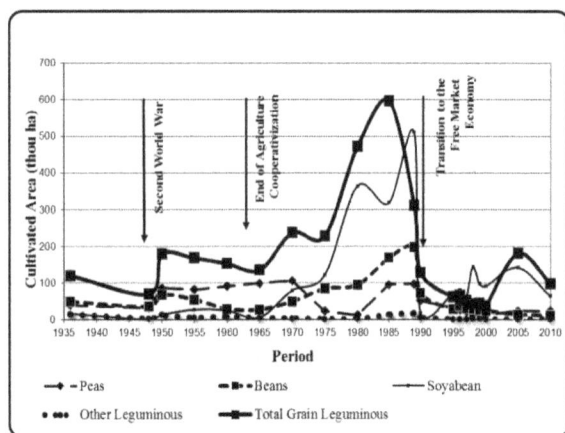

Figure 1. Evolution of Grain Legumes grown acreage in Romania, in 1935-2010 (thou ha) (Gh.V. Roman, 2015)

Soybeans presents a more special situation, in the sense that there were only 5.6 thou ha in 1935, reached over 300 thou ha during 1980-1985 and 500 thou ha in 1989; in that period,

Romania was the most important soybean grower in Europe.

After 1990, the soybean crop has shrunk to 58.1-73.3 thou ha, with a short period with growing areas nearly 200 thou ha, between 2002-2006, when it was permitted growing of GM-modified varieties.

In recent years soybean was grown on 42 thou ha in 2012, 67 thou ha in 2013, 72 thou ha in 2014 and over 100 thou ha in 2015.

European studies have shown that a soybean crop is competitive with wheat at a production least 3.4 tons/ha and with maize at least 4.3 tons/ha. From own research carried out in South Romania, resulted that, under intensive cultivation technology a soybean crop becomes profitable at a production of 3,000 kg grains/ha.

Table 6. Economic efficiency of soybean crop in Dobrudja region (2010-2014)

Index	2010 (very favorable year)	2014 (less favorable year)	Average (2010-2014)
Grain yields (kg/ha)	4,285.2	2.508.1	3,291.1
Market price (lei/t)	1,450	1,450	1,450
Total production value (lei/ha)	6,187.1	3,636.7	4,772.1
Total expenses (lei/ha)	3,773.7	3,610.8	3,680.5
Profit (lei/ha)	2,413.4	25.9	1,091.6
Rentability rate (%)	63.9	0.7	22.9
Production cost (lei/t)	880.6	1,439.7	1,118.3

Donau Soja Initiative and prospects of soybean cultivation in the Danube Basin.
In 2012, it founded Donau Soja (Danube Soya) Association, based in Vienna, whose declaration of constitution was signed so far by 18 European ministers of agriculture (Figure 2).

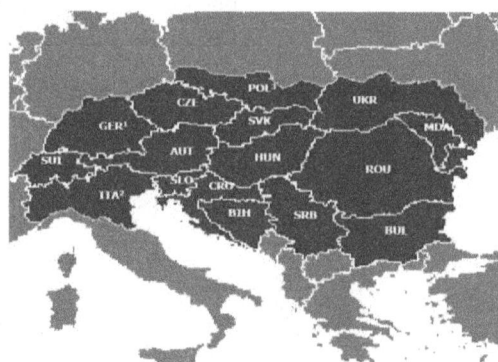

Figure 2. Region of "Donau Soja" Association ("Danube Soya") activity
(*"2nd International Danube Soya Congress"*, Augsburg, 2013)

The organization aims mainly: promoting of GMO-free soybean varieties growing; building a reliable soybean grain and meat supply and a value chain as a contribution to European protein supply; to establish an international breeding-research and control program for GMO-free soybean seeds along the Danube and it will establish concepts for improved crop growing.

The strategies developed in this respect stipulate that potential, among other things, increasing area growing with soybean up to 700 thou ha in Romania (which it can produced about 0.51 mil. tons of proteins), 125 thou ha for Bulgaria, 1 mil. ha for Ukraine, and 75.5 thou ha for Republic of Moldova (Dima, 2015).

CONCLUSIONS

The development of livestock in Europe and the need to balance feed rations based on cereals and the drastic reduction of areas grown with grain legumes crops led to reliance on Europe's imports, which reached 70% for agricultural protein products and over 95% for soybean grains and meal.

Diminishing area with grain legumes was due to the reduction of the use in human food, but also to deficiencies of these crops, including lower productivity, sensitivity to stressors and productions stability.

The consequences of these processes are more important if are considering restricting diversity of agricultural crops in crop rotations, which are dominated by cereals, lack of improving soil fertility that have legumes crops, and high dependence of Europe on imported protein products, in soybean grains and meal, majority GMO.

In recent decades the EU stimulus measures were initiated to extend protein crops growing, and since 2013 this has become a priority objective of the Common Agricultural Policy.

Assortment of legumes (protein crops group) from Europe consists mainly of pea, soybean, field beans, chick pea, lentils, broad beans, and lupine (and alfalfa). As protein sources can be included the oilseed crops important for Europe - rapeseed (a potential of 0.25 mil. tons of protein) and sunflower (a potential of 0.40 mil. tons of protein).

European studies have shown that a soybean crop is competitive with wheat at a production least 3.4 tons/ha and with maize at least 4.3 tons/ha. From own research carried out in South Romania, resulted that, under intensive cultivation technology a soybean crop becomes profitable at a production of 3,000 kg grains/ha. Stimulating farmers to expand the protein crops refers to coupled payments - for example, for non-GM (conventional) soybean in Romania were provided 345 €/ha, by provisions the use of certified seed and harvesting at least 1,300 kg ha (plus current subsidies per ha).

Donau Soja initiative will support farmers in order to increase soybean grown area, will intent to help for establishing harvest processing companies and to develop the trade, accompanied by the certification Donau Soja, allowing production recovery on specific channels. Following this program, provides for Romania, increasing soybean areas up to 700 thou ha, which would ensure a production of about 0.51 mil. tons of proteins.

REFERENCES

Dima D., 2015. Country Reports: Romania, Bulgaria, Republic of Moldova. Danube Soya East-West Protein Report Berlin.

Djordjevic V., Roman Gh.V., 2013. Soybean – Handbook for Danube region. Institute for Vegetables and Field Crops, Novi Sad, Serbia.

Lombardi A.R., 2015. Research on the Field Crops Assortment diversification through Grain Legumes Crops growing in the North Dobrudja conditions. PhD Thesis. Faculty of Agriculture, Bucharest.

Lombardi A.R., Epure L.I., Roman, Gh.V., 2013. Studies on Soybean growing in Tulcea County. Scientific Papers, Series A, Agronomy. Vol. LVI, USAMV, Faculty of Agriculture, Bucharest.

Roman Gh.V., Bucata L. I., Ion V., 2004. Evolution of the Field Crops Production in Romania. Proceedings, International Conference Ruse, Bulgaria.

Roman Gh.V. et al., 2015. Field Crops — Cereals and Grain Legumes. „Universitatea" Publishing House, Bucharest, Romania.

Schreuder R., de Visser C. de, 2014. Raport EIP-AGRI Focus Group on Protein Crops. Bruxelles.

***European Parliament, Directorate General for Internal Policies. Agricultural and Rural Development, 2013. The Environmental Role of Protein Crops in the New, Common Agricultural Policy. Strasbourg.

23

COMPARATIVE STUDY OF *IN VITRO, EX VITRO* AND *IN VIVO* PROPAGATED *Salvia hispanica* (CHIA) PLANTS: MORPHOMETRIC ANALYSIS AND ANTIOXIDANT ACTIVITY

Ely ZAYOVA[1], **Milena NIKOLOVA**[2], **Ludmila DIMITROVA**[1], **Maria PETROVA**[1]

[1]Institute of Plant Physiology and Genetics, 21, Acad. G. Bonchev, Str., 1113 Sofia, Bulgaria
[2]Institute of Biodiversity and Ecosystem Research 23, Acad. G. Bonchev, Str., 1113 Sofia, Bulgaria

Corresponding author email: mtihomirova@gmail.com

Abstract

A tissue culture technique for micropropagation of chia (Salvia hispanica L.) was established. High percentage of seed germination (100%) was recorded on Murashige and Shoog medium enriched with 0.4 mg l⁻¹ gibberellic acid and 10 mg l⁻¹ ascorbic acid after one week of culture. The maximum number of shoots per explant *(2.7) gave MS medium with 2 mg l⁻¹ 6-benzylaminopurine after five weeks of culture. The best plant rooting was achieved on half-strength MS medium with 0.1 mg l⁻¹ indole-3-butyric acid after four weeks of culture. The multiple plants were successfully ex vitro adapted with 95% survival. After six months under field conditions, important morphometric traits of ex vitro and in vivo derived plants were evaluated. Some morphological characteristics as plant height, leaf size, number of branches and leaves per plant were influenced by the propagation way. Differences between extracts of in vitro, ex vitro and in vivo developed plants in regard to free radical scavenging activity, flavonoid and lipid composition were not established. This is the first comparative study of micropropagated and in vivo seed derived chia plants. The study presents also for the first time data for antioxidant potential and lipid components of leaves of S. hispanica.*

Key words: DPPH, GC/MS, fatty acids, fatty alcohols, flavonoids, micropropagation.

INTRODUCTION

Salvia hispanica L. (chia) Lamiaceae is an annual herbaceous plant native to southern Mexico and northern Guatemala but it is cultivated in Argentina, Australia, Bolivia, Colombia, Guatemala, Mexico, Peru, Southeast Asia and naturalized in the Caribbean (Jamboonsri et al., 2012). High nutritional and pharmacological value of chia seed (Muñoz et al., 2013) generate interest to be explored opportunities the species to be grown outside these areas. The important problem is that at higher latitudes the plant cannot produce seeds (Jamboonsri et al., 2012). Preliminary our study has shown that a chia plant successfully grown in Bulgaria, but the main limitation is seed production. However, the plants produce a large amount of leaves, which could be used as a source of biologically active substances. In contrast to the seeds the leaves of *Salvia hispanica* have been poorly studied for chemical composition and pharmacological activities. The major leaf essential oil components and leaf metabolites including flavonoids and phenolic acids of *S. hispanica*

have been reported (Ahmed et al., 1994; Peiretti and Gai, 2009; Amato et al., 2015). As it is known, the phenolic compounds have capacity to act as antioxidants. Although chia seeds are widely studied for antioxidant activity (Taga et al., 1984; Muñoz et al., 2013) to the best of our knowledge there are no data on the antioxidant action of the leaves.

Plant micropropagation is one of the techniques to obtain large number of plants with constant biosubstances, irrespective of season. The efficacy of propagation depending on the explant and the type and concentration of used plant growth regulators. A higher proliferation of leaves/plant were observed with 0.75 and 1 µM of BA (Bueno et al., 2010). The possibility of establishing Chia plant and callus cultures for production of fatty acids were studied (Marconi et al., 2013). The authors reported that the fatty acid content of calli is significantly lower than that of seeds after two years in culture. The available literature concerning *S. hispanica in vitro* cultures is quite limited.

The aim of present study is to establish *in vitro* propagation protocol and compare morphometric traits, antioxidant activity, flavonoid and lipid components of leaves of *ex vitro* and *in vivo* derived *S. hispanica* plants.

MATERIALS AND METHODS

Initial plant material and surface sterilization

Chia (*S. hispanica*) seeds were obtained from the commercial seed source (Chia Seed Company, Mexico). The seeds were soaked in 70% ethanol for 30 s and agitated in 15% bleach solution (ACE Procter & Gamble Co., USA) containing 5% active chloride for 5 min. Then they were washed three times each for 5 min in sterilized distilled water. Each treatment consisted of 2 replicates with 20 seeds. The disinfected seeds were germinated on Murashige and Skoog, 1962 basal medium (MS) supplemented with 3% sucrose, 7% agar, 0.4 mg l^{-1} gibberellic acid (GA$_3$) and 10 mg l^{-1} ascorbic acid (AA). As control medium was used MS basal medium without GA$_3$ and AA. The percentage of germination was determinated after two weeks of culture. Initial explants were taken from four weeks old seedlings for further *in vitro* cultivation.

Shoot multiplication and *in vitro* rooting

For shoot multiplication, all tips from three weeks old seedling were aseptically excised and cultured on full strength MS media with vitamins, containing 6-benzylaminopurine (BAP), zeatin, thidiazuron (TDZ) or 6-(γ,γ-dimethylallylamino) purine (2-iP) alone (Table 2). The explants were placed in the culture vessels (tubes 140 x 20 mm), one explant per 10 ml medium. Twenty explants per nutrient medium were utilized. The frequency of shoot formation, number of developed shoots and their length was determined after five weeks of culture. For root induction, well developed shoots were cultivated on half strength MS medium with 2% sucrose, 7% agar and auxin indole-3-butyric acid (IBA) at concentrations 0; 0.1; 0.5 and 1.0 mg l^{-1}. Twenty shoots per medium were used. Data were recorded on percentage rooting, mean number of roots per plant, and root length after four weeks of culture.

Each experiment was repeated two times. pH of the medium was adjusted to 5.8 before autoclaving at 121°C for 20 min at a pressure of 1.1 kg cm^{-2}. All chemicals used for preparing the media were purchased from Duchefa Biochemie B.V., the Netherlands. The *in vitro* cultures were maintained under growth room conditions at a temperature of 24 ± 1 °C, relative humidity of 70% and a 16 h photoperiod under 40 μmol m^{-2} s^{-1} illumination provided by Philips 36W cool white fluorescent tubes.

Ex vitro acclimatization

Plants with a well developed root system were carefully removed from the culture vessels and their roots were washed to remove agar. Then they were transferred to the small pots (12 cm in diameter, 10 cm in height) containing a mixture: soil, sand and perlite in the volume ratio 1:1:1, v/v/v. The plants were grown in a culture room at a temperature of 25 ± 1 °C under 16 h photoperiod and 50 μmol m^{-2} s^{-1} light intensity. The potted plants were covered with a transparent polythene membrane to ensure high humidity (90%). The polythene membrane was opened after two weeks. The survival efficiency, defined as the percentage of plants survived the transfer from *in vitro* to *ex vitro* growth conditions, was determined after six weeks of acclimatization. After two months of adaptation, fully acclimatized plants were transferred to the field.

In vivo seed propagation

The seeds were sown 0.4 cm deep in small pots (12 cm diameter) filled with a mixture of soil, sand and perlite 1:1:1, v/v/v (20 seeds per pot, five pots per treatment). The pots were placed under the same *ex vitro* conditions. The seed germination percentage was determined after two weeks of planting. The young seedlings were transferred separately into the pots with the same soil mixture. The *ex vitro* and *in vivo* propagated plants were planted at the experiment field of the Institute of Plant Physiology and Genetics, Sofia, Bulgaria (42°50′N, 23°00′E, and altitude of 595 m). The soil type at Sofia, classified according to physical characteristics was sandy loam (0-15 cm, top soil). Each trial was laid out in a randomized block design with two replications. The *ex vitro* and *in vivo* plants were planting by hand on 1.5 m x 1.2 m plots, with 50 cm spacing between rows and 40 cm spacing between plants. Standard agronomic practices used included regular watering and hand weeding. In the temperate climate of Bulgaria, the planting were performed in early spring, the beginning of April. The leaves and flowers from plants, propagated by the both ways were collected at the end of September.

Morphometric analysis

Five plants from each plot were tagged randomly for data recording (total 10 selected individuals from each *ex vitro* and *in vivo* origin). The *ex vitro* and *in vivo* derived plants of the same age were evaluated and compared on the basis of selected morphological parameters (Table 3).

Antioxidant activity

Metanolic extracts: Air-dried, ground plant material (1 g) was extracted with 80% (3 x 30 ml) methanol by classical maceration for 24 h. After evaporation of the solvent the crude extract was subject to subsequent analysis.

DPPH radical scavenging activity: The effect of methanolic extracts on DPPH radicals was estimated according to Stanojević et al. (2009). The IC_{50} values were calculated by Software Prizm 3.00. All of the experiments were carried out in triplicate.

Flavonoid analysis

Thin layer chromatographic analysis (TLC): Two TLC sorbents and two mobile phases were used for the analysis of the methanolic extracts for flavonoids. Ethyl acetate/formic acid/acetic acid/ methyl ethyl ketone/water (50:7:3:30:10 v/v/v/v/v) was used for the development of the extracts on silica gel plates Kiselgel 60 F_{254} (5554). Acetic acid/water (15:85 v/v) was used for Cellulose F (5574) plates. Chromatograms were viewed under UV = 366 nm light before and after spraying with „Naturstoffreagenz A": 1% solution of diphenylboric acid 2-aminoethyl ester complex in methanol.

Lipid fraction

100 μg of plant material as well as internal standards of 50 μg of nonadecanoic acid were placed in 2 ml Ependorf tubes and extracted with 1 ml of MeOH for 2 h at room temperature assisted by an ultrasonic bath for 15 min at 70°C every 30 min, after that the sample was centrifuged. Aliquot of 800 μl was transferred in other Ependorf tubes and was added of 500 μL H_2O and 500 μl of $CHCl_3$, vortexing for 2 min, and the mixture was centrifuged. The chloroform fraction was separated, evaporated and transmethylated with

2% of H_2SO_4 in MeOH at 60°C for 18 h, than lipids were extracted with *n*-hexane (2x500 μl) which was dried with anhydrous Na_2SO_4 and evaporated to obtain lipid fraction.

Metabolite analysis

The GC-MS spectra were recorded on a Termo Scientific Focus GC coupled with Termo Scientific DSQ mass detector operating in EI mode at 70 eV. ADB-5MS column (30 m x 0.25 mm x 0.25 μm) was used. The temperature program was: 100-180 °C at 15 °C x min^{-1}, 180-300 20 at 5 °C x min^{-1} and 10 min hold at 300 °C. The injector temperature was 250 °C. The flow rate of carrier gas (Helium) was 0.8 ml x min^{-1}. The split ratio was 1:10 1 μl of the solution was injected.

The metabolites were identified as TMSi derivatives comparing their mass spectra and Kovats Indexes (RI) with those of an on-line available plant specific database (The Golm Metabolome Database, the NIST 05 database and mass spectra available in the on-line lipid library.The measured mass spectra were deconvoluted by the Automated Mass Spectral Deconvolution and Identification System (AMDIS), before comparison with the databases. RI of the compounds were recorded with standard n-hydrocarbon calibration mixture (C9-C36) (Restek, Cat no. 31614, supplied by Teknokroma, Spain) using AMDIS 3.6 software.

Statistical analysis

The data were subjected to one-way ANOVA for comparison of means, and significant differences were calculated according to Fisher's least significance difference (LSD) test at the 5% significance level using a statistical software package (Statgraphics Plus, version 5.1 for Windows). Data were presented as means ± standard error.

RESULTS AND DISCUSSIONS

In vitro seed germination

Due to the successful sterilization no bacterial and fungal contamination of the *in vitro* cultured seeds of *S. hispanica* was observed. Usually after applying of sterilizing agents, the initial plant material washed three times each 15 min in sterile distilled water (Zayova et al., 2013) but these small chia seeds easily were coated with a sticky gel when they were soaked in water, so duration time was reduced to 5 min. The results showed that seed germination depended on nutrient medium composition. Most efficient proved to be MS basal medium supplemented with 0.4 mg l^{-1} GA$_3$ and 10 mg l^{-1} AA ensured 100% seed germination 7 days after sowing. On the MS control medium without GA$_3$ and AA the percentage of germinated seeds was 70% after 14 days of cultivation.

It might suggest that favorable influence of GA$_3$ and AA resulted in better seed germination compared to MS control medium. This combination was effective for increasing the speed of germination and good development of the seedlings (Figure 1a).

Figure 1. Micropropagation of *S. hispanica*: a) *in vitro* seedlings on MS basal medium with 0.4 mg l^{-1} GA$_3$ and 10 mg l^{-1} AA; b) *in vitro* propagated plant on MS medium with 2 mg l^{-1} BAP; c) *in vitro* rooted plant on ½ MS medium with 0.1 mg l^{-1} IBA; d) *ex vitro* acclimatized plant; e) *ex vitro* plant, grown under field conditions; f) plant with flowers

The seed germination of *S. officinalis* was also improved, when 1 mg l^{-1} GA$_3$ was added in culture medium (Grzegorczyk et al., 2005). Dadasoglu and Özer (2014) reported that the germination is strongly influenced by gibberellic acid application for five wild growing Salvia species.

Shoot multiplication and *in vitro* rooting
The data on shoot formation of tip explants isolated from four weeks old seedling cultured on MS medium supplemented with different concentrations of BAP, Zeatin, TDZ or 2-iP alone are presented in Table 1. The number of

induced shoots per explant varied depending on the type and concentration of cytokinin.

Table 1. Effect of different cytokinin types on shoot multiplication of *S. hispanica*

Cytokinin types	mg l^{-1}	Shoot formation, (%)	Number of shoots per explant	Shoot height (cm)
BAP	0.5	NR	NR	NR
	1.0	45	1. ±0.08^b	1.7±0.15^{ab}
	2.0	65	2.7±0.20 ^{d}	1.4±0.07^a
Zeatin	0.5	NR	NR	NR
	1.0	40	1.2±0.06^{ab}	1.6±0.12^{ab}
	2.0	50	1.8±0.15^c	1.5±0.10^{ab}
TDZ	0.5	NR	NR	NR
	1.0	NR	NR	NR
	2.0	30	1.0±0.06 ^{a}	1.8±0.10^b
2-iP	0.5	NR	NR	NR
	1.0	NR	NR	NR
	2.0	35	1.2±0.09^{ab}	2.2±0.10^c
LSD	-		0.37	0.31

The data are presented as means of 20 individuals per treatment ± standard error. Different letters indicate significant differences assessed by Fisher LSD test (P≤0.05) after performing ANOVA multifactor analysis. NR: Not Responded

The shoot induction and multiplication occurred in the presence of cytokinins BAP and zeatin at higher concentration. The maximum number of shoots per explant (2.7 with 1.4 cm average height) gave MS medium supplemented with 2 mg l^{-1} BAP followed by MS medium with 2 mg l^{-1} zeatin after five weeks of culture (Table 1). The new shoots induced on MS medium, containing 2 mg l^{-1} BAP were characterized by normal leaves and less average shoot height (Figure 1b) compared to those grown on MS media with the other tested cytokinins. The cytokinins TDZ or 2-iP positively influenced the plant growth and development, but not stimulated shoot formation. The results show that cytokinins, particularly BAP or zeatin in the MS culture medium are important for multiplication of *S. hispanica* shoots under *in vitro* conditions).

Bueno et al. (2010) also reported that higher concentration BA increased the shoots formation of *S. hispanica*. A similar response has been demonstrated in *Salvia santolinifolia* where addition of BA at 2.0 and 3.0 mg l^{-1} produced maximum number of shoots (Tour and Khatoon, 2014). The same stimulatory effect of BAP in MS medium on the shoot formation in *S. officinalis* was described by other authors (Weielgus et al., 2011; Cristea et al., 2014). For shoot induction and plant growth of *Salvia* species often high concentrations of cytokinins from adenine type are necessary (Kintzios, 2000).

The multiple shoots were separated from one another and transferred to half strength MS root induction medium with different concentrations of IBA (Table 2).

Table 2. Effect of auxin IBA on the root induction of S. hispanica micropropagated plants

IBA ($mg\ l^{-1}$)	Rooting (%)	Number of roots per plant	Root length (cm)
0	15	1.0 ± 0.12^a	0.7 ± 0.09^a
0.1	100	8.6 ± 0.44^d	1.5 ± 0.17^b
0.5	80	3.2 ± 0.31^c	1.4 ± 0.14^b
1.0	65	2.1 ± 0.23^b	1.6 ± 0.16^b
LSD		0.54	0.26

The data are presented as means of 20 individuals per treatment ± standard error. Different letters indicate significant differences assessed by Fisher LSD test (P≤0.05) after performing ANOVA multifactor analysis

Initial root formation was observed 14 days after transferring to the rooting medium. The best rhizogenesis was observed on ½ MS medium supplemented with 0.1 mg l^{-1} IBA, where 100% of plants produced high number of roots (average 8.6 roots per plant with 1.5 cm root length) after four weeks of culture. The formed roots were short and firmly fixed to the plant (Figure 1c).

The content of 0.5 mg l^{-1} IBA in ½ MS medium was found to be effective for root induction, too.

Treatment with IBA has been previously reported for *in vitro* rooting of *Salvia blancoana* and *Salvia valentine* (Cuenca and Amo-Marco, 2000). MS medium with 2.7 µM IBA was efficient for root induction of *Salvia fruticosa* (Arikat et al., 2004). In contrast, NAA were found to have a better effect on *rhizogenesis* of *S. officinalis* than IBA (Cristea et al., 2014).

The rooted plants of *S. hispanica* were successfully *ex vitro* acclimatized with 95% survival.

Ex vitro acclimatization

The existence of well developed root system and controlled reduction of humidity are important factors for the success of *ex vitro* acclimatization.

An appropriate potting substrate was facilitated the successful transfer of plants from *in vitro* to *ex vitro* conditions. In our case, well drained mixture (soil: sand and perlite 1:1:1, v/v/v) was found to be the best for hardening (Figure 1d) and its use ensures high survival rate (95%) of micropropagated plants. The plants were successfully transferred to field conditions, where they continued to grow and develop (Figure 1e).

In vivo seed propagation

The results showed that the seed germination on the used mixture (soil, sand and perlite 1:1:1, v/v/v) was reached 80% after two weeks of sowing. The seedlings in the pots developed normally (Figure 2a).

Figure 2. *In vivo* propagation of *S. hispanica*: a) *in vivo* plants on mix of soil, sand and perlite (1:1:1 v/v/v); b) *in vivo* plant, grown under field conditions; c) plant with flowers

The plants were ready to be transplanted in the field in the beginning of April within eight weeks. All plants were healthy and fully vigorous after planting in the soil (Figure 2b). It was not observed diseases and pests of plants.

It should be noted that the plants produced a prolific amount of leaves, but only a few plants bloomed with a small number of flowers (Figures 1f and 2c). Seeds were not obtained until the end of culturing under the field conditions.

Morphometric characteristics

A comparative study was conducted between *ex vitro* and *in vivo* grown plants after six months of cultivation using morphometric parameters. Morphological differences were observed (Table 3). The plant height, number of branches and leaves per plant and leaf size were influenced by the propagation way. The *ex vitro* plants grown under field conditions were higher (average 160.4±1.52 cm) than those propagated *in vivo* (average 125.3±1.34 cm). They exhibited more branches production when compared to the *in vivo* plants.

Table 3. Morphometric characteristic of *ex vitro* and *in vivo* propagated *S. hispanica* plants grown in the field (six months)

Morphological character	Ex vitro plants	In vivo plants
Plant height, cm	160.4±1.52[a]	125.3±1.34[b]
Number of branches per plant	19.0±0.55[b]	15.0±0.48[a]
Number of leaves per plant	305.0±1.84[b]	256.0±1.65[a]
Leaf length, cm	5.4±0.31[b]	4.6±0.24[a]
Leaf width, cm	4.2±0.23[b]	3.4±0.21[a]
Internode distance, cm	5.9±0.30[a]	5.3±0.28[a]
LSD	1.31	1.10

The data are presented as means of 10 individual's ± standard error

Each plant derived from *ex vitro* conditions produced 19.0 branches, while those from *in vivo* seeds derived plants - 15.0 branches, which were fragile and easily break. Similar tendency was observed for the number of leaves. Both *ex vitro* and *in vivo* chia leaves looked similar with normal leaf shape, but differed in their size. The *ex vitro* plants grow rapidly and had larger leaves and thicker stems compared with *in vivo* plants. Moreover, the flowers of both *ex vitro* and *in vivo* plants also showed similar morphology. On the other hands, there were morphometric differences among obtained *in vivo* plants; some generally had higher plant stature, whereas others had shorter ones. Similar differences were also observed for number of branches per plant, while all *in vitro* propagated and *ex vitro* adapted plants were morphologically homogeneous. The better performance of *in vitro* derived plants than the conventionally propagated plants was reported by other authors (Gustavsson and Stanys 2008; Singh et al., 2013).

Antioxidant activity
Methanolic extracts of leaves of *in vitro*, *ex vitro* and *in vivo* plants of *S. hispanica* were

analyzed for their antioxidant potential assessed by scavenging of DPPH radicals. Results presented as IC_{50} values (µg ml^{-1}) - extract concentration providing 50% inhibition of the DPPH solution.

Differences between extracts of *in vitro*, *ex vitro* and *in vivo* propagated plants were not found. The studied extracts showed good antioxidant activity with IC_{50} values about 100 µg ml^{-1} (Figure 3).

This study presents for the first time evidence for antioxidant potential of leaves of *S. hispanica*.

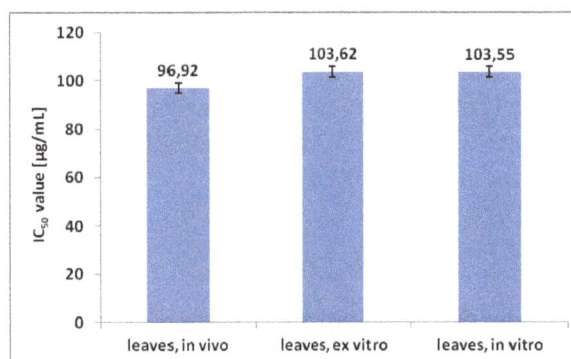

Figure 3. DPPH radical scavenging activity of methanolic extracts of leaves and flowers of *in vitro*, *ex vitro* and *in vivo* propagated *S. hispanica* plants

Flavonoid composition
The methanolic extracts were examined for occurrence of flavonoids by TLC analysis. Differences of the flavonoid composition between *in vitro*, *ex vitro* and *in vivo* developed plants were not established. Several components with TLC behavior (Rf-values and color) of flavonoids were detected. Some of them were identified with comparative analysis with reference compounds as luteolin-7-0-glycoside, apigenin-7-0-glycoside, orientin (luteolin-6-C-glucoside), kaempferol-3-0-glucoside and *vitexin* derivative.

Other components remained unidentified. The received results showed that the leaves of *S. hispanica* are rich in flavonoid compounds, mainly derivatives of flavones.

This is in accordance with reported data about flavonoid compounds in *Salvia* species (Lu and Foo 2002; Cvetkovikj et al., 2013).

Reported here flavonoid components coincide with those established by Amato et al. (2015) as leaf metabolites of *S. hispanica* plants, grown in Basilicata, Italy.

Lipid composition

Lipid fraction of methanolic extract from leaves of *ex vitro* and *in vivo* developed plants of *S. hispanica* were analyzed by GC/MS. In the fractions of the samples four main peak signals were detected.

They were identified as representatives of fatty alcohols and fatty acid. Three fatty alcohols were identified: 1-dodecanol, 1-hexadecanol, 1-octadecanol. Among them the 1-dodecanol was the most abundant.

Palmitic acid was detected as major fatty acid in the lipid fraction, approximately 40% of all lipid components.

Difference in the lipid composition between *ex vitro* and *in vivo* developed plants was not established. In the leaves α-linolenic acid was not found in contrast to the seed where it is the major acid (Amato et al., 2015).

CONCLUSIONS

An effective protocol for micropropagation of *Salvia hispanica* was developed. The described *in vitro* propagation technique could be applied to produce elite planting material within a short period of time (five months).

Multiplication of chia was highly dependent on the type of cytokinins used.

Comparison between both types of propagated plants revealed that some morphometric traits (plant height, number of branches per plant leaves number and leaf size) were influenced by the propagation way.

A micropropagation method may reduce the variation inherent in seed population.

The lack of significant differences in antioxidant properties, flavonoid and lipid composition between extracts of *in vitro*, *ex vitro* and *in vivo* developed plants of *S. hispanica* found in present study indicates that biotechnological tools may be applied for production of homogeneous plants with profuse leaves as a rich source of natural antioxidants.

ACKNOWLEDGEMENTS

The authors are grateful to Vasil Velev, manager of a private company "Tevio" Ltd., Sofia, Bulgaria for financial support of present study.

REFERENCES

Amato M., Caruso M.C., Guzzo F., Commisso M., Bochicchio R., Galgano F., Labella R., Favati F., 2015. Nutritional quality of seeds and leaf metabolites of Chia (*Salvia hispanica* L.) from Southern Italy. European Food Research and Technology, 241, p. 615-625.

Arikat N.A., Jawad F.M., Daram N.S., Shibli R.A., 2004. Micropropagation and accumulation of essential oils in wild sage (*Salvia fruticosa* Mill). Scientia Horticulturae, 100, p. 193-202.

Ahmed M., Ting I.P., Scora R.W., 1994. Leaf oil composition of *Salvia hispanica* L. from three geographical areas. Journal of Essential Oil Research, 6, p. 223-228.

Bueno M., Sapio D., Barolo M., Villalonga E.M., Busilacchi H., Severin C., 2010. *In vitro* response of different *Salvia hispanica* L. (Lamiaceae) explants. Molecular Medicinal Chemistry, 21, p. 125-126.

Cristea O.T., Prisecaru M., Ambarus S., Calin M., Brezeanu C., Brezeanu M., Florin G., 2014. Vegetative multiplication of *Salvia officinalis* L. for the obtaining of true-to-type plants. Biologie, 23, p. 104-107.

Cvetkovikj I., Stefkov G., Acevska J., Petreska J., Karapandzova M., Stefova M., Dimitrovska A., Kulevanova S., 2013. Polyphenolic characterization and chromatographic methods for fast assessment of culinary *Salvia* species from South East Europe. Journal of Chromatography A, 1282, p. 38-45.

Cuenca S., Amo-Marco J.B., 2000. *In vitro* propagation of two Spanish endemic species of *Salvia* throught bud proliferation. In Vitro Cellular & Developmental Biology - Plant, 36, p. 225-229.

Dadasoglu E., Özer H., 2014. Effects of different temperatures and gibberellic acid (GA₃) doses on germination of Salvia species. Journal of Medicinal and Spice Plants, 19, p. 47-51.

Grzegorczyk I., Bilichowski I., Mikiciuk-Olasik E., Wysokińska H., 2005. *In vitro* cultures of *Salvia officinalis* L. as a source of antioxidant compounds. Acta Societatis Botanicorum Poloniae, 74, p. 17-21.

Gustavsson B.A., Stanys V., 2008. Field performance of sanna lingonberry derived by micropropagation vs stem cutting. Hortic. Science, 35, p. 742-744.

Jamboonsri W., Phillips T.D., Geneve R.L., Cahill J.P., Hildebran D.F., 2012. Extending the range of an ancient crop, *Salvia hispanica* L., a new & ω-3 source. Genetic Resources and Crop Evolution, 59, p. 171-178.

Kintzios S., 2000. *Salvia* sp.: Tissue culture, somatic embryogenesis, micropropagation and biotransformation. In Kintzios S. (ed.), Medicinal and Aromatic Plants-Industrial Approaches: The genus *Salvia*. Harwood Publishers, Amsterdam, p. 243-250.

Lu Y., Foo L., 2002. Polyphenolics of *Salvia* - a review. Phytochemistry, 59, p. 117-140.

Marconi P.L., López M.C., Jürgen D.M., Bovjin C., Alvarez M.A., 2013. *In vitro* establishment of *Salvia hispanica* L. plants and callus. Biotecnología Vegetal, 13, p. 203-207.

Muñoz L.A., Cobos A., Diaz O., Aguilera J.M., 2013. Chia Seed (*Salvia hispanica*): An ancient grain and a new functional food. Food Reviews International, 29, p. 394-408.

Peiretti P.G. and Gai F., 2009. Fatty acid and nutritive quality of Chia (*Salvia hispanica* L.) seeds and plant during growth. Animal Feed Science and Technology, 148, p. 267-275.

Singh T.D., Singh C.H., Nongalleima K., Moirangthem S., Devi H.S., 2013. Analysis of growth, yield potential and horticultural performance of conventional vs. micropropagated plants of *Curcuma longa* var. Lakadong. African Journal of Biotechnology, 12, p. 1604-1608.

Stanojević L., Stanković M., Nikolić V., Nikolić L., Ristić D., Čanadanovic-Brunet J., Tumbas V., 2009. Antioxidant activity and total phenolic and flavonoid contents of *Hieracium pilosella* L. extracts. Sensors, 9, p. 5702-5714.

Taga M.S., Miller E.E., Pratt D.E., 1984. Chia seeds as a source of natural lipid antioxidants. Journal of the American Oil Chemists' Society, 61, p. 928-931.

Tour J., Khatoon K., 2014. *In vitro* regeneration of *Salvia santolinifolia*. Pakistan Journal of Botany, 46, p. 325-328.

Wielgus K., Luwanska A., Szalata M., Mielcarek S., Gryszczynska A., Lipinski D., Slomski R., 2011. Phytochemical estimation of Sage (*Salvia officinalis* L.) cultivated *in vitro* - flavonoids and phenolic acids. Acta Fytotechnica et Zootechnica, Sp. Number, p. 8-11.

Zayova E., Stancheva I., Geneva M., Petrova M., Dimitrova L., 2013. Antioxidant activity of *in vitro* propagated *Stevia rebaudiana* Bertoni plants of different origins. Turkish Journal of Biology, 37, p. 106-113.

24

CAUSES AND SOLUTIONS FOR THE REMEDIATION OF THE POOR ALLOCATION OF P AND K TO WHEAT CROPS IN ROMANIA

Florin SALA, Ciprian RUJESCU, Cristian CONSTANTINESCU

Banat University of Agricultural Sciences and Veterinary Medicine "King Michael I of Romania", Timisoara, Calea Aradului 119, 300645, Timişoara, Romania, Email: florin_sala@usab-tm.ro, rujescu_ciprian@yahoo.com

Corresponding author, email: florin_sala@usab-tm.ro

Abstract

The aim of the present study was to establish the main cause of the poor allocation of PK fertilizers to wheat crops, as this phenomenon has been observed in Romania. The experimental research consisted in differentiated allocation of fertilizers with N and the PK complex in order to create controlled nutrition deficits. It was conducted within the Didactic Station in Timisoara, from 2012 to 2014. Based on the yields obtained and on the associated economic elements, two scenarios were used: variation of wheat selling price and the price of fertilizers; each scenario had several variants. For each variant, the study assessed the optimal dose of N and PK respectively that resulted in the maximum profit. The model employed for this purpose was a model given by the production function verified by experimental data. The results were analysed and processed with SPSS and the graphic representations with Wolfram Alpha. The study discovered that the main cause of the Romanian farmers' low interest in using PK fertilizers is the dissonance between the cost of PK fertilizers and the low market price of wheat. Given the current price of wheat - between 0.5 - 0.6 lei/ kg[-1], the authorities have to adopt incentive pricing for PK fertilizers in order to stimulate the use of such fertilizers by Romanian farmers. These measures will result not only in better quality and quantity of the yield, but also in medium and long-term improvement of soil fertility.

Key words: fertilization, models, phosphorus, potassium, wheat.

INTRODUCTION

Wheat is one of the most widely used cereals in the world, together with rice and corn. On average, the world produces over 600 million tons of wheat annually (Shewry, 2009).

Wheat production is mainly concentrated in a few large areas: the European Union is responsible for around 21% of the entire production of wheat in the world (Eurostat Database). Over 50% of the wheat production is obtained in France (about 41 Mio. tons), Germany (approx. 24 Mio. tons) and the UK (approx. 15 Mio. tons). Other countries with large wheat yields are Poland (approx. 10 Mio. tons), Italy (approx.6 Mio. tons), Denmark, Romania and Spain (approx. 5 Mio. tons each), Bulgaria and Hungary (almost 4 Mio. tons each). The development of better crop technologies and the improvement of the biologic material in recent years have led to a significant increase in wheat production worldwide (Vigani et al., 2013). Romania is among the biggest 6 producers of wheat in Europe: here, wheat is cultivated on an area of approx. 2.135 million ha, with an average yield of 3479 kg ha[-1] and a total production of 7.428 million tons (INS, 2014).

Fertilizers are necessary for increasing the productivity of agricultural crops, especially wheat crops. Nevertheless, fertilization has to be well-balanced in terms of nutrients and soil fertility, the consumption needs of plants and the estimated yields (Otiman and Creț, 2002; Havlin et al., 2005; Sala, 2011). Addition of macro- and micro-elements and balanced fertilization are essential for ensuring good quality and quantity of the wheat yield (Calderini and Ortiz-Monasterio, 2003; Malakouti, 2008; Habib, 2009; Zeidan et al., 2010; Pîslea et al., 2013; Hamzeh and Sala, 2015; Wang et al., 2016).

The basic principles of plant nutrition are known (Marschner, 1995; Mengel and Kirkby, 2001), as are a series of principles and methods for establishing fertilization programs and fertilizer doses in relation to different agricultural systems (Borlan and Hera, 1982;

Buresh and Witt, 2007; Cui et al., 2008; Chuan et al., 2013). Nevertheless, certain limitations of technical and economical nature have generated imbalance in the optimal allocation of fertilizers in Romania, affecting soil fertility over large areas, the yields of different crops and the productivity of agricultural exploitations (Otiman, 1999).

It is well-known that, after 1989, in Romania, the allocation of fertilizing substances has been unevenly distributed in relation to the types of agricultural exploitations, being generally insufficient and generating nutritive imbalance on extensive areas (Dumitru, 2002; Hera, 2010).

Eurostat statistics show that, when compared to countries such as Germany or France, where on average 7-8 kilos P are added per hectare, as reported to UAA, or Croatia, where on average the allocation is 13 kg P/ha reported to UAA, in Romania the average allocation is 3 kilos P/ha, as reported to UAA. Certainly, if we look at the area cultivated with cereals (from UAA), which in Romania is approximately 37%, the quantity of phosphorus fertilizers allotted to cereal crops will be higher than the average mentioned above, but from the point of view of the biological potential of the cultivated plants and the necessary support for soil fertility in relation to the estimated yields, it is insufficient (Hera, Otiman et al. 2015). Eurostat statistics show that things are similar in respect to K fertilization, as well: almost 2 kg K/ha reported to UAA in Romania, while in France there are 13 kg K/ha reported to UAA, and in Germany there are 20 kg K/ha reported to UAA (Eurostat Database).

Optimization of wheat crop fertilization has always been of interest. Thus, a series of studies have developed models focusing on the interdependence among fertilizers, yield and quality in relation to the soil and climate conditions and to the cultivars (Fu et al., 2014; Swain et al., 2014; Rawashdeh et al., 2015; Sala et al., 2015; Nutall et al., 2016). Others have focused on optimizing the fertilization process with insufficient quantities of fertilizers (Sala and Boldea, 2011; Boldea and Sala, 2013). Recent measures for financing agriculture, based on the use of an initial fix capital, have introduced a different approach to

the optimization issue (in this case abandoning the idea of increasing the profits by minimizing the costs - concept that often involves fertilization deficit). In this sense, some studies have been made on particular cases, regarding the determination of optimal productions for a reference area, given a certain initial capital (Rujescu et al., 2014, 2015).

Taking into consideration the demand for quality wheat for human consumption and also some aspects that have been identified in the Romanian agricultural practice, the aim of the present study was to establish, through adequate mathematical analysis, the fact that the main cause of poor allocation of PK fertilizers is given by the lack of correlation between the market price of wheat and the prices of PK fertilizers, as well as to indicate threshold values for economic profitability.

MATERIALS AND METHODS

The starting point of the study is represented by direct results obtained under experimental conditions at the Didactic Station within BUASVM Timişoara, regarding the influence that different doses of N fertilizers and PK fertilizers, respectively, have on wheat yield. The variety of wheat cultivated was Alex. The studies were done on cambic chernozem with medium fertility, neutral reaction (pH = 6.7-6.8), good humus supply (H = 3.2%), nitrogen index IN = 2.8, high base saturation (88-90%), poor phosphorus supply (P = 11.4 ppm) and medium potassium supply (K = 130.5 ppm). The experiments were run between 2012 and 2014, and the climate conditions were favourable for wheat crops.

The fertilization variants were made based on the PK complex in rates between 0-150 kg ha^{-1} and N in rates between 0-200 kg ha^{-1}.

Variants and combinations of variables. The study was focussed on two variables and the phosphorus and potassium fertilization was determined/motivated in relation to them. These variables are the price of wheat and the price of PK fertilizers. Price variations of the two variables were simulated in order to find the justification for fertilizer application at farmer level, Table 1.

Table 1. Schematic representation of the simulation variants – input data

V 1
Fixed data: p_{wheat} = 0.5 lei/kg; p_N = 4.1 lei; fixed costs per 1 hectare c= 1900 lei
Variable: p_{PK} = 6.1 lei (current price)

	Variables:		
V 1A	V 1B	V 1C	V 1D
p_{PK}=4.1 lei	p_{PK}=4.6 lei	p_{PK}=5.1 lei	p_{PK}=5.6 lei

V 2
Fixed data: p_{wheat} = 0.6 lei/kg; p_N = 4.1 lei; fixed costs per 1 hectare c= 1900 lei
Variable: p_{PK} = 6.1 lei

	Variables:	
V 2A		V 2B
p_{PK}=5.1 lei		p_{PK}=5.6 lei

Fixed data: p_N = 4.1 lei; p_{PK} = 6.1 lei; Fixed costs per 1 hectare: c= 1900 lei

	Variables	
V3	V4	V5
p_{wheat} = 0.7 lei/kg	p_{wheat} = 0.8 lei/kg	p_{wheat} = 0.9 lei/kg

The aim of the simulation was to indicate, for each variant, the optimal quantity of N and PK respectively, to be allocated for obtaining maximum profit. The negative values of the production factors PK indicate that the optimal point is not located in the first quadrant of the rectangular coordinate system $O\,N\,PK$, but in a different one, which translates into economic inefficiency under the respective initial conditions. The threshold values of the input were observed, which led to positioning the optimal point in the positive region of the coordinate system.

The mathematical model and functions. The first stage consisted in determining the production function Q = Q(N, PK), which immediately led to the determination of the technical maximum, with the help of SPSS software. The expression of Q is given by a real-valued function with two real variables, N and PK, respectively, and real coefficients $a_1,..., a_6$ of the form:

$$Q(N,PK)=a_1 N^2 + a_2 PK^2 + a_3 N \cdot PK + a_4 N + a_5 PK + a_6$$

expressions which give an objective description of the yield variation in relation to certain production factors, and on which one can easily apply the techniques for determining the extreme points (Intriligator, 2002).

The issue was then discussed from an economic perspective, with prices for fertilizers and wheat that are currently on the market. After that, a simulation was made for various hypothetical prices of wheat and PK fertilizers, respectively, which resulted in the identification of variants/combinations which could indicate an increase in the interest to allocate P and K fertilizers in higher rates. The graphical representations in the present paper were made by using the application Wolfram Alpha.

RESULTS AND DISCUSSIONS

The differentiated conditions of plant nutrition ensured by fertilization with the two types of fertilizers, N and the PK complex, determined a specific development of the wheat crop in the experimental variants in relation to the nutrients supplied. Table 2 presents the yield results obtained.

Starting from the results obtained in real conditions, through the SPSS software, the corresponding production function was determined, represented by expression (1) and Figure 1.

$$Q(N,PK) = -0.0643N^2 - 0.0237PK^2 + 0.02362N \cdot PK + \quad (1)$$
$$26.333N + 5.361PK + 3300.53$$

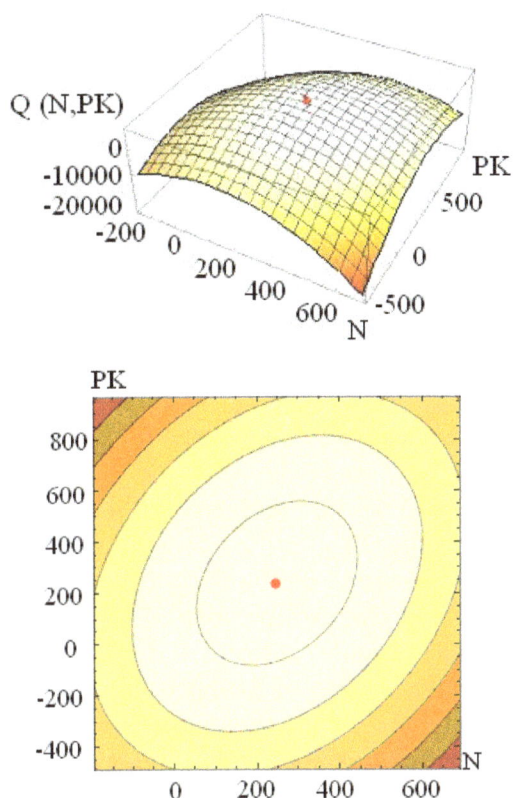

Figure 1. Graphical representations of the production function (graph and isoquants)

The high value (r = 0.995) of the correlation coefficient indicates closeness of the chosen functional model to the real phenomenon. By annulling the partial derivatives of function Q(N, PK) one can get to the N and PK quantities that lead to the technical maximum of the function, namely 7204 kg ha^{-1}.

$$N = 248 \, kg/ha, \quad PK = 236 \, kg/ha$$

Table 2. The influence of the rates of N and PK fertilizers respectively on the wheat yield

N	PK	Q
0	0	3290
100	0	5180
200	0	6045
50	50	4765
100	50	5845
200	50	6385
100	100	5650
150	100	6445
200	100	6740
150	150	6655
200	150	6995

Taking into account the aim of the present paper, the variations in the prices of wheat and

PK fertilizers were simulated, in order to find the economic moment/conditions where PK fertilization is profitable and motivating for farmers.

The results obtained through simulation are described below, together with their graphical distribution in accordance with each variant.

Simulation variant 1 (V 1)

The input considered in variant 1 was: p_{wheat}= 0.5 lei/kg; p_N=4.1 lei; p_{PK}=6.1 lei; fixed costs per 1 hectare c=1900 lei and output: N 126 kg ha^{-1}, PK=81 kg ha^{-1}.

It is easy to note that the result obtained through simulation does not justify the use of PK fertilizers, for two reasons: the low price of wheat and the high price of PK fertilizers. This aspect will be justified in the following results, for other values of the market price of wheat and of PK fertilizers, respectively.

Simulation variant 1A (V 1A) considered the following input: p_{wheat} =0.5 lei kg^{-1}; p_N=4.1 lei ka^{-1}; p_{PK}=4.1 lei kg^{-1}; fixed costs per 1 hectare c=1900 lei, and output: N=143 kg ha^{-1}, PK=11 kg ha^{-1}, profit 390 lei ha^{-1}. The cost of 4.1 lei/kg PK induces an economic optimum which is realized for a positive value of PK rates. This is realized even if the wheat is sold for a low price, 0.5 lei kg^{-1}.

Simulation variant 1B (V 1B) considered the input: p_{wheat}=0.5 lei kg^{-1}; p_N=4.1 lei kg^{-1}; p_{PK}=4.6 lei kg^{-1}; fixed costs per 1 hectare c=1900; Output: N=138 kg ha^{-1}, PK-11 kg ha^{-1}, profit 390 lei ha^{-1}. With these values of the prices for wheat and PK fertilizers, respectively, the results obtained through simulation do not justify the use of PK fertilizers.

Simulation variant 1C (V 1C) considered the input: p_{wheat} =0.5 lei kg^{-1}; p_N=4.1 lei ka^{-1}; p_{PK}=5.1 lei ka^{-1}; fixed costs per 1 hectare c=1900; Output: N=134 kg ha^{-1}, PK-35 kg ha^{-1}, profit=402 lei ha^{-1}.

With these values of the prices for wheat and PK fertilizers, respectively, the results obtained through simulation do not justify the use of PK fertilizers.

Simulation variant 1D (V 1 D) considered the input: p_{wheat}=0.5 lei kg^{-1}; p_N=4.1 lei kg^{-1}; p_{PK}=5.6 lei kg^{-1}; fixed costs per 1 hectare c=1900; Output: N=130 kg ha^{-1}, PK=-58 kg ha^{-1}, profit=426 lei ha^{-1}.

With these values of the prices for wheat and

PK fertilizers, respectively, the results obtained through simulation do not justify the use of PK fertilizers.

Simulation variant 2 (V 2) considered the input: p_{wheat}=0.6 lei kg^{-1}; p_N=4.1 lei ka^{-1}; p_{PK}=6.1 lei ka^{-1}; fixed costs per 1 hectare c=1900; Output: N = 46 kg ha^{-1}, PK=-28 kg ha^{-1}, profit=977 lei ha^{-1}. With these values of the prices for wheat and PK fertilizers, respectively, the results obtained through simulation do not justify the use of PK fertilizers.

Simulation variant 2A (V 2A) considered the input: p p_{wheat}=0.6 lei kg^{-1}; p_N=4.1 lei ka^{-1}; p_{PK}=5.1 lei kg^{-1}; fixed costs per 1 hectare c=1900; Output: N=153 kg ha^{-1}, PK=10 kg ha^{-1}, profit=968 lei ha^{-1}.

The cost of 5.1 lei/kg PK induces an economic optimum which is realized for a positive value of PK rates. This is realized even if the wheat is sold for a low price, 0.6 lei kg^{-1}.

Simulation variant 2B (V 2B) considered the input: p_{wheat}=0.6 lei kg^{-1}; p_N=4.1 lei ka^{-1}; p_{PK}=5.6 lei ka^{-1}; fixed costs per 1 hectare c 1900; Output: N=149 kg ha^{-1}, PK=-9 kg ha^{-1}, profit=968 lei ha^{-1}.

With these values of the prices for wheat and PK fertilizers, respectively, the results obtained through simulation do not justify the use of PK fertilizers.

Simulation variant 3 (V 3) considered the input: p_{wheat}=0.7 lei kg^{-1}; p_N=4.1 lei ka^{-1}; p_{PK}=6.1 lei ka^{-1}; fixed costs per 1 hectare c=1900; Output: N=160 kg ha^{-1}, PK=9 kg ha^{-1}, profit=1552 lei ha^{-1}.

A higher market price for wheat, 0.7 lei kg^{-1}, induces an economic optimum which is realized for a positive value of PK rates, even if the cost of fertilizers is 6.1 lei kg^{-1} PK.

Simulation variant 4 (V 4) considered the input: p_{wheat}=0.8 lei kg^{-1}; p_N=4.1 lei kg^{-1}; p_{PK}=6.1 lei ka^{-1}; fixed costs per 1 hectare c=1900; Output: N=171 kg ha^{-1}, PK=37 kg ha^{-1}, profit=2164 lei ha^{-1}.

A higher market price for wheat, 0.8 lei kg^{-1}, induces an economic optimum which is realized for a positive value of PK rates, even if the cost of fertilizers is 6.1 lei kg^{-1} PK. Moreover, the profit will be much higher than in the previous cases.

Simulation variant 5 (V 5) considered the input: p_{wheat}=0.9 lei kg^{-1}, p_N=4.1 lei kg^{-1}, p_{PK}=6.1 lei kg^{-1}, fixed costs per 1 hectare c=1900; Output: N=180 kg ha^{-1}, PK 59 kg ha^{-1}, profit=2799 lei ha^{-1}. In this case, as well, a higher market price of wheat, 0.9 lei kg^{-1}, induces an economic optimum which is realized for a positive value of PK rates, even if the cost of fertilizers is 6.1 lei kg^{-1} PK. The profit will be much higher than in the previous cases. Figure 2 presents the graphical distribution of the values in the form of isoquants, obtained from the simulations.

After comparing the results generated in simulation variants V1, V1A-D, where the price of wheat was considered to be constant, 0.5 lei kg^{-1}, and the price of the nitrogen fertilizer 4.1 lei kg^{-1} respectively, the threshold value of the PK fertilizers over which there appears economic inefficiency was 4.1 lei kg^{-1}, suggested in simulation variant V1A.

Practically, higher costs of the PK fertilizers lead to a relocation of the optimum point to an area that indicates economic inefficiency when using PK fertilizers for wheat crops.

Now we can observe what would happen if the market price of wheat rose. Obviously, the threshold value for efficiency of buying PK would grow, so it would be more easily supported by subsidies. Thus, for simulation variants V2, V2A-B, where the price of wheat is 0.6 lei kg^{-1} and the price of nitrogen is kept constant, the threshold value for profitability when buying PK fertilizers is 5.1 lei kg^{-1}.

In the hypothetical situations given by the simulation variants V3-5 where the price of wheat was 0.7 lei kg^{-1} or higher, this type of fertilization would be attractive from an economic point of view, even with the current cost of PK of approximately 6.1 lei kg^{-1}. Therefore, it is safe to say that, if a fixed buying price is kept of 6.1 lei kg^{-1} for PK fertilizers and if all other given conditions are met, the threshold value for the price of wheat should be 0.77 lei kg^{-1} so that PK application can be profitable.

Taking into consideration the modification of soil fertility due to the lack of application of PK, as well as secondary acidification generated by unilateral application of N, PK application should be stimulated through direct measures (lower prices for fertilizers) and the costs could be recovered from selling the wheat, through improvements in both the quality and the quantity of the yield.

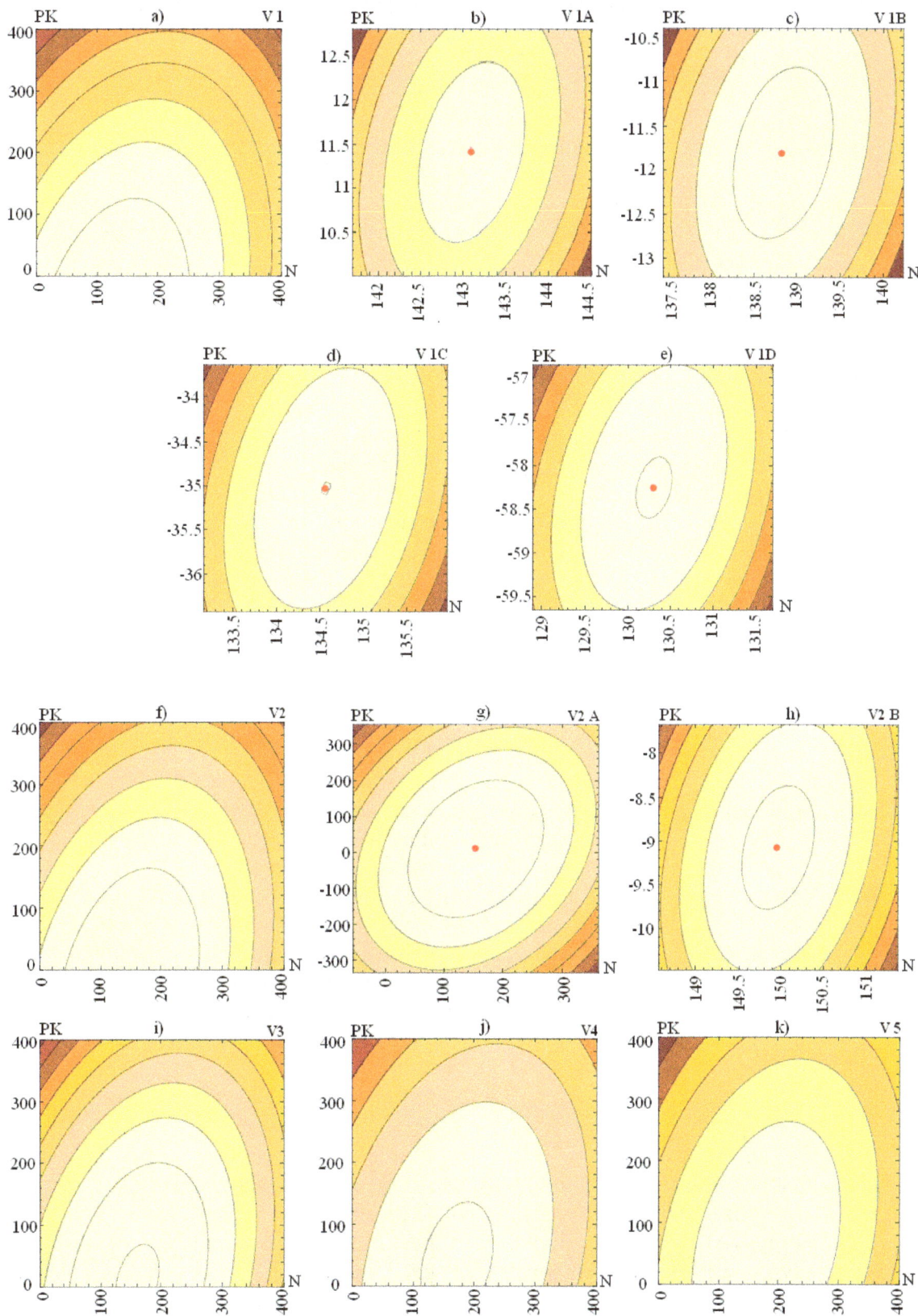

Figure 2. Graphical distribution, in the form of isoquants, of the results regarding the profitability of the wheat crop as the balance between the price of PK fertilizers and the price of wheat; a – V1, b – V1A, c – V1B, d – V1C, e – V1D, f – V2, g – V2A, h – V2B, i – V3, j – V4, k – V5.

Econometric studies of the optimization of wheat crops in relation to the cultivated area, the level of fertilization, the estimated yields and the types of agricultural systems, have been done in various geographical areas, with specific pedoclimatic and socio-economic conditions.

Efficient management of fertilizers in the agronomic practice, for ensuring the sustainability of the wheat crop, has proven necessary in relation to the biological material and germplasm potential (Hawkesford, 2014), agricultural system, pedoclimatic conditions, technological conditions and economic efficiency (Ghosh et al., 2003; Jat et al., 2012; Velasco et al., 2012; Poruțiu et al., 2015).

Due to the fact that nitrogen has a greater contribution to the wheat yield (Spiertz and De Vos, 1983; Marschner, 1995; Ortiz-Monasterio et al., 1997; Mengel and Kirkby, 2001), but at the same time a higher mobility, with the risk of loss on the soil profile, with unfavourable effects on the environment (Lehmann and Schroth, 2003; Liu et al., 2003; Beaudoin et al., 2005), many studies have focused on the optimization of the rates of nitrogen on wheat crops (Delogu et al., 1998; Merle, 2007; Gandorfer and Rajsik, 2008; Abedi et al., 2011; Hawkesford, 2014; Khalid et al., 2014).

Gandorfer and Rajsik (2008) addressed economic modelling of the optimum rate of nitrogen for the wheat crop, if the input affects the production and the price of the yield. However, phosphorus and potassium represented the object of the study, together with nitrogen, for the optimization of fertilizing treatments on the wheat crop; the economic aspects were dealt with in relation to the three nutrients.

Yadav (2003) made experiments at farm level, in different pedoclimatic conditions, in order to study the production of cereals (wheat and rice) in relation to the fertilizer unit used through agronomic efficiency (AE) and the partial factor productivity (PFP) through marginal analyses.

Niamatullah et al. (2010) communicated the results of studies performed over the course of 30 years, regarding the evolution of the areas covered by wheat crops, the evolution of support prices and of fertilizer use in Pakistan. They measured the significant contribution of the price factor (support price) and the price of fertilizers on the production of rice and wheat. They found that the support price had a strong influence on the production of rice ($p<0.05$) and wheat ($p<0.10$) and that the use of fertilizers has a significant relation with the production ($p<0.10$). These results are similar to those communicated in the present study and they highlight the necessity of support prices for enhancing the efficiency of phosphorus and potassium application for the final purpose of optimizing the fertilization process of wheat crops.

Jat et al. (2012) assessed the response of wheat crops (in the conditions of the Indo-Gangetic Plains) to the application of phosphorus fertilizers and the profitability of applying phosphorus according to some models/ equations that took into account the rates of phosphorus applied in the context of NPK fertilization, the yield obtained, the return on investment when applying P, and the costs of P fertilizers. Based on two scenarios, they assessed the profitability of phosphorus application on wheat crops (together with rice and corn) and concluded that, for the profitability of phosphorus application, besides other factors such as application rates, crop response to P application, and the cost of P fertilizers, there is another very important factor: the minimum support price for wheat crops.

These results are similar with the ones obtained in the present study. Taking into account the response of the wheat crop, the flexibility of the support prices had an important medium-term and long-term role. The flexibility of the support prices was in slight contradiction with the fertilizer price levels.

Therefore, the problem of the nutrient deficit in wheat crops (and for rice and corn crops, as well) was solved through balanced and flexible use of the market prices of wheat and the costs of fertilizers, of NPK especially, which played a crucial role in obtaining the yields for the respective crops.

Other studies have focused on the analysis of wheat crop performance as a response to different fertilizer rates (Aujla et al., 2010; Ali et al., 2015) or to differentiated N fertilization on gradual PK backgrounds (Sala et al., 2015). Singh and Sharma (2014) used a Fuzzy basic expert system for optimizing the fertilization rate of the wheat crop. In order to do this, they considered 24 Fuzzy rules for describing the relations among the three inputs NPK, soil and wheat yield.

This study has highlighted the balance and flexibility of prices of fertilizers and wheat production, for an incentive and also to determine farmers to use fertilizers with phosphorus and potassium for balanced wheat crops. The results of this study are consistent with results obtained in other research and studies from different countries, to which reference has been made. These results were obtained in certain climatic conditions and biological material (Alex cultivar) specific but it is possible that the equilibrium level become other value according to other climatic conditions, or biological material. Thus, the dynamic and flexible nature of support prices for crop fertilization can solve the problem together with other financial incentives for farmers. Thereby the soil fertility will undergo a rebalancing as a result of fertilization with phosphorus and potassium, which will contribute to the sustainable use of agricultural land.

CONCLUSIONS

The main cause of Romanian farmers' low interest in using PK fertilizers is the low market price of wheat, correlated with the high cost of PK fertilizers. Thus, the results obtained from simulations indicate that only starting from a market price of wheat of 0.7 lei kg^{-1}, even with the current cost of PK fertilizer, which is 6.1 lei kg^{-1}, it would be profitable to use PK fertilizers. With the current price of wheat (between 0.5 and 0.6 lei kg^{-1}), it is mandatory to change the policy in such a way as to make it possible to buy PK fertilizers at a lower price. More exactly, for the price of 0.5 lei kg^{-1} for wheat, the maximum acquisition price of PK fertilizers is approximately 4.1 lei kg^{-1}. If the price of wheat is 0.6 lei kg^{-1}, the results indicate the maximum acquisition price of PK fertilizers should be approximately 5.1 lei kg^{-1} in order for their use to be profitable.

ACKNOWLEDGEMENTS

The authors would like to thank the management of the Didactic and Experimental Station of BUASVM Timişoara for facilitating the setting up of the experiment and research activity. The biological material (wheat seed, Alex cultivar) was provided by the Agricultural Research and Development Station, Lovrin, Romania.

REFERENCES

Abedi T., Alemzadeh A., Kazemeini S.A., 2011. Wheat yield and grain protein response to nitrogen amount and timing. Australian Journal of Crop Science 5(3): p. 330-336.

Ali M., Ali L., Din M., Sattar M., Waqar M.Q., Alim M.A., Khalid L., 2015. Comparative performance of wheat in response to different phosphatic fertilizers. International Journal of Research in Agriculture and Forestry 2(6): p. 1-9.

Aujla K.M., Taj S., Mahmood K., Akmal N., 2010. Wheat management practices and factors of yield decline in the Punjab province of Pakistan. Pakistan Journal of Agricultural Research 23(1-2): p. 17-24.

Beaudoin N., Saad J.K., Van Laethem C., Machet J.M., Maucorps J., Mary B., 2005. Nitrate leaching in intensive agriculture in Northern France: Effect of farming practices, soils and crop rotations. Agriculture, Ecosystems & Environment 111(1-4): p. 292-310.

Boldea M., Sala F., 2013. Optimizing the area fertilized with nitrogen-based chemical fertilizers for wheat crops. Applied Mathematics 3(3): p. 93-97.

Borlan Z., Hera Cr., 1982. Tabele si nomograme agrochimice. Ed. Ceres, Bucuresti.

Buresh R.J., Witt C., 2007. Site-specific nutrient management. In: Proceedings of the IFA International Workshop on Fertilizer Best Management Practices, 7-9 March 2007, Brussels, Belgium. International Fertilizer Industry Association, Paris, p. 47-55.

Calderini D.F., Ortiz-Monasterio I., 2003. Grain position affects grain macronutrient and micronutrient concentrations in wheat. Crop Science 43(1): p. 141-151.

Chuan L., He P., Pampolino M.F., Johnston A.M., Jin J., Xu X., Zhao S., Qiu S., Zhou W., 2013. Establishing a scientific basis for fertilizer recommendations for wheat in China: Yield response and agronomic efficiency. Field Crop Research 140: p. 1-8.

Cui Z.L., Zhang F.S., Chen X.P., Miao Y.X., Li J.L., Shi L.W., Xu J.F., Ye Y.L., Liu C.S., Yang Z.P., Zhang Q., Huang S.M., Bao D.J., 2008. On-farm evaluation of an in season nitrogen management strategy based on soil N$_{min}$ test. Field Crops Research 105: p. 48-55.

Delogu G., Cattivelli L., Pecchioni N., De Falcis D., Maggiore T., Stanca A.M., 1998. Uptake and agronomic efficiency of nitrogen in winter barley and winter wheat. European Journal of Agronomy 9(1): p. 11-20.

Dumitru M., 2002. Factors that influenced fertilizers consumption in post revolutionary Romania. Fertilizarea echilibrată a principalelor culturi în România, p. 23-46.

Fu Q.P., Wang Q.J., Shen X.L., Fan J., 2014. Optimizing water and nitrogen inputs for winter wheat cropping system on the Loess Plateau, China. Journal of Arid Land 6(2): p. 230-242.

Gandorfer M., Rajsik P., 2008. Modeling economic optimum nitrogen rates for winter wheat when inputs affect yield and output-price. Agricultural Economics Review 9(2): p. 54-64.

Ghosh P.K., Dayal D., Mandal K.G., Wanjari R.H., Hati K.M., 2003. Optimization of fertilizer schedules in fallow and groundnut-based cropping systems and an assessment of system sustainability. Field Crop Research 80(2): p. 83-98.

Habib M., 2009. Effect of foliar application of Zn and Fe on wheat yield and quality. African Journal of Biotechnology 8(24): p. 6795-6798.

Havlin J.L., Beaton J.D., Tisdale S.L., Nelson W.L., 2005. Soil fertility and fertilizers, an introduction to nutrient management. Seventh Edition, Pearson Education Inc. New Jersey, USA, 515 p.

Hawkesford M.J., 2014. Reducing the reliance on nitrogen fertilizer for wheat production. Journal of Cereal Science 59: p. 276-283.

Hera C., 2010. CIEC - 77 years dedicated to enhancing soil fertility for the benefit of sustainable performing agriculture. Proceedings of the 15th World Fertilizer Congress of the International Scientific Centre for Fertilizer (CIEC) 29 Aug-2 Sept. Romania, p. 17-29.

Hera C. (Coord), Otiman P.I. (Coord), Alexandri C., Luca L., Băneş A., Feher A., Ionel I., Peticila A., Rujescu C., Rusu M., 2015. Securitate şi siguranţă alimentară. In: Vlad I.V. (Coord) 2015. Strategia de dezvoltare a României în următorii 20 de ani. Ed. Academiei Române, Vol. I: p. 169-234.

Intriligator M.D., 2002. Mathematical optimization and economic theory. 1st Edition, Soc. for industrial and applied mathematics (SIAM), Philadelphia, 508 pp.

Jat M.L., Kumar D., Majumdar K., Kumar A., Shahi V., Satyanarayana T., Pampolino M., Gupta N., Singh V., Dwivedi B.S., Singh V.K., Singh V., Kamboj B.R., Sidhu H.S., Johnston A., 2012. Crop response and economics of phosphorus fertiliser application in rice, wheat and maize in the Indo-Gangetic Plains. Indian Journal of Fertilisers 8(6): p. 62-72.

Khalid M.S., Saleem M.F., Ali S., Pervez M.W., Rehman M., Hussain S., Rehman K., 2014. Optimization of nitrogen fertilizer level for newly evolved wheat (Triticum aestivum) cultivars. Applied Scientific Reports 7(2): p. 83-87.

Lehmann J., Schroth G., 2003. Nutrient leaching. In: (Schroth G., Sinclair F.L., Eds.) Trees, Crops and Soil Fertility. CAB International, p. 151-166.

Liu X., Ju X., Zhang F., Pan J., Christie P., 2003. Nitrogen dynamics and budgets in a winter wheat-maize cropping system in the North China Plain. Field Crops Research 83(2): p. 111-124.

Malakouti M.J., 2008. The effect of micronutrients in ensuring efficient use of macronutrients. Turkish Journal of Agriculture and Forestry 32: p. 215-220.

Marschner H., 1995. Mineral Nutrition of Higher Plants. Second Edition, Academic Press, Elsevier, p. 889.

Mengel K., Kirkby E.A., 2001. Principles of Plant Nutrition. Kluwer Academic Publisher, Dordrecht/Boston/London, 849 p.

Merle F., 2007. Optimal nitrogen fertilization of dry land

wheat. High Plains Ag Lab PHREC 07-30: 20-23.

Niamatullah M., Zaman K.U., Khan M.A., 2010. Impact of support price and fertilizer off take on rice production and wheat acreage in NWFP, Pakistan. The J. of Animal & Plant Sciences 20(1): p. 28-33.

Nuttall J.G., O'Leary G.J., Panozzo J.F., Walker C.K., Barlow K.M., Fitzgerald G.J., 2015. Models of grain quality in wheat-A review. Field Crop Research xxx: xxx-xxx.

Ortiz-Monasterio R.J.I., Sayre K.D., Rajaram S., McMahon M., 1997. Genetic progress in wheat yield and nitrogen use efficiency under four nitrogen rates. Crop Science 37(3): p. 898-904.

Otiman P.I., 1999. Economie rurală. Ed. Agroprint, Timişoara, p. 153-160, 440.

Otiman P.I., Creţ F., 2002. Elemente de matematici aplicate în economia agroalimentară. Ed. Agroprint, Timişoara, p. 333-383.

Pislea D., Boldea M., Sala F., 2013. Assessing the gluten content in wheat as an expression of the nutrition deficit, through beta distribution. AIP Conference Proceedings 1558: p. 1571-1574.

Poruţiu A., Arion F., Mureşan I., Sălăgean T., Fărcaş R., 2015. Economic considerations on wheat crops cultivated following soy crops on an argic chernozem in the Transylvanian Plain. Agriculture - Science and Practice 3-4(95-96): p. 99-103.

Rawashdeh H., Sala F., 2015. Effect of some micronutrients on growth and yield of wheat and its leaves and grain content of iron and boron. Bulletin USAMV series Agriculture 72(2): p. 503-508.

Rawashdeh H., Sala F., Boldea M., 2015. Mathematical and statistical analysis of the effect of boron on yield parameters of wheat. AIP Conference Proceedings 1646: p. 670010-670014.

Rujescu C.I., Ciolac M.R., Martin S.C., Butur M., 2014. The use of a two-factor model for the fertilization of agricultural crops in order to determine the optimal production area. Lucrări ştiinţifice, Management Agricol, Seria I, XVI(1): p. 9-12.

Rujescu C.I., Mateia A.N., Martin S., Ciolac R., 2015. Determination of the optimum productions surface monofactorial model having an initial fixed capital. Quaestus Multidisciplinary Res. J. 7: p. 149-156.

Sala F., 2011. Agrochimie. Ed. Eurobit, Timişoara, p. 493-502.

Sala F., Boldea M., 2011. On the optimization of the doses of chemical fertilizers for crops. AIP Conference Proceedings 1389: p. 1297-1300.

Sala F., Boldea M., Rawashdeh H., Nemet I., 2015. Mathematical model for determining the optimal doses of mineral fertilizers for wheat crops. Pakistan Journal of Agricultural Sciences 52: p. 609-617.

Sala F., Rawashdeh H., Boldea M., 2015. Differentiated contribution of minerals through soil and foliar fertilization to the winter wheat yield. American Journal of Experimental Agriculture 6(3): p. 158-167.

Shewry P.R., 2009. Wheat, Darwin review. Journal of Experimental Botany 60(6): p. 1537-1553.

Singh H., Sharma N., 2014. Optimization of fertilizer rates for wheat crop using fuzzy expert system. International Journal of Computer Applications 100(1): p. 36-40.

Spiertz J.H.J., De Vos N.M., 1983. Agronomical and physiological aspects of the role of nitrogen in yield formation of cereals. Plant and Soil 75(3): p. 379-391.

Swain E.Y., Rempelos L., Orr C.H., Hall G., Chapman R., Almadni M., Stockdale E.A., Kidd J., Leifert C., Cooper J.M., 2014. Optimizing nitrogen use efficiency in wheat and potatoes: interactions between genotypes and agronomic practices. Euphytica 199(1): p. 119-136.

Velasco J.L., Rozas H.S., Echeverría H.E., Barbieri P.A., 2012. Optimizing fertilizer nitrogen use efficiency by intensively managed spring wheat in humid regions: Effect of split application. Canadian Journal of Plant Science 92: p. 847-856.

Vigani M., Dillen K., Cerezo E.R., 2013. Proceedings of a workshop on "Wheat productivity in the EU: determinants and challenges for food security and for climate change". European Commission, EUR 25934 EN: p. 1-36.

Wang F., Wang Z., Kou C., Ma Z., Zhao D., 2016. Responses of wheat yield, macro- and micro-nutrients, and heavy metals in soil and wheat following the application of manure compost on the North China Plain. PLoS ONE 11(1): e0146453.

Yadav R.L., 2003. Assessing on-farm efficiency and economics of fertilizer N, P and K in rice wheat systems of India. Field Crop Res. 81(1): p. 39-51.

Zeidan M.S., Mohamed M.F., Hamouda H.A., 2010. Effect of foliar fertilization of Fe, Mn and Zn on wheat yield and quality in low sandy soils fertility. World Journal of Agricultural Sciences 6 (6): p. 696-699.

***INS, 2014. Crop production for the main crops in 2013. National institute of statistics, No. 75, March 31, 2014, 5 p.

***http://statistici.insse.ro (acces date: 15.06.2015).

***ec.europa.eu/eurostat (acces date: 15.06.2015).

***http://www.wolframalpha.com/ (acces date: 01.10.2015).

25

THE EFFECT OF PLANTING DATE AND CLIMATIC CONDITION ON OIL CONTENT AND FATTY ACID COMPOSITION IN SOME ROMANIAN SUNFLOWER HYBRIDS

Mihaela POPA[1], Gabriel Florin ANTON[2], Luxiţa RÎŞNOVEANU[3], Elena PETCU[2], Narcisa BĂBEANU[1]

[1]University of Agronomic Sciences and Veterinary Medicine of Bucharest, 59 Mărăşti Blvd., District 1, Bucharest, Romania
[2]National Agricultural Research and Development Institute Fundulea, 1 Nicolae Titulescu Street, Romania
[3]Agricultural Research and Development Station from Brăila, Romania
Corresponding author email: ionita_mihaela84@yahoo.com

Abstract

Grain oil content and fatty acid composition are very important traits in sunflower. A research was carried on in 2014 and 2016 to determine planting date and climatic conditions effects on oil content and fatty acid composition, using three sunflower hybrids (F 708, F 911, FD 15 C 44).
Samples of seeds were harvested at maturity and provided from experimental field conducted in two years at NARDI Fundulea, Romania. The oil content and fatty acid composition of sunflower seeds were determined and analyzed using a Soxhlet apparatus and gas chromatography according to the conventional method. The results of experiment and analyses described in this paper showed that the planting date interact with climatic conditions and affect both the quantity and the quality of the seeds yield. Climatic conditions was the main source of variance for the yield. The oil content in sunflower seeds was very significantly affected by year, planting date and hybrids, as well as by most interactions between these factors. The early planting date in both years led to an increase of grain oil content in all studied sunflower hybrids. The results showed that there was a significant negative effect of 2014 conditions (low rainfall during seed maturation) on the oleic acid concentration in all studied sunflower hybrids. The delay in planting decreased the concentration of oleic acid and increased linoleic acid concentration in all sunflower hybrids, except hybrid F 708, which is more stable in this regard. It is concluded that rainfall, genotype and planting date influenced yield and fatty acid composition from seeds.

Key words: sunflower, planting date, climatic conditions, yield, fatty acid composition.

INTRODUCTION

Sunflower (*Helianthus annuus* L.) is an important oilseed crop whose oil content varies from 25 to 50% of seed content (Sahari Khoufi et al., 2014). Sunflower oil quality is determined mainly by fatty acid composition. This consists of different types of saturated (palmitic acid, stearic acid) and unsaturated fatty acids (linoleic acid, oleic acid) (Kowalski 2007). Traditional sunflower oil rich in linoleic acid is used in the food industry and in various commercial products while oil with high proportion of oleic acid is more stable than others and is desirable for improved quality of life (Onemli, 2012).
The genotype is the most important factor that defines the fatty acid composition (Petcu et al., 2010) but also the environmental factors during seed-filling period can widely affect the oil

percentage and fatty acid composition of oil (Atanasi et al., 2010). Thus, the oleic acid/linoleic acid sunflower ratio increases at high temperatures occuring during seed maturation and on contrary, decrease at lower temperatures conditions (Chalermkwan Sukkasem et al., 2013). The water stress increase of oleic acid in the high oleic sunflower hybrids but in the standard hybrids the water stress caused a significant reduction of the concentration of oleic acid (Baldini et al., 2002; Petcu et al., 2001).
The productivity of sunflower is largely determined by the prevailing weather conditions throughout its life cycle and imposed cultural practices (Vrânceanu, 2000; Oshundiya et al., 2014). Of this planting date is one of the most important cultural practices to be considered in sunflower production, as it is in all crops.

The main objective of this present work was to study the influence of genotypes, planting date and climatic conditions on some agronomic traits and fatty acid composition of several news Romanian sunflower hybrids.

MATERIALS AND METHODS

The three standard sunflower hybrids provided by Fundulea Research Institute (F 708, F 911 and FD 15 C44) were used in this study.

The seeds for analysis were produced during two vegetation periods (2014, 2016) in the experimental field of National Agricultural Research and Development Institute at Fundulea. The hybrids were sown at two different planting dates (beginning of April and May).

Evaluation of agronomic traits

Days to flowering, days to maturity, height of plant, head diameter were measured as agronomic traits. Height of plant was measured (in centimeters) at the completion of flowering. Five plants were selected at random from each plot and their heights were measured from the soil surface to top of flower. Five heads were taken randomly from each plot and diameter of each head was measured using measuring tape.

Determination of seed oil content and fatty acid composition

The dry seeds were ground with a Waring blender. Four grams of dried sunflower seeds were extracted with petroleum ether for 4 hours in a Soxhlet system (Buchi B-811, Germany) according to the SR-EN_ISO 659/2003 method. The oil extract was evaporated by distillation at a reduced pressure in a rotary evaporator at 40°C until the solvent was totally removed. The oil was extracted 2 times from a 2 g air dried seed sample by homogenization with the same solvent. Oil content was calculated with the formula: $W_0 = (m_1/m_0) \times 100$, where m_1 is the weight (in grams) of total seed sample and m_0 is the weight (in grams) of air dried seed sample.

The fatty acids were analysed by gas cromatography (GS) according to the conventional method. Thus, the transesterify of triglycerides to fatty acid methyl esters was performed with trimethylsulfoniumhydroxid (TMSH). The capillary column (BP x 70) by 25 m lengths on a DELSI gas cromatography

with FID detector was used. Injector and detector temperature were kept at 270 and 280°C. The carrier gas was helium, with a flow rate of 20 ml/min.

RESULTS AND DISCUSSIONS

Average temperatures in the experimental years were higher than normal of the zone. In both years of experimentation, July was above multi-annual average by 2.3°C in 2014 respectively 1.4°C in 2016. The mean temperature increased from sowing period (April to May) to flowering period (August), while it decreased a litlle during maturity period (from August to October) (Table 1). The years of experimentation were totally different from the viewpoint of quantity and monthly repartitions of rainfall. In 2016, the cumulated rainfall from sowing to maturity stage was 341.1 mm, as compared with 436 mm recorded in 2014 year (Table 1). In 2014, the cumulated rainfall during May-June exceeded with 101.4 mm the normal of the zone (135.4 mm), suggesting favorable conditions for sunflower crop, but rains were unevenly distributed along the sunflower vegetation period.

Thus, July and August registered a moisture deficit of 41.05 mm, vs. multi-annual average (Table 1). This moisture deficits increased unfavorable conditions during reproductive organs appearance and grain formation, determining relatively smaller yields of 2395 kg/ha (F 911, late planting date) to 3300 kg/ha (F 708, early planting date) (Figure 1) as compared to those in 2016.

Table 1. Average temperature (°C) and monthly distribution of rainfall (mm) during the sunflower vegetation period. Fundulea 2014, 2016

Month	April	May	June	July	Aug.	Sept
Temperature 2014	11	18.3	21.2	25.0	23.9	19.6
Temperature 2016	13.7	16.1	22.9	24.1	23.4	19.1
Multi-annual average	*11.2*	*17.0*	*20.8*	*22.7*	*22.3*	*17.4*
Rainfall 2014	82.8	100.6	136.2	52.1	27.3	37
Rainfall 2016	73.7	81.2	43.7	31.3	64.6	46.6
Multi-annual average	*45.1*	*61.84*	*73.59*	*70*	*50.5*	*50.22*

The weather conditions of 2016 led to yields between 3300 kg/ha (F 708, late planting data) and 4100 kg/ha (F D 15 C 44, early planting data) (Figure 1).

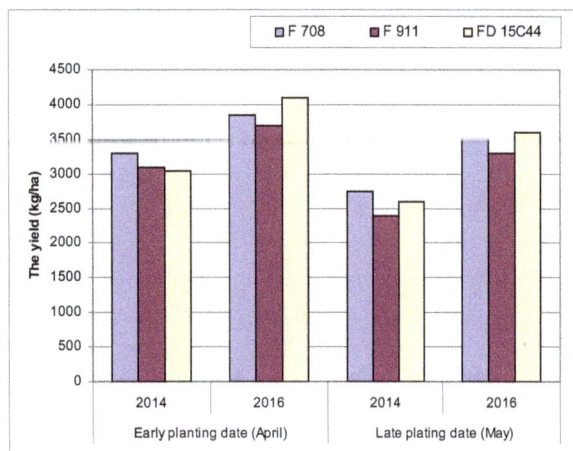

Figure 1. Yields obtained in sunflower hybrids under experimental conditions. Fundulea, 2014 and 2016

The results of the analysis of variance showed that the sunflower yield was affected by the weather conditions of the two years (79.9%), by planting date (5.9%), hybrids (10.10%) and as well as by interactions between these factors (Table 2). Except interactions, these influences were statistically very significant.

Table 2. Analysis of variance for yield

Sources of variation	DF	MS	F_e value and significations
Years (A)	1	1834670 (79.9%)	123.67***
Error (A)	2	14835.5 (0.6%)	-
Planting date (B)	1	134801.8 (5.9%)	9.71**
Years* Planting date	1	12967.09 (0.6%)	0.93
Error (B)	4	13884.47 (0.6%)	-
Hybrids (C)	2	222128 (9.7%)	10.10***
Years*Hybrids	2	30590.25 (1.3)	1.39
Planting date *Hybrids	2	6977.77 (0.3%)	0.32
Year*Planting date *Hybrids	2	2540 (0.3%)	0.12
Error (C)	16	22003 (1%)	-

In this study, all agronomic traits studied were found to be significantly different. All hybrids have a greater height in 2014, which may be

explained by excess of rainfall in May-June, which that favored the growth of plants (Table 3). The later planting date and higher temperatures during 2014 led to an acceleration of the processes of development, flowering and maturity respectively (Table 3). In terms of head diameter, it had higher values in 2016 and early sowing. Which explains sunflower productions obtained (Table 3).

Table 3. Effect of planting date on agronomic traits of sunflower genotypes studied

Planting date	Hybrid	Height of plants	Days to flowering	Days to maturity	Head diameter
2014					
Early	F 708	178	83	128	22
Late		189	75	125	17
Early	F 911	176	66	105	19
Late		180	58	100	16
Early	FD 15 C44	165	77	124	20
Late		168	62	121	18
2016					
Early	F 708	175	86	130	25
Late		176	76	125	20
Early	F 911	170	67	105	22.5
Late		174	60	100	19.3
Early	FD 15 C44	162	78	127	23
Late		162	63	125	20

The results of the analysis of variance showed that the oil content was affected very significant by the planting date and hybrids (Table 4).

Table 4. Analysis of variance for oil content from sunflower seeds

Sources of variation	DF	MS	F_e value and significations
Years (A)	1	140.42	9.74
Error (A)	2	14.42	
Planting date (B)	1	617.52	54***
A x B	1	20.40	1.78
Error (B)	4	11.43	
Hybrids (C)	2	58.42	19.61***
A x C	2	0.05	0.02
B x C	2	13.52	0.54
A x B x C	2	13.52	0.54
Error (C)	16	2.98	

The earlier planting date was shown to give better oil content than later planting date for all hybrids and both climatic conditions (Figure 2). Oil content varied widely from 39% in the hybrid F 708 planted late in 2016 to 53% in the hybrid FD 15 C 44 planted early, in 2014. This may be due to shortening of grain filling and decrease of intercepted radiation per plant during a critical period observed at late planting. Vega and Hall (2002) showed a significant reduction in oil concentration associated with a strong reduction in the duration of grain filling observed at late planting, it was due to changes in kernel oil proportion, rather than to changes in kernel percentage. Late sowing decreases not only grain yield but also grain oil concentration in high-oil sunflower hybrids. For three locations in Argentina, simulated grain oil concentration decreased when sowing was delayed (Pereyra-Irujo and Aguirrezábal, 2007).

Oil content was a little bigger in the first year of the experiment (2014) than in the second (2016) for both planting dates (Figure 2).

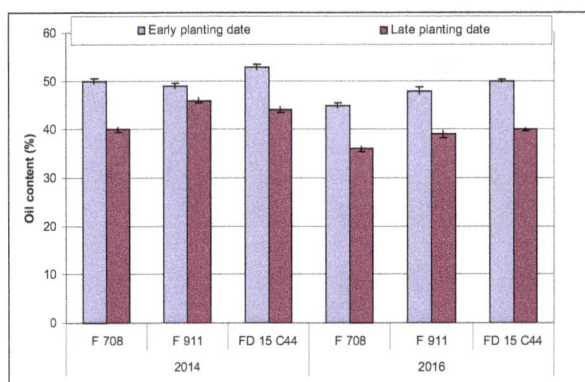

Figure 2. Oil content of sunflower hybrids studied. Fundulea, 2014 and 2016

This was due to the longer period of drought during August-September in 2016. Other results, shows that an increase in water deficit is associated with an increase in oil content and a low water deficit is associated with a decrease in oil content (Flagella et al., 2002; Anastasi et al., 2010). The correlation between oil content and watering regime cannot sometimes be determined in some hybrids (Kaya and Kolsarici, 2011). Erdemoglu et al. (2003) showed that genotypes and climatic conditions such as temperature, altitude and soil structure affected oil content of sunflower more than irrigation.

On the contrary, Baldini et al. (2002) noted a decrease in oil content in severe hydric stress, while a moderate water deficit increased oil content, showing the great adaptability of sunflower to early hydric stress applied during

flowering to seed filling. The authors attributed this result to abscisic acid produced in leaves of stressed plants and then translocated to the seed, thus contributing to the decline in seed oil content.

Oil yield was highly significant and positively correlated with seed yield obtained in 2014 (r=0.72**) and 2016 (r=0.80**) (Figure 3). Other researchers (Teklewold et al., 1999; Anandhan et al., 2010) indicated a positive relationship of different intensity between grain and oil yield.

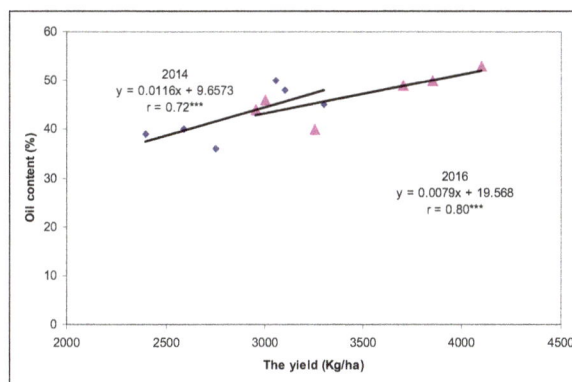

Figure 3. Correlation between the yield and oil content of sunflower hybrids studied

The characters plant height, day to flowering, days to maturity and head diameter had non-significant association with oil yield (r=0.068; 0.35; 0.068, 0.55). In literature results are conflicting. Some have found positive correlations between days to maturity and oil content (Anto Mijic et al., 2009) or negative correlation with head diameter (Hladni et al., 2006).

The saturated fatty acid (palmitic and stearic acid) contents were insignificant affected by weather conditions and planting date. The palmitic acid concentration increase in 2014 conditions (from 0.57% for F 911 to 0.74% for F 708 hybrid) and stearic acid concentration decreased in the same conditions (from 0.14% for F 708 to 0.78% for FD 15 C27 hybrid respectively) (Table 5).

There was a significant negative effect of climatic condition of 2014 on the oleic acid concentration in all studied sunflower hybrids. The low oleic content was recorded by F 911 hybrid (22.95% for early planting date and 21.10% for late planting date). The reason for this can be explained by low rainfall during seed maturation (August, September) in 2014 (Table 1). The relation between fatty acid composition and drought stress in sunflower remains poorly explained and studies on this topic are contradictory. The research work of Petcu et al. (2001, 2011) revealed that water stress causes a significant reduction of about 8-14% in the concentration of oleic acid in standard hybrid while Baldini et al. (2002) and Flagella et al. (2000) found that water stress increased oleic acid content and decreased linoleic acid content in both standard and high oleic genotypes.

The linoleic acid concentration increases in 2014 year conditions. The proportion of linoleic acid in fatty acid of the sunflower was from between 59.02-64.11% for early planting date up to 66.85% for late planting date (Table 5).

The delay in planting decreased the concentration of oleic acid and increased linoleic acid concentration in all sunflower hybrids, except hybrid F 708, which is more stable in this regard. The modifications were more obviously in 2014 conditions that in 2016 (almost normal conditions) (Table 5). This could be in correlation with high drought resistance of this hybrid. The lowest oleic acid content caused by late planting was found in the hybrid F 911 in 2014 (21.10%).

Table 5. The effect of climatic conditions on fatty acid compositions in sunflower seeds

Fatty acids	Planting date	F 708		F 911		FD 15 C27	
		2014	2016	2014	2016	2014	2016
Palmitic acid (C 16:0)	Early	7.10	6.36	6.6	6.03	6.06	5.47
	Late	6.10	5.95	6.35	6.10	6.10	5.91
Stearic acid (C 18:0)	Early	6.10	6.24	3.85	4.05	4.29	5.07
	Late	6.15	6.52	3.90	3.85	4.30	4.20
Oleic acid (C 18:1)	Early	23.5	37.6	22.9	33.0	27.4	35.6
	Late	24.1	37,1	21.1	32.1	26.1	33.2
Linoleic acid (C 18:2)	Early	60.5	46.4	64.1	54.6	59.1	51.7
	Late	61.2	46.9	65.8	55.1	63.1	49.2

Previous literatures reported increasing intercepted solar radiation per plant during seed filling raised the oleic acid percentage and reduced that of linoleic acid (Izquierdo et al., 2009).

CONCLUSIONS

Agronomic traits such as plant height, head diameter, days to maturity, oil content of sunflower had influenced by climatic conditions and planting date. The early planting date in both year conditions led to an increase of grain oil content in all studied sunflower hybrids.

In the normal sunflower hybrids, like the hybrids from this study, in addition with the reduction in oleic acid concentration was a simultaneous increase in the content of linoleic acid due to drought conditions during seed maturation and late planting date.

In conclusion apart from rainfall during seed filling, genotype remains the most important factor controlling the variability in oil and fatty acid composition, mainly oleic and linoleic acids.

Based on these results, chose of genotypes and management practices that modify temperature, soil water content during seed filling period (e.g. planting date, plant density and so on) could contribute to obtain high oleic acid content of sunflower.

REFERENCES

Anastasi U., Santonoceto C., Giuffrè A.M., Sortino O., Gresta F., Abbate V., 2010. Yield performance and grain lipid composition of standard and oleic sunflower as affected by water supply. Field Crops Research 119: p. 145–153.

Anandhan T., Manivannan N., Vindhiyavarman P. and Jeyakumar P., 2010. Correlation for oil yield in sunflower (*Helianthus annuus*. L). Electronic Journal of Plant Breeding, 1(4): p. 869-871.

Anto Mijic, Ivica Liovic, Zvonimir Zdunic, Sonja Maric, Ana Marjanovic Jeromela, Mirjana Jankulovska, 2009. Quantitative analysis of oil yield and its components in sunflower (*Helianthus annuus* L.). Romanian Agric. Res., 26: p. 41-46.

Baldini M., Giovanardi R., Tahmasebi-enferadi S., Vannozzi G., 2002. Effects of water regime on fatty acid accumulation and final fatty acid composition in the oil of standard and high oleic sunflower hybrids. Italian Journal of Agronomy 6: p. 119–126.

Chalermkwan S., Laosuwan P., Wonprasaid S. and Thitiporn Machikowa, 2013. Effects of Environmental Conditions on Oleic Acid of Sunflower Seeds. International Journal of Chemical, Environmental & Biological Sciences (IJCEBS) Vol. 1, Issue 2 (2013) ISSN 2320–4087 (Online),

Erdemoglu N., Kusmenoglu S., Yenice N., 2003. Effect of irrigation on the oil content and fatty acid composition of some sunflower seeds. Chemistry of Natural Compounds 39: p. 1-4.

Flagella Z., Rotunno T., Di Caterina R., De Simone G., De Caro A., 2000. Effect of environmental crop conditions to produce useful sunflower oil components. Meteorology 143: p. 252–265.

Kowalski R., 2007. GC analysis of changes in the fatty acid composition of sunflower and olive oil. Acta chromatographica, no. 18, p. 15-22.

Izquierdo N.G., Aguirrezabal A.N., and F. Andrade., 2009. Intercepted solar radiation affects oil fatty acid composition in crop species. Field Crop. Res., Vol. 114, p. 66-74.

Kaya M.D., Kolsarici O., 2011. Seed yield and oil content of some sunflower (*Helianthus annuus* L.) hybrids irrigated at different growth stages. African Journal of Biotechnology 10: p. 4591–4595.

Hladni N., Škorić D., Kraljević-Balalić M., Sakač Z., Jovanović D., 2006. Combining ability for oil content and its correlations with other yield components in sunflower (*Helianthus annuus* L.). HELIA, 29, Nr. 44, p. 101-110.

Onemli F., 2012. Changes in oil fatty acid composition during seed development of sunflower. Asian Journal of Plant Sciences, ISSN 1682-3974, p. 1-7.

Oshundiya F.O., Olowe V.I.O., Sowemimo F.A., Odedina J.N., 2014. Seed yield and quality of sunflower (*Helianthus annuus* L.) as Influenced by Staggered Sowing and Organic Fertilizer Application in the Humid Tropics.

Petcu E., Babeanu N., Popa O., Partal E., Pricop S.M., 2010. Effect of planting date, plant population and genotype on oil content and fatty acid composition in sunflower. Romanian Agricultural Research 27: p. 53–57.

Petcu E., Arsintescu A., Stanciu D., 2001. The effect of drought stress on fatty acid composition in some Romanian sunflower hybrids. Rom. Agr. Res., 15: p. 39-42.

Pereyra-Irujo G.A., Aguirrezábal L.A.N., 2007. Sunflower yield and oil quality interactions and variability: Analysis through a simple simulation model. Agric. For. Meteorol. 143, p. 252–265.

Sahari Khoufi, Khalil Khamasi, Jaime A., Teixeira da Silva, Salah Rezcui, Faycal Ben Jeddi, 2014. Watering regime affects oil content and fatty acid composition of six sunflower lines. Journal of New Sciences Vol. 7 (1), p. 1-9.

Teklewold A., Jayaramaiah H., Ramesh S., 1999. Genetic variability studies in sunflower (*Helianthus annuus* L.). Crop Impro., 26 (2): p. 236-240.

Vega A.J. and Hall A., 2002. Effect of planting date, genotype and their interactions on sunflower yield: II. Components of oil yield. Crop Sci., 42: p. 1202-1210.

Vrânceanu A.V., 2000. Floarea-soarelui hibridă. Ed. Ceres, Bucharest.

26

INFLUENCE OF SEED RATE AND FERTILIZATION ON YIELD AND YIELD COMPONENTS OF *Nigella sativa* L. CULTIVATED UNDER MEDITERRANEAN SEMI-ARID CONDITIONS

Ioannis ROUSSIS, Ilias TRAVLOS, Dimitrios BILALIS, Ioanna KAKABOUKI

Agricultural University of Athens, School of Agriculture, Engineering and Environmental Sciences, 75 Iera Odos Street, 118 55 Athens, Greece

Corresponding author email: bilalisdimitrios@yahoo.gr

Abstract

A field experiment was conducted to determine the effect of seed rate and fertilization on yield and yield components of Nigella sativa crop. The experiment was laid out according to a split-plot design with three replicates, two main plots (seed rates: 50 kg ha^{-1} and 60 kg ha^{-1}) and four sub-plots (fertilization treatments: untreated, compost, sheep manure, inorganic fertilizer). Plants were higher in plots sown at a rate of 60 kg ha^{-1} (18.2-22.7 cm). The highest number of capsules per plant (5.0-5.8) were found in sub-plots subjected to inorganic fertilization. Moreover, there were significant differences between fertilization treatment regarding seed yield and biological yield. The highest seed yield (911-1066 kg ha^{-1}) and biological yield (3864-4063 kg ha^{-1}) were found in inorganic treatments. The number of branches per plant, number of seeds per capsule, thousand-seed weight, and Harvest Index was not affected neither by seed rate nor by fertilization. Finally, there was not any significant interaction between seed rate and fertilization.

Key words: compost, inorganic fertilizer, Nigella sativa, seed rates, yield components.

INTRODUCTION

Medicinal plants are used for treating many disorders that are either non-curable or rarely cured by modern systems of medicine (Abadi et al., 2015). Over the last three decades, the use of herbal medicinal products and supplements have grown to such an extent that about 80% of the world population relying on them for some part of primary healthcare. (Ekor, 2014). *Nigella sativa* L. is an annual medicinal plant belonging to the Ranunculaceae family. It is native to southern Europe, North Africa, South and West Asia (Tuncturk et al., 2012) and widely cultivated throughout Syria, Egypt, Saudi Arabia, Iran, Pakistan, India and Turkeyfor seed yield and oil production (Riaz et al., 1996).

The height of plant ranges from 20 to 30 cm and its leaves are finely divided and linear with grayish-green color. The flowers are of pale blue and white color with 5 to 10 pedals (2-2.5 cm). The fruit is an inflated capsule composed of 3 to 7 united follicles. Each capsule contains a various number of dark grey or black small sized (1-5 mg) seeds (Valadabadi, 2013).

Seeds of *N. sativa* are characterized for their unique chemical properties that may contribute to the improvement of human health. Various surveys have shown that the whole seeds or their extracts have diuretic, antihypertensive, antidiabetic, anticancer, immunomodulatory, anthelmintic, analgesic, antimicrobial, anti-inflammatory, spasmolytic, bronchodilator, hepatoprotective, gastroprotective, nephron-protective, antihypertensive and antioxidant effects (Al-Jassir, 1992; Riaz et al., 1996; Ahmad et al., 2013). Moreover, the seeds are rich in fats, fiber, minerals such as Fe, Na, Cu, Zn, P, Ca and vitamins such as ascorbic acid, thiamin, niacin, pyridoxine, and folic acid (Takruri and Dameh, 1998). *N. sativa* seeds contain 30-35% oil and 0.5-1.5% essential oil which have several uses for pharmaceutical and food industries (Üstun et al., 1990; Ashraf et al., 2006). One of the most important constituents of essential oil is thymoquinone that belongs to the chemical class of terpenoids and imports the plant under investigation about the ability to influence on important human diseases such as cancer (Banerjee et al., 2010) or the metabolic syndrome (Razavi and Hosseinzadeh, 2014).

The effective use of *N. sativa* for different therapeutic purposes is highly dependent on the yield and quality (Datta et al., 2012). Seed rate is one of the main key factors for obtaining high yield and quality in the production of crops. Several studies carried out in countries where systematically cultivated, have demonstrated that suitable seed rate can increase the growth and yield of *N. sativa* (Toncer and Kizil, 2004; Tuncturk et al., 2004). Talafih et al. (2007) reported that under Mediterranean semi-arid environment the highest seed yield (801.2 kg ha^{-1}) was obtain under 35 kg seed ha^{-1}.

The appropriate use of fertilizers increases the growth and quality of the medicinal plants (Mohamed et al., 2014). Many experiments have been conducted to investigate the effect of different amounts of nitrogen (Ashraf et al., 2006; Tuncturk et al., 2012) and phosphate (Kizil et al., 2008) fertilizers on different agronomic characteristics, yield and yield components in *N. sativa*. According to Rana et al. (2012), the maximum values of agronomic characteristic such as plant height and number of branches and the highest yield of seed were observed at a ratio of 60:120 kg NP ha^{-1}.

Because *N. sativa* is a plant that has been systematically studied only during last two decades, the available literature on plant growth and yield under organic fertilization is still quite limited. Data by other researchers (Efthimiadou et al., 2009; Bilalis et al., 2012) clearly proved the beneficial effects of organic fertilization on the yields ofs everal crops.

Limited data are available regarding the performance of *N. sativa* growth under Mediterranean semi-arid conditions and organic cropping system. Therefore, the aim of this study was to determine the effects of seed rate and organic and inorganic fertilization on growth and yield parameters of nigella crop.

MATERIALS AND METHODS

A field experiment was carried out in the organic experimental field of the Agricultural University of Athens (Latitude: 37°59′ 1.70″ N, Longitude: 23°42′ 7.04″ E, Altitude: 29 m above sea level) from April to July 2016. The soil was clay loam (29.8% clay, 34.3% silt and 35.9% sand) (Bouyoucos, 1962) with pH (1:1 H$_2$O) 7.29, nitrate-nitrogen (NO$_3$-N) 12.4 mg kg^{-1} soil, available phosphorus (P) 13.2 mg kg^{-1} soil, available potassium (K) 201 mg kg^{-1} soil,

15.99% CaCO$_3$ and 1.47% organic matter (Wakley and Black, 1934). The site was managed according to organic agricultural guidelines (EC 834/2007). Meteorological data (mean monthly air temperature and precipitation) during the experimental period as recorded by the weather station of Agricultural University of Athens are presented in Figure 1. The mean temperature during the crop cycle (April-July) was higher (24.5°C) than that observed in 35-year average (22.7°C). Accumulated precipitation of cultivation period was 26.8 mm, which means it was approximately two-thirds lower than 35-year average (62.1 mm).

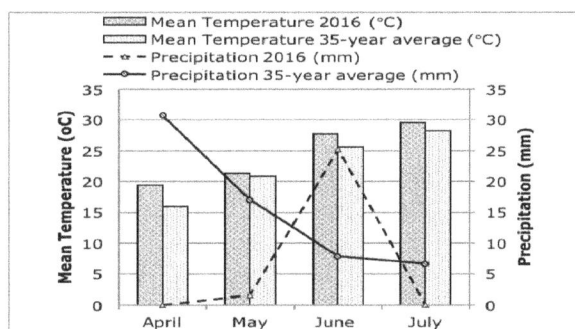

Figure 1. Meteorological data (mean monthly air temperature and precipitation) during the experimental period (April-July, 2016) and the 35-year average (1981-2015) in Athens, Greece

The experiment was set up on an area of 302 m^2 according to the split-plot design with three replicates, two main plots (seed rates: 50 kg ha^{-1} and 60 kg ha^{-1}) and four sub-plots [fertilization treatments: control (untreated), compost (2000 kg/ha), sheep manure (2750 kg/ha), inorganic fertilizer (15-15-15+5 S, 400 kg/ha)]. The main and sub-plot sizes were 42.25 m^2 (6.5 m X 6.5 m) and 9 m^2 (3 m X 3 m), respectively. There was a space of 0.5 m between sub-plots, 1 m between plots and 1 m between replications. Soil was prepared by ploughing at a depth of about 0.25 m. Fertilizers were applied by hand on the soil surface and then harrowed. *N. sativa* seeds were sown on 19th April by hand in rows 30 cm apart, at a depth of 1 cm. Drip irrigation system was installed in experimental plots. The first irrigation was applied immediately after sowing and the second irrigation was done two days later for uniform emergence. After that, plots were irrigated to field capacity every 3-4 days until the 10th day before harvest. The total quantity of water applied during the experiment was 735.2 mm. Throughout the experimental period, there was not observed any incidence of pest or disease on *N. sativa* crop. Weeds were controlled by hand hoeing when it was necessary. Finally, harvesting was done manually on 30th July 2016 after the stage of full maturity. Data were recorded on some growth and yield parameters including plant weight, number of branches and number of capsules per plant, number of seeds per capsule and thousand seed weight using five randomly selected plants from each sub-plot. Moreover, seed yield, biological yield and harvest index were assessed. For the determination of these parameters, the middle sub-plot area of 4 m^2 (2 m X 2 m) was used.

Plant weight was measured in centimeters at the stage of physiological maturity from the ground level to the tip of plant. Number of branches per plant was computed by counting the number of primary branches of five middle plants of the four middle rows. After that, the number of capsule per plant and number of seed per capsule on individual plant and capsule basis, respectively, were measured. For the computation of thousand seed weight, the seeds obtained from each of the ten selected plants were counted, weighed by analytical balance and expressed in grams. Biological yield was recorded by measuring the whole weight of plants from middle sub-plot areas after harvesting and drying for three days. Seed yield was determined by measuring the seed weight of plants used for biological yield. Harvest Index was calculated by divining seed yield by the biological yield.

The experimental data were checked for normality and subjected to statistical analysis according to the split-plot design. The statistical analysis was performed with JMP 9statistical software (SAS Institute Inc., Cary, NC, USA). The differences between means were compared using Least Significance Difference (LSD) test. All comparisons were made at the 5% level of significance.

RESULTS AND DISCUSSIONS

Results of variance analysis showed that the plant weight was significantly ($p<0.05$) affected by the different seed rates. In particular, the tallest plant (20.0 cm) recorded at the high seed rate (60 kg ha^{-1}), while, the lowest values (18.1 cm) were obtained from the low seed rate (50 kg ha^{-1}). These data are in accordance with previous studies reporting that increasing seed rates noticeably increased plant weight but with values lower than those of literature (Toncer and Kizil, 2004; Talafih et al., 2007).

These differences between our results and other studies could be explained by different variations in environmental conditions and the genetic potential of populations. Fertilization had not significant effect on plant weight. Despite the differences among fertilization treatments were not important, generally inorganic fertilizer gave slightly higher plant height (20.8 cm) and the minimum height obtained from control treatment (17.2 cm). Concerning number of branches per plant, the effect of different seed rates was found not to be statistically significant for the high and low seed rates. However, it was observed that the highest value (3.7) was obtained in the case of the low seed rate. This can be explained by the interplant competition in higher plant density which could induce plant growth reduction and produced the highest number of plants. Similar results were also reported by Tunctur et al. (2005). Fertilization treatments showed higher values than control in number of branches per plant but values were not statistically significant.

During this study, the highest number of branches per plant was achieved in inorganic fertilization treatment. Ali et al. (2015) mentioned that the number of primary branches per plant was significantly increased in the case of high nitrogen treatment, whereas number of secondary and tertiary branches per plant were not influenced.

The number of capsules per plant was only affected by different fertilization treatments. The highest value (5.4) was observed in inorganic fertilization treatment and the lowest was obtained in the untreated plots. Similar results concerning the positive response of nigella crop to inorganic fertilization were also recorded by other researchers (Rana et al., 2012; Khalid and Shedeed, 2015; Yiman et al., 2015).

Table 1. Plant height, number of branches per plant and number of capsules per plant and number of seeds per capsule of *Nigella sativa* as affected by different seed rates and fertilization

Fertilization	Seed Rate							
	Plant height (cm)		Number of branches per plant		Number of capsules per plant		Number of seeds per capsule	
	50 kg ha⁻¹	60 kg ha⁻¹	50 kg ha⁻¹	60 kg ha⁻¹	50 kg ha⁻¹	60 kg ha⁻¹	50 kg ha⁻¹	60 kg ha⁻¹
Control	16.2	18.2	3.0	2.2	3.0	1.7	37.0	36.6
Compost	18.7	19.4	3.5	3.3	5.4	4.3	42.5	40.8
Manure	18.5	19.5	3.7	2.3	4.7	4.6	41.6	37.9
Inorganic	18.9	22.7	4.6	2.7	5.8	5.0	42.6	42.0
$F_{seed\ rate}$	19.2417^* (LSD$_{5\%}$= 0.4274)		5.4516^{ns}		0.7557^{ns}		1.0852^{ns}	
$F_{fertilization}$	1.7297^{ns}		1.8958^{ns}		4.8728^* (LSD$_{5\%}$= 0.8672)		1.8025^{ns}	
$F_{seed\ rate\ X\ fertilization}$	0.3881^{ns}		1.2292^{ns}		0.9020^{ns}		0.1703^{ns}	

F-test ratios are from ANOVA. Significant at *p=0.05, **p=0.01, ***p=0.001, ns: not significant The LSD (p=0.05) for seed rate and fertilization are also presented.

Regarding number of seeds per capsule, the results have revealed that different seed rates had no significant effect on this trait. The results of number of seeds in capsule were lower than Toncer and Kizil (2004) demonstrated. Moreover, fertilization treatments resulted in values not statistically higher than control. However, the maximum number of seeds per capsule (42.3) was recorded after application of inorganic fertilizer. In addition, Hadi et al. (2015) found that application of manure in *N. sativa* crop did not affect the seed number per capsule.

Thousand-seed weight was not influenced neither by seed rate nor by fertilization. Despite the seed rates had no significant effect on thousand seed weight the high seed rate produced higher results (2.70 g). Talafih et al. (2007) supported that the increase in plant density resulted in an increase in the thousand-seed weight that could be related to the competition between plants on available soil moisture, nutrients and light. This result is in agreement with Ali et al. (2015) and Yimam et al. (2015) who found that the increase in levels of inorganic fertilizer did not affect the thousand-seed weight.

The results of experiment indicated that the effect of different seed rates on seed yield was insignificant. Despite that, the study data showed that the application of high seed rate led to declines in seed yield. Toncer and Kizil (2004) and Tuncturk et al. (2005) explained that increasing seed rate over 20 kg ha⁻¹, plant population density increased, which might cause higher interplant competition, thus reducing seed yield. Fertilization effect on this trait was statistically significant. The highest value (988 kg ha⁻¹) was recorded from inorganic treatment. This result is in agreement with Tuncturk et al. (2012) reported that yield components such as branches and capsules affects seed yield. In this study as mentioned above, the highest values of number of branches and capsules noted in inorganic treatment. Moreover, Yimam (2015) stated the maximum seed yield (1337 kg ha⁻¹) obtained from application of 60:40 kg NP ha⁻¹ and Seyyedi et al. (2015) reported than increase in amount of phosphorus might significantly increase seed and biological yield.

The biological yield was only influenced by fertilization. The maximum yield (3963 kg ha⁻¹) was achieved in the inorganic treatment followed by compost (3347 kg ha⁻¹) and manure (3263 kg ha⁻¹). Biological yield was actually affected by the combined supply of nitrogen and phosphorus.

The increase in these nutrients might be resulted in a more vigorous plant growth with greater plant height, number of branches, number of leaves and number of capsules producing a greater total plant biomass, thereby resulting in higher biological yield (Rana et al., 2012; Ali et al., 2015). This result is in full accordance with Yimam et al. (2015) reported that the highest biological yield was observed after application of 60 and 40 kg of N and P ha⁻¹, respectively.

The results obtained from variance analysis indicated that Harvest Index was not affected by different seed rates and fertilization.

Table 2. Thousand seed weight, seed yield, biological yield and Harvest Index of *Nigella sativa* crop
as affected by different seed rates and fertilization

Fertilization	Seed Rate							
	Thousand-seed weight (g)		Seed yield (kg ha^{-1})		Biological yield (kg ha^{-1})		Harvest Index (%)	
	50 kg ha^{-1}	60 kg ha^{-1}	50 kg ha^{-1}	60 kg ha^{-1}	50 kg ha^{-1}	60 kg ha^{-1}	50 kg ha^{-1}	60 kg ha^{-1}
Control	2.65	2.50	601	497	2471	2387	24.53	20.83
Compost	2.67	2.73	881	828	3455	3239.	25.50	25.57
Manure	2.58	2.79	820	773	3279	3247	25.00	23.81
Inorganic	2.68	2.76	1066	911	4063	3864	26.23	23.57
$F_{seed\ rate}$	0.1555^{ns}		3.5316^{ns}		0.5280^{ns}		3.9431^{ns}	
$F_{fertilization}$	1.1637^{ns}		6.7781** (LSD$_{5\%}$= 100.57)		9.3441*** (LSD$_{5\%}$= 291.73)		1.2908^{ns}	
$F_{seed\ rate\ X\ fertilization}$	1.5043^{ns}		0.3414		0.0465^{ns}		1.1118^{ns}	

F-test ratios are from ANOVA. Significant at *p=0.05, **p=0.01, ***p=0.001, ns: not significant The LSD (p=0.05) for seed rate and fertilization are also presented.

Despite the no significant effect of fertilization, the highest value (25.54%) was found in compost treatment followed by inorganic (24.90%) and manure (24.41%) treatments. Harvest Index values in our study were higher than Yimam et al. (2015) and lower than Rana et al. (2012) who studied the influence of various nitrogen and phosphorus rates in *N. sativa*.

CONCLUSIONS

The results of the present study indicate that nigella crop was not affected neither by seed rate nor by fertilization. The highest plant height was found under 60 kg ha^{-1} seed rate. Moreover, it was observed that plants received the inorganic fertilizer had higher number of capsules per plant which is a key yield component. Finally, there were significant differences between fertilization treatments regarding the seed yield and biological yield with highest values found under inorganic treatment.

REFERENCES

Abadi B.H.M., Ganjali H.R., Mobasser H.R., 2015. Effect of mycorrhiza and phosphorus fertilizer on some characteristics of black cumin. Biological Forum - An International Journal, 7(1): p. 1115-1120.

Ahmad A., Husain A., Mujeeb M., Khan S.A., Najmi A.K., Siddique N.A. et al., 2013. A review on therapeutic potential of *Nigella sativa*: A miracle herb. Asian Pacific Journal of Tropical Biomedicine. 3: p. 337-352.

Al-Jassir S.M., 1992. Chemical composition and microflora of black cumin (*Nigella sativa*) seeds growing in Saudi Arabia. Food Chemistry, 45: p. 239-242.

Ali M.M.K., Hasan M.A., Islam M.R., 2015. Influence of fertilizer levels on the growth and yield of black cumin *(Nigella sativa* L.). The Agriculturists, 13(2): p. 97-104.

Ashraf M., Ali Q., Iqbal Z., 2006. Effect of nitrogen application rate on the content and composition of oil, essential oil and minerals in black cumin (*Nigella sativa* L.) seeds. Journal of the Science of Food and Agriculture, 86: p. 871-876.

Banerjee S., Padhye S., Azmi A., Wang Z., Philip P.A., Kucuk O., Sarkar F.H., Mohammad R.M., 2010. Review on molecular and therapeutic potential of thy- moquinone in cancer. Nutrition and Cancer 62(7): p. 938-946.

Bilalis D., Kakabouki I., Karkanis A., Travlos I., Triantafyllidis V., Hela D., 2012. Seed and saponin production of organic quinoa (*Chenopodium quinoa* Willd.) for different tillage and fertilization. Notulae Botanicae Horti Agrobotanici Cluj-Napoca, 40(1): p. 42-46.

Bouyoucos G.J., 1962. Hydrometer method improved for making particle size analyses of soils. Agronomy Journal 54:464–65.

Datta A.K., Saha A., Bhattacharya A., Mandal A., Paul R., Sengupta, S., 2012. Black cumin (*Nigella sativa* L.) - A review. Journal of Plant Development Sciences, 4(1): p. 1-43.

Ekor M., 2014. The growing use of herbal medicines: issues relating to adverse reactions and challenges in monitoring safety. Frontiers in Pharmacology. 4 JAN, Article 177 (doi: 10.3389/fphar.2013.00177).

Efthimiadou A., Bilalis D., Karkanis A., Froud-Williams B., Eleftherochorinos I., 2009. Effects of cultural system (organic and conventional) on growth, photosynthesis and yield components of sweet corn (*Zea mays* L.), under semi-arid environment. Notulae Botanicae Horti AgrobotaniciCluj-Napoca, 37(2): p. 104-111.

Hadi M.R.H.S., Ghanepasand F., Darzi M.T., 2015. Evaluation of biofertilizer and manure effects on quantitative yield of *Nigella sativa* L., International

Journal of Biological, Biomolecular, Agricultural, Food and Biotechnological Engineering, 9(8): p. 799-802.

Khalid K.A., Shedeed M.R., 2015. Effect of NPK and foliar nutrition on growth, yield and chemical constituents in *Nigella sativa* L. Journal of Materials and Environmental Science, 6(6): p. 1709-1714.

Kizil S., Kirici S., Cakmak O., Khawar K.M., 2008. Effects of sowing periods and P application rates on yield and oil composition of black cumin (*Nigella sativa* L.). Journal of Food, Agriculture and Environment, 6(2): p. 242-246.

Mohamed N.M., Helmy A.M., Shiha A.A., Khalil M.N.I., 2014. Response of black cumin (*Nigella sativa* L.) to fertilization with chicken manure, mineral N fertilizer and varying K doses under different soil moisture contents. Zagazig Journal of Agricultural Research. 41(5): p. 1003-1019.

Rana S., Singh P.P., Naruka I.S., Rathore S.S., 2012. Effect of nitrogen and phosphorus on growth, yield and quality of black cumin (*Nigella sativa* L.). International Journal of Seed Spices, 2(2): p. 5-8.

Razavi B.M., Hosseinzadeh H., 2014. A review of the effects of *Nigella sativa* L. and its constituent, thymoquinone, in metabolic syndrome. Journal of Endocrinological Investigation, 37(11): p. 1031-1040.

Riaz M., Syed M., Chaudhary F.M., 1996. Chemistry of the medicinal plants of the genus *Nigella*. Hamdard Medicus, 39: p. 40-45.

Seyyedi S.M., Moghaddam P.R., Khajeh-Hosseini M., Shahandeh H., 2015. Influence of phosphorus and soil amendments on black seed (*Nigella sativa* L.) oil yield and nutrient uptake. Industrial Crop and Products, 77: p. 167-174.

Takruri H.R.H., Dameh M.A.F., 1998. Study of the nutritional value of black cumin seeds (*Nigella sativa*).Journal of the Science of Food and Agriculture, 76(3): p. 404-410.

Talafih K.A., Haddad N.I., Hattar B.I., Kharallah K., 2007. Effect of some agricultural practices on the productivity of black cumin (Nigella sativa L.) grown under rainfed semi-arid conditions. Jordan Journal of Agricultural Sciences 3(4): p. 385-397.

Toncer O., Kizil S., 2004. Effect of seed rate on agronomic and technologic characters of *Nigella sativa* L. International Journal of Agriculture and Biology, 6(3): p. 529-532.

Tuncturk M., Ekin Z., Turkozu D., 2005. Response of black cumin (*Nigella sativa* L.) to different seed rate growth, yield components and essential oil content. Journal of Agronomy, 4(3): p. 216-219.

Tuncturk R., Tuncturk M., Ciftci V., 2012. The effects of varying nitrogen doses on yield and some yield components of black cumin (*Nigella sativa* L.). Advances in Environmental Biology, 6(2): p. 855-858.

Ustun G., Kent L., Cekin N., Civelekoglu H., 1990. Investigation of the technological properties of *Nigella sativa*, L. (black cumin) seed oil. Journal of the American Oil Chemists' Society,67 (12): p. 71-86.

Valadabadi S.A., Farahani H.A., 2013. Influence of biofertilizer on essential oil, harvest index and productivity effort of black cumin (*Nigella sativa* L.). International Journal for Biotechnology and Molecular Research, 4(2): p. 24-27.

Wakley A., Black I.A., 1934. An examination of the Degtiareff methods for determining soil organic matter and a proposed modification of chromic acid titration method. Soil Science, 37: p. 29-38.

Yimam E., Nebiyu A., Mohammed A., Getachew M., 2015. Effects of nitrogen and phosphorus fertilizers on growth, yield and yield components of black cumin (*Nigella sativa* L.) at Konta district, South West Ethiopia. Journal of Agronomy, 14(3): p. 112-120.

27

FRACTAL ANALYSIS AS A TOOL FOR POMOLOGY STUDIES: CASE STUDY IN APPLE

Florin SALA, Olimpia IORDĂNESCU, Alin DOBREI

Banat University of Agricultural Sciences and Veterinary Medicine, "King Michael I of Romania", from Timisoara, Str. Calea Aradului 119, 300645, Timişoara, Romania

Emails: florin_sala@usab-tm.ro; alin1969tmro@yahoo.com; olimpia.iordanescu@yahoo.com

Corresponding author email: florin_sala@usab-tm.ro

Abstract

This study evaluated the possibility of characterization and discrimination of apple cultivars by fractal analysis of the shape of leaves. The geometry of the leaf lamina from five apple cultivars (Delicios de Voineşti, Florina, Generos, Jonathan and Pionier) was studied by fractal analysis, box-counting method, for obtaining fractal dimensions (D). Leaf lamina descriptor parameters (length - L, width – W, perimeter – P, and leaf area - LA) were determined for the comparative analysis with fractal dimensions in order to characterize the cultivars studied. Based on fractal dimensions (D) obtained from the analysis of the geometry of the leaf, the five apple cultivars studied were characterized and discriminated safely statistics (Cophenetic coefficient = 0.967). Regression analysis revealed the relationship of interdependence between leaf area and fractal dimensions (D) for each apple cultivar. From comparing the coefficient of variation of foliar parameters studied (L, W, P, LA), that the fractal dimensions (D) had the smallest variation within each apple cultivar, which shows that the fractal dimension (D) is a more stable parameter for characterization of the apple cultivars studied compared with other elements of the leaf lamina. According to the values of the correlation coefficient R^2 and safety parameters (p and RMSE), the fractal dimensions (D) facilitated the prediction of leaf area certainly higher than in the classics descriptors parameters of the leaves L, W and P ($R_D^2 = 0.991$ for Florina; $R_D^2 = 0.985$ for Generos and Pionier; $R_D^2 = 0.982$ for Delicios de Voineşti and $R_D^2 = 0.979$ for Jonathan). Fractal analysis has proven to be a tool with great power for the analysis, characterization and discrimination of the five apple cultivars studied, and predicting of the leaf area based on fractal dimension (D), which recommends it as a tool in pomology studies.

Key words: apple, fractal geometry, leaf shape, image analysis, texture.

INTRODUCTION

The description and characterization of plant leaves based on Euclidean geometry is difficult or impossible in most cases, so that the fractal geometry proposed by Mandelbrot (1983) was found to be a suitable tool to it. Fractal analysis has become a tool to study very accurate and useful for different fields, box-counting method showing a greater interest (Liebovitch and Toth 1989; Buchníček et al., 2000; Nežádal et al., 2001). Numerous studies have addressed by fractal analysis, various shapes and textures to develop models of representation in respect of the fractal dimensions (Backes and Bruno, 2010; Xu et al., 2012; Florindo and Bruno, 2013; Zhao et al., 2013).

Given the power of detection of the analyzed elements, and high precision in geometry study of various objects, shapes and textures, fractal analysis was used in the study of plants and vegetation. For such studies, estimating the complexity of leaves shape, can be made based on fractal dimension, by the method box-counting (Tricot, 1995; Buczkowski et al., 1998) or multiscale fractal dimension, which has developed based on Minkowski method (Costa and Cesar, 2000).

Bruno et al. (2008) and Backes and Bruno (2009) used fractal analysis, as multi-scale fractal dimension, for identification of plant species based on the complexity of leaf. De Araujo Mariath et al. (2010) based on fractal analysis have studied and characterized the leaves of several *Relbunium* species and have obtained a clear deceleration of species analyzed based on fractal dimensions. Characterization of lamina leaves based on fractal dimension for identification and recognition of different plant species, was carried out in other studies (Backes et al., 2009; Cope et al., 2012; Sun et al., 2013; Plotze et al., 2015).

Da Silva et al. (2015) used fractal analysis to characterize the over 50 species of Brazilian flora, based on the images of the section of median rib of leaves, considering that in biological studies such method is important because it is linked to the evolutionary aspects.

In this study it was used box-counting method for the analysis of fractal geometry of leaves for five apple cultivars, in order to characterization of these cultivars and development of study models based on fractal analysis in order to be used in pomology studies.

MATERIALS AND METHODS

The purpose of the study was to analyze the fractal geometry of the leaves of apple to characterize the five cultivars studied based on fractal dimensions and promote the method of fractal analysis as a tool in studies of pomology.

Biological material. The biological material was represented by five apple cultivars: Generous, Florina, Delicious de Voineşti, Jonathan and Pionier. Age plantation was 12 years old, of medium vigor rootstock (MM 110), and semi-intensive type of exploitation.

Sampling leaves. For the analysis of fractal geometry of leaves at the five apple cultivars studied, were randomly sampled 50 typical leaves at every cultivar, from the middle of the annual shoots during fruit growth phenophase; 50-70% final size of fruit, 75-77 BBCH code (Meier et al., 1994). The leaves were collected randomly from the crown of the tree, placed in plastic bags in cool box and transported to the laboratory for tests.

Measurements and determinations of leaves parameters. Each leaf was analyzed individually by measuring and scanning to obtain foliar parameter values (length, width, perimeter, leaf area). Leaf dimensions (length, width) obtained by measuring ruler with an accuracy of ± 0.5 mm. The perimeter of leaves and leaf area were obtained by the measuring method (Measured leaf area - MLA) and the scanning method (scanned leaf area - SLA) based on the model proposed by Sala et al. (2015).

Fractal analysis of leaves geometry. Based on binarized images (Figure 1), fractal geometry of each leaf was analyzed with ImageJ software (Rasband, 1997). Fractal dimensions (D) were obtained using box-counting, technique developed by Voss (1985) and commonly used to determine the fractal dimension based on the analysis of images (Li et al., 2009; Annadhason, 2012; Long and Peng, 2013). Fractal dimensions averages, equation (1), were calculated on the basis of intermediate values determined for each leaf equation (2).

$$Mean\ D = \sum (D) / GRIDS \qquad (1)$$

$$D = m\left[\frac{\ln(F)}{\ln \varepsilon}\right] \qquad (2)$$

where: D – fractal dimension;
  m – slope to regression line, from equation (3);
  F – number of new part;
  ε – scale applied to an object.

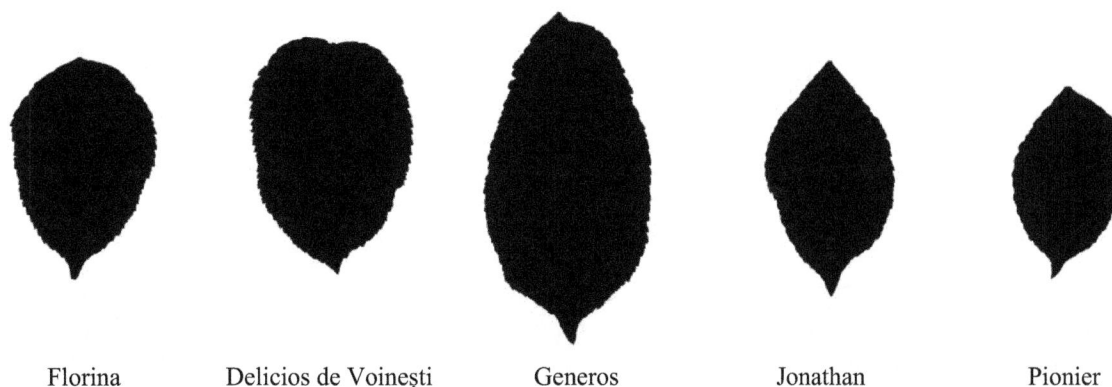

Florina   Delicios de Voineşti   Generos   Jonathan   Pionier

Figure 1. Binarized representation of the types of leaves in apple studied cultivars

$$m=(n\sum SC-\sum S\sum C)/(n\sum S^2-(\sum S)^2) \qquad (3)$$

where: m – slope of the regression line;
        S – log of scale or size;
        C – log of count;
        n – number of size;

Statistical analysis. The experimental data were processed by analysis of variance (ANOVA), regression analysis and multivariate analysis based on Euclidean distances (Cluster analysis). For this were used PAST software (Hammer et al., 2001) and Excel in Office 2007 suite. Statistical safety of the results, for descriptors parameters (L, W, P), leaf area (LA) and fractal dimensions (D), was evaluated based on Standard Error for regression line (SE) relationship (4), correlation coefficient $R2$, and safety parameters p and RMSE, equation (5).

$$SE=\sqrt{\frac{\sum C^2-b\sum C-m\sum SC}{n-2}} \qquad (4)$$

where: S = log of scale or size;
        C = log of count;
        n = number of size;
        b = y, intercept of the regression line;
        m = slope of the regression line.

$$RMSE=\sqrt{\frac{1}{n}\sum_{j=1}^{n}(y_j-\hat{y}_j)^2} \qquad (5)$$

RESULTS AND DISCUSSIONS

From the determinations of the leaves of apple tree, were obtained values for the leaves dimensions (length and width), and from the scanning method, were obtained values for the perimeter, and the leaf area of the leaves. Fractal geometry of each leaf was analyzed by Box-counting method, which facilitated the obtaining of the fractal dimensions (D), the results being presented in Table 1. Analysis of variance (ANOVA single factor type), showed statistical safety of experimental results ($p \ll 0.001$; $F_{teoretic} \ll F_{calculat}$, for Alpha = 0.001), Table 2.

Fractal dimensions, that describe the geometry leaves in the five apple cultivars studied, had similar values but through multivariate analysis based on Euclidean distances (Cluster analysis method), were differentiated two distinct clusters, safety statistics (Cophenetic coefficient = 0.967). A first cluster comprised two subgroups, one with the cultivar Florina, that has highest value of fractal dimension (D = 1.891) and the second with the cultivars Delicios de Voineşti (D = 1.884) and Generous (D = 1.885), with similar values of fractal dimensions. The second cluster included the cultivars Jonathan (D = 1.864) and Pionier (D = 1.865) with the lowest values of fractal dimensions.

Table 1. Descriptors parameters and fractal dimensions at apple leaves

Apple cultivars	L	W	P	LA	D
Florina	10.01±0.17	5.61±0.11	307.11±6.04	42.53±1.41	1.891±0.037
Delicios de Voineşti	9.68±0.25	5.49±0.13	306.50±8.91	42.38±1.97	1.884±0.041
Generos	10.77±0.25	5.51±0.14	322.63±9.55	43.25±2.07	1.885±0.039
Jonathan	9.11±0.16	4.82±0.08	262.91±5.41	32.21±1.01	1.864±0.045
Pionier	9.48±0.16	4.72±0.08	264.95±5.38	32.67±1.14	1.865±0.044

L – leaf length; W – leaf width; P – leaf perimeter; LA – leaf area; D – fractal dimension

Table 2. Variance analysis by ANOVA test

Source of Variation	SS	df	MS	F	P-value	F crit
Between Groups	15773359	4	3943340	5871.081	3.8E-229	4.647362
Within Groups	836210.2	1245	671.6548			
Total	16609569	1249				

Alpha = 0.001

The coefficient of variation of the fractal dimension (CV_D), revealed a differentiated degree of variability in the fractal geometry of leaves at the five apple cultivars studied. The highest variation of fractal dimensions was registered for the cultivar Generous (CV_{DG} = 2.3388) and the lowest variability of fractal dimensions was recorded for the cultivar Jonathan (CV_{DJ}) = 1.1299), the other cultivars having intermediate values of the coefficient of variation of fractal dimensions, Delicios de Voineşti (CV_{DDV} = 1.8163), Florina

(CV_{DF} = 1.2411), Pionier (CV_{DP} = 1.1491).
The others parameters descriptors of the leaves (L, W, P, LA) had the coefficients of variation with a larger range of expressions, their graphical distribution compared to the fractal dimensions (D), being shown in Figure 2. The coefficient of variation for leaf length (CV_L) had values between CV_{LP} = 12.1960 at Pioneer cultivar, and CV_{LDV} = 18.2677 at Delicious de Voineşti cultivar. Leaf width had coefficient of variation (CV_W) of between CV_{WJ} = 11.6519 at Jonathan cultivar and CV_{WG} = 18.5510 Generos cultivar. For leaf perimeter, coefficient of variation (CV_P) recorded values between

CV_{PF} = 13.9179 at Florina cultivar and CV_{PG} = 20.9393 at Generous cultivar. Leaf area had coefficient of variation (CV_{LA}) between CV_{LAJ} = 22.1478 at Jonathan cultivar and CV_{LAG} = 33.9624 at Generos cultivar.

From comparing the coefficient of variation of foliar parameters studied (Figure 2), it was observed that the fractal dimensions (D) had the smallest variation within each apple cultivar, which shows that the fractal dimension (D) is a more stable parameter for characterization of the apple cultivars studied compared with other elements of the leaf lamina.

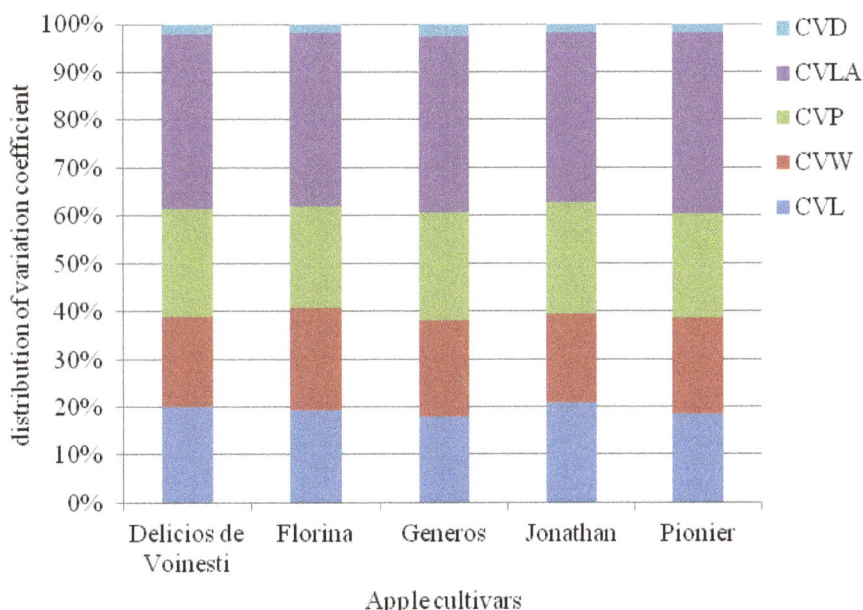

Figure 2. Graphical representation of the amplitude of coefficient of variation for descriptors parameters and fractal dimensions of leaves at the apple cultivars studied; (CV_L – coefficient of variation for length of the leaves; CV_W – coefficient of variation for width of the leaves; CV_P – coefficient of variation for perimeter of the leaves; CV_{LA} – coefficient of variation for the leaf area; CV_D – coefficient of variation for fractal dimensions)

Correlative analysis of fractal dimensions with descriptors parameters of leaves had shown any positive correlation with variable statistical certainty, Table 3. The fractal dimensions (D) had high correlation with the leaf area (R^2 = 0.960 – 0.985), and to the perimeter of the leaves (R^2 = 0.937 – 0.951), and lower values of correlation were recorded with the length of the leaves (R^2 = 0.829 – 0.858) and with the width of the leaves (R^2 = 0.885 – 0.922).

Among descriptor parameters of leaves, with the greatest practical utility is the leaf area. It is used to evaluate interception of sunlight (Wünsche and Lasko, 2000; Pandey and Singh, 2011; Fascella et al., 2013), the rate of photo-

synthesis and primary production (Guo and Sun, 2001; Bălan, 2010), nutritional status of plants and fruits quality (Fallahi et al., 2001; Nachtigall and Dechen, 2006; Jivan and Sala, 2014), tolerance to certain stress factors (Davies and Lasko, 1979; Wen et al., 2007). Also leaf area is useful for determining derivatives indices: leaf area index (LAI), leaf area duration (LAD), specific leaf area (SLA), net assimilation rate (NAR) used to characterize the status of plant and crops vegetation, in relation to technological or environmental factors (Williams, 1987; Williams and Ayards, 2005; Delalieux et al., 2008; Mokhtarpour et al., 2010; Poblete-Echeverría et al., 2015).

Table 3. Matrix table of correlations between descriptors parameters of leaves, and fractal dimensions of apple varieties studied

	L	W	P	LA	D
			Delicios		
L		9.02E-07	1.50E-22	5.89E-21	1.94E-15
W	0.631		4.11E-13	2.18E-15	1.97E-18
P	0.930	0.818		8.21E-31	4.69E-24
LA	0.918	0.856	0.969		3.57E-28
D	0.857	0.895	0.940	0.960	
			Florina		
L		2.47E-07	2.99E-16	1.50E-16	1.07E-13
W	0.655		3.60E-15	2.43E-20	2.03E-21
P	0.869	0.853		2.26E-28	2.53E-24
LA	0.873	0.913	0.961		1.04E-34
D	0.829	0.922	0.942	0.979	
			Generos		
L		4.14E-12	6.43E-20	1.54E-17	1.73E-15
W	0.798		1.59E-15	1.39E-21	5.34E-21
P	0.909	0.858		3.04E-26	1.28E-23
LA	0.885	0.923	0.952		3.99E-30
D	0.858	0.919	0.937	0.967	
			Jonathan		
L		6.76E-07	2.41E-19	8.02E-16	4.59E-14
W	0.636		1.29E-12	1.07E-19	1.61E-18
P	0.904	0.808		1.76E-27	3.97E-24
LA	0.863	0.907	0.957		2.03E-36
D	0.835	0.896	0.940	0.982	
			Pionier		
L		3.11E-06	1.72E-20	3.04E-16	1.43E-14
W	0.606		6.62E-12	4.01E-17	1.64E-17
P	0.914	0.793		1.48E-30	3.70E-26
LA	0.869	0.880	0.968		2.26E-38
D	0.844	0.885	0.951	0.985	

L – leaf length; W – leaf width; P – leaf perimeter; LA – leaf area; D – fractal dimension

Given the correlation between fractal dimension and the parameters descriptors of leaves that were identified in this study, regression analysis was used to predict leaf area based on fractal dimensions. From regression analysis were obtained prediction models of leaf area in the form of polynomial equations of order 2, relations (6) - (15) in Table 4, under which it was possible to predict leaf area with a high degree of statistical certainty according to the correlation coefficient R^2 and the parameters p and RMSE. Fractal dimensions ensured a safer prediction of leaf area compared with that achieved under the parameters descriptors (perimeter of the leaves - P, and leaf dimensions - L and W). According to the values of the correlation coefficient R^2 and parameter RMSE, the safest prediction of leaf area based on fractal dimension (D) was obtained at cultivar Florina ($R_D^2 = 0.991$), followed by cultivars Generous and Pioneer ($R_D^2 = 0.985$), the cultivar Delicios de Voineşti ($R_D^2 = 0.982$) and the cultivar Jonathan ($R_D^2 = 0.979$).

The perimeter of the leaves, facilitated predicting leaf area with lower level of statistical certainty compared to the fractal dimensions ($R_P^2 = 0.931 - 0.970$), the distribution of particular values of leaf area predicted based on fractal dimension (D) and on the perimeter of the leaves (P) being shown in figures 3-12. The length and width of leaves (L, W) have facilitated the prediction of leaf area certainly lower than the fractal dimensions (D) and perimeter (P) of the leaves ($R_W^2 = 0.753 - 0.895$, $R_L^2 = 0.672 - 0.879$).

Table 4. Prediction of leaf area in apple cultivars based on fractal dimensions and perimeter of leaves

Cultivar	Parameter for PLA	Equation	Equation number	R^2	P	RMSE
Delicios	P	$PLA_P = 2864.5x^2 - 10372x + 9411.5$	(6)	0.962	<<0.001	5.2405
	D	$PLA_D = 1624.1x^2 - 5707.1x + 5027.9$	(7)	0.970	<<0.001	2.1837
Florina	P	$PLA_P = -0.0003x^2 + 0.4373x + 0.9412$	(8)	0.941	<<0.001	5.3668
	D	$PLA_D = 808.42x^2 - 2628.8x + 2122.3$	(9)	0.991	<<0.001	1.3003
Generos	P	$PLA_P = -0.0005x^2 + 0.5319x + 73.471$	(10)	0.968	<<0.001	4.4593
	D	$PLA_D = 860.33x^2 - 2890.1x + 2432.9$	(11)	0.985	<<0.001	2.6492
Jonathan	P	$PLA_P = -0.0002x^2 + 0.269x - 26.483$	(12)	0.931	<<0.001	2.9974
	D	$PLA_D = 38.384x^2 + 189.6x - 454.49$	(13)	0.979	<<0.001	1.3277
Pionier	P	$PLA_P = 6E - 07x^2 + 0.2058x - 21.9$	(14)	0.952	<<0.001	2.0055
	D	$PLA_D = -82.4x^2 + 681.14x - 950.87$	(15)	0.985	<<0.001	1.3796

PLA – predicted leafa rea; P – perimeter of leaves; D – fractal dimension

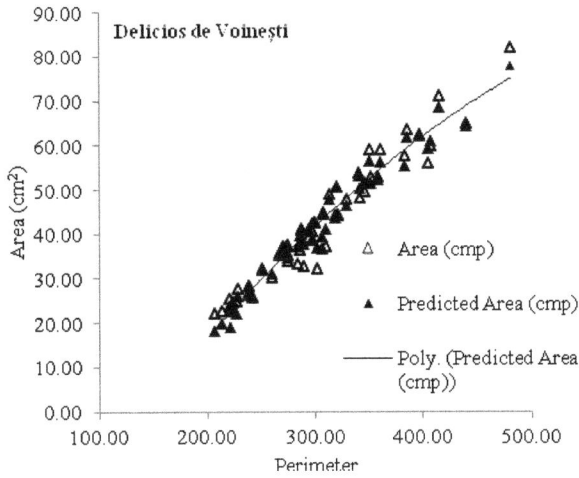

Figure 3. The particular distribution of leaf area values, depending on perimeter of leaves at the cultivar Delicios de Voineşti

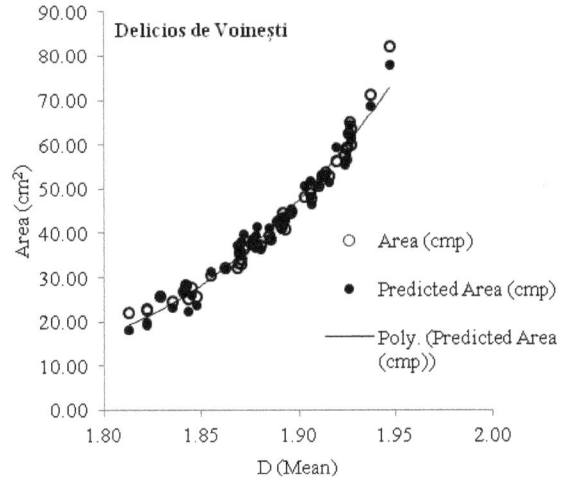

Figure 4. The particular distribution of leaf area values, depending on fractal dimension at the cultivar Delicios de Voineşti

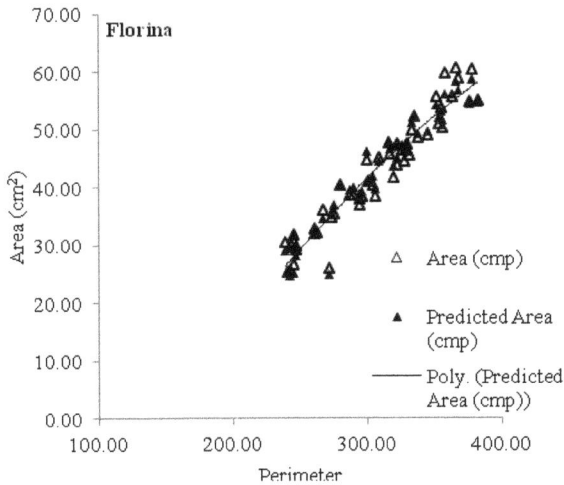

Figure 5. The particular distribution of leaf area values, depending on perimeter of leaves at the cultivar Florina

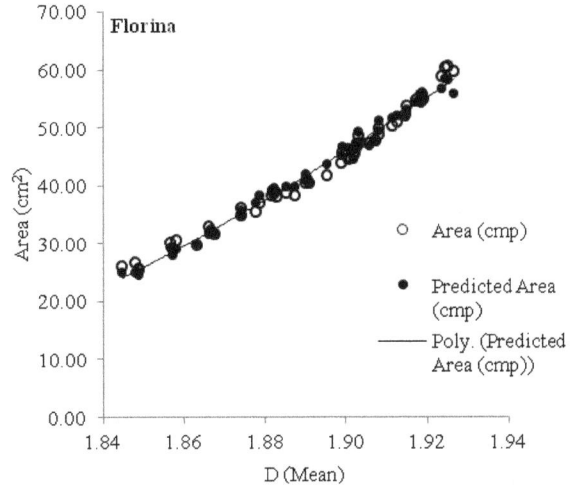

Figure 6. The particular distribution of leaf area values, depending on fractal dimension at the cultivar Florina

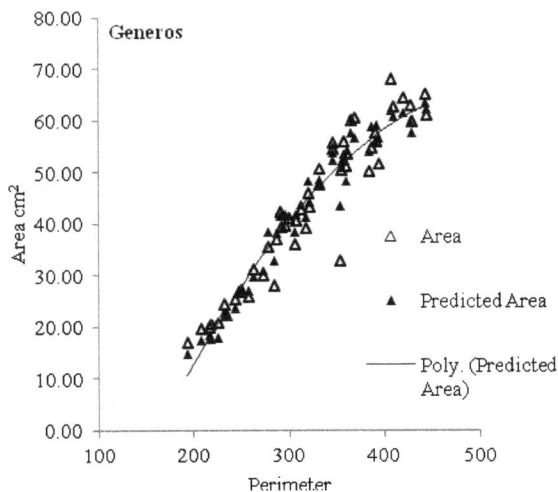

Figure 7. The particular distribution of leaf area values, depending on perimeter of leaves at the cultivar Generos

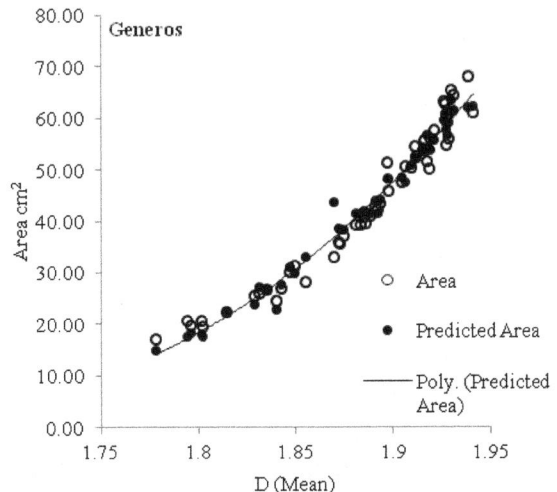

Figure 8. The particular distribution of leaf area values, depending on fractal dimension at the cultivar Generos

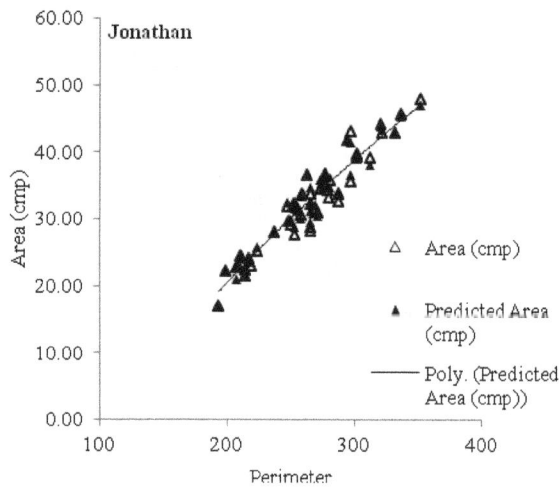

Figure 9. The particular distribution of leaf area values, depending on perimeter of leaves at the cultivar Jonathan

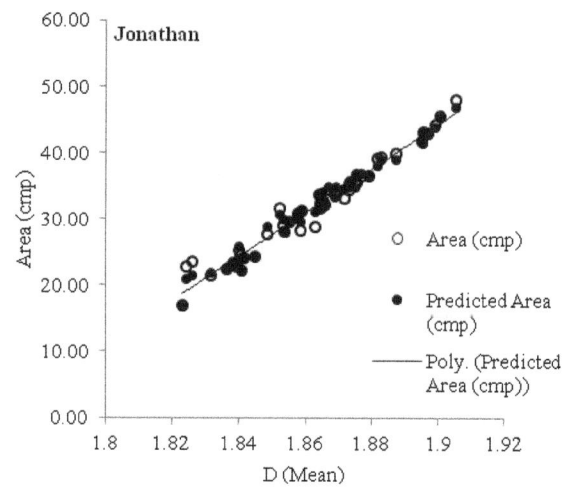

Figure 10. The particular distribution of leaf area values, depending on fractal dimension at the cultivar Jonathan

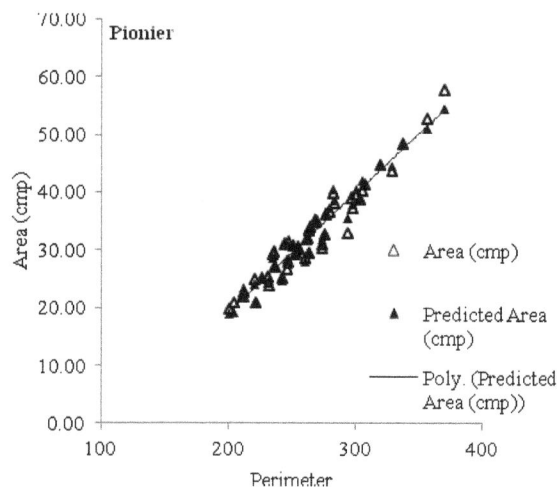

Figure 11. The particular distribution of leaf area values, depending on perimeter of leaves at the cultivar Pionier

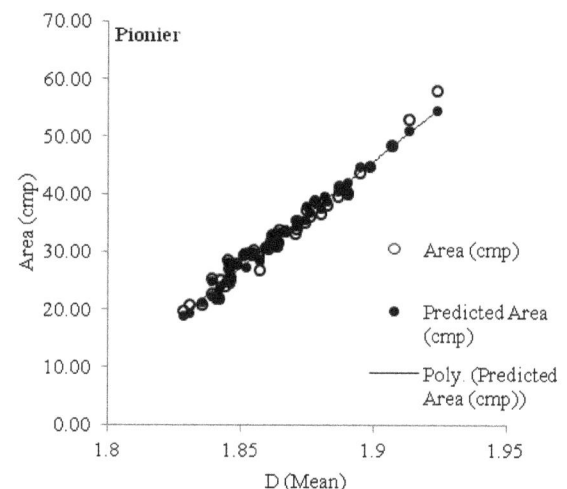

Figure 12. The particular distribution of leaf area values, depending on fractal dimension at the cultivar Pionier

Similar levels of safety in predicting leaf area based on L and W were reported and by Blanco and Folegatti (2003), Mokhtarpour et al. (2010) and Rouphael et al. (2010a, b), but their results of were not compared with fractal analysis.

In various studies, fractal analysis was used as an appropriate tool in order to characterize the structural complexity of plant species studied. Vlcek and Cheng (1986) have used the fractal analysis to study blade of the leaves at six species of trees, and the results have shown that the fractal dimension is sensitive to variations of the leaf within a species. They found the fractal dimension variation from average at red oak, white oak and maple sugar, depending on the W/L and the degree of jaggedness of the leaf.

Borkowski (1999) used the fractal analysis to characterize the leaves at 10 species of trees, and found that the fractal dimension appropriately defined can be used to discriminate between species with accuracy greater than 90%, especially when it is used together with other measurements. He summed that fractal analysis can be used in informatics systems for identifying species and taxonomic purposes only. Mancuso (1999, 2001) used the fractal dimension as an appropriate tool in ampelographic research, to characterization of over 10 vine genotypes, similar studies being conducted and by Oancea (2007) who has used fractal analysis to characterize six varieties for grapes table and wine. Jonckheere et al. (2006) used the fractal dimension (D) as a correction parameter for the evaluation of leaf area index (LAI) in the studies of some forest species in relation to the theoretical distribution of the foliage.

Based on morphological and fractals parameters, Mugnai et al. (2008) studied the possibility of identification of 25 genotypes at the species *Camellia japonica* L. The results highlighted the potential of those parameters to discriminate genotypes studied, having on hand a relatively accessible technology based on an optical scanner and a personal computer.

Rosatto et al. (2011) used fractal analysis to characterize the leaves, and as a tool for taxonomic analysis, at the different species of the Melastomataceae family. With all the difficulties of identifying species of that family without the presence of reproductive organs, analysis of fractal geometry leaves on both surfaces (adaxial and abaxial), facilitated parameterization and identifying those species with a precision much higher than other traditional methods. They concluded that the fractal approach may be useful for identifying species in the absence of reproductive material, and it is a quick and powerful tool for discrimination and identification of plant species. Gazda (2013), studying blackberry leaves through fractal analysis found that the fractal dimension is more related to the complexity of shape than the leaf size. And in the case of this study was found an interdependent relationship between the fractal dimension and leaf area and leaf descriptor parameters (L, W and P) for the five apple cultivars studied.

Systems of branching of the tree, the geometry of the stem bark in relation to the species and age of the trees, the root system or the leaves structure were studied by fractal analysis to characterize different species and genotypes of spontaneous or cultivated plants (Hartvigsen, 2000; Bruno et al., 2008; de Araujo Mariath et al., 2010; Cope et al., 2012; Du et al., 2013; Sala and Boldea, 2014). Zhang et al. (2011) have used fractal analysis to characterize the configuration of crown of apple tree "Fuji" after winter pruning, and have found the correlation of fractal dimensions (D) with the architecture of tree crown and fruit production. Fractal dimensions variation was mainly determined by branches distribution on the trunk and crown, and they concluded that the fractal dimensions can be used as an accurate

indicator for the evaluation of the crown and production at apple depending on the cutting trees maintenance.

Safe differentiation obtained from the analysis of three species of *Relbuniun* (Rubiaceae) by fractal analysis (based on fractal dimensions) was the argument which the Mariath Araujo et al. (2010) have considered that the fractal measure (fractal dimension), can be regarded as a taxonomic character in differentiating species, based on analysis of fractal properties at the leaf lamina. Da Silva et al. (2015), based on the study of more than 50 plant species, have found that fractal analysis is an accurate tool for the characterization and discrimination between different biological structures, the fractals descriptors (D) being accurate and reliable in the taxonomic identification of plant species.

Most studies used fractal analysis for the characterization of leaf lamina at various plants, for the characterization and discrimination of species or biotypes, for association of fractal properties with plants age, or biological material behavior evaluation to certain environmental conditions. Do not have found studies in which the fractal dimensions to be related with the foliar surface, usually the fractal dimensions are associated with the shape and complexity of the leaf lamina. In this study, the fractal dimensions were correlated also with leaf area, along with other foliar parameters (L, W, P), and facilitated the safe discrimination between apple cultivars studied, for which can be used as a precise tool for analysis in studies of pomology.

CONCLUSIONS

Based on fractal dimensions obtained from the analysis of fractal geometry at the apple leaves, box-counting method, the five apple cultivars studied were discriminated safely statistics (Cophenetic coefficient = 0.967) by multivariate analysis. Compared with the leaves descriptor parameters (L, W, P), the fractal dimension (D) offered a prediction of leaf area certainly higher, they being closely related to the fractal geometry of the leaf lamina. From the overall analysis of the results obtained (CV, prediction of leaf area based on L, W), the fractal dimension was more related to the complexity of leaves shape than of their size at the apple cultivars studied. Fractal analysis has proven to be a tool with great power to separation of the results, and discrimination of the ones five apple cultivars studied, which recommends her as a tool in studies of pomology.

ACKNOWLEDGEMENTS

The authors express thanks to the leaders and staff of the Didactic and Experimental Station of the Banat University of Agricultural Sciences and Veterinary Medicine "King Michael I of Romania" from Timişoara, Romania, Fruit-Vine Research Center, for facilitating the setup of the experimental field for this research.

REFERENCES

Annadhason A., 2012. Methods of fractal dimension computation. International Journal of Computer Science and Information Technology & Security (IJCSITS), 2(1): p. 166-169.

Backes A.R., Bruno O.M., 2009. Shape classification using complex network and Multi-scale fractal dimension. Image Analysis and Processing, 5716: p. 143-150.

Backes A.R., Casanova D., Bruno O.M., 2009. Plant leaf identification based on volumetric fractal dimension. International Journal of Pattern Recognition and Artificial Intelligence (IJPRAI), 23: p. 1145–1160.

Backes A.R., Bruno O.M., 2010. Shape classification using complex network and multi-scale fractal dimension. Pattern Recognition Letters, 31: p. 44–51.

Balan V., 2010. Methods of Foliar Surface Determination along the ontogenetic cycle on apple trees. Notulae Botanicae Horti Agrobotanici Cluj-Napoca 38(3): p. 219-222.

Blanco F.F., Folegatti M.V., 2003. A new method for estimating the leaf area index of cucumber and tomato plants. Horticultura Brasileira, Brasília, 21(4): p. 666-669.

Borkowski W., 1999. Fractal dimension based features are useful descriptors of leaf complexity and shape. Canadian Journal of Forest Research, 29(9): p. 1301-1310.

Bruno O.M., de Oliveira Plotze R., Falvo M., de Castro M., 2008. Fractal dimension applied to plant identification. Inform. Sciences 178: p. 2722–2733.

Buchníček M., Nežádal M., Zmeškal O., 2000. Numeric Calculation of Fractal Dimension. 3rd Prediction Conference, Faculty of Technology Zlin, BUT Brno, October 2 - 3: p. 10-15.

Buczkowski S., Kyriacos S., Nekka F., Cartilier L., 1998. The modified box-counting method: analysis of some characteristic parameters Pattern Recognition. 31 (4): p. 411–418.

Cope J.S., Corney D., Clark J.Y., Remagnino P., Wilkin P. 2012. Plant species identification using digital morphometrics: A review. Expert Systems with Applications, 39: p. 7562–7573.

Costa L.F., Cesar R.M., 2000. Shape Analysis and Classification: Theory and Practice. CRC Press, Pennsylvania.

Da Silva N.R., Florindo J.B., Gómez M.C., Rossatto D.R., Kolb R.M., Bruno O.M., 2015. Plant identification based on leaf midrib cross-section images using fractal descriptors. PLoS ONE, 10(6): e0130014.

Davies F.S., Lasko A.N., 1979. Diurnal and seasonal changes in leaf water potential components and elastic properties in response to water stress in apple trees. Physiologia Plantarum, 46(2): p. 109-114.

De Araujo Mariath J.E., dos Santos R.P., dos Santos R.P., 2010. Fractal dimension of the leaf vascular system of three *Relbunium* species (Rubiaceae). Brazilian Journal of Biosciences, 8(1): p. 30-33.

Delalieux S., Somers B., Hereijgers S., Verstraeten W.W., Keulemans W., Coppin P., 2008. A near-infrared narrow-waveband ratio to determine Leaf Area Index in orchards. Remote Sensing of Environment, 112(10): p. 3762-3772.

Du J., Zhai C.M., Wang Q.-P., 2013. Recognition of plant leaf image based on fractal dimension features. Neurocomputing, 116: p. 150–156.

Fallahi E., Chun I.-J., Neilsen G.H., Colt W.M., 2001. Effects of three rootstocks on photosynthesis, leaf mineral nutrition, and vegetative growth of "Bc-2 Fuji" apple trees. Journal of Plant Nutrition, 24(6): p. 827-834.

Fascella G., Darwich S., Rouphael Y., 2013. Validation of a leaf area prediction model proposed for rose. Chilean Journal of Agricultural Research, 73(1): p. 73-76.

Florindo J.B., Bruno O.M., 2012. Fractal descriptors based on Fourier spectrum applied to texture analysis. Physica A., 391: p. 4909–4922.

Gazda A., 2013. Fractal analysis of leaves: are all leaves self-similar along the cane? Ekologia, 32(1): p. 104-110.

Guo D.P., Sun Y.Z., 2001. Estimation of leaf area of stem lettuce (*Lactuca sativa* var angustana) from linear measurements. Indian Journal of Agricultural Sciences, 71(7): p. 483-486.

Hammer Ø., Harper D.A.T., Ryan P.D., 2001. PAST: paleontological statistics software package for education and data analysis. Palaeontologia Electronica, 4(1): p. 1-9.

Hartvigsen G., 2000. The analysis of leaf shape using fractal geometry. The American Biology Teacher, 62(9): p. 664-669.

Jivan C., Sala F., 2014. Relationship between tree nutritional status and apple quality. Horticultural Science, 41(1): p. 1-9.

Jonckheere I., Nackaerts K., Muys B., van Aardt J., Coppin P., 2006. A fractal dimension-based modelling approach for studying the effect of leaf distribution on LAI retrieval in forest canopies. Ecological Modelling, 197: p. 179-195.

Li J., Du Q., Sun C., 2009. An improved box-counting

method for image fractal dimension estimation. Pattern Recognition, 42: p. 2460-2469.

Liebovitch L.S., Toth T., 1989. A fast algorithm to determine fractal dimensions by box counting. Physics Letters A., 141(8–9): p. 386–390.

Long M., Peng F., 2013. A box-counting method with adaptable box height for measuring the fractal feature of images. Radioengineering, 22(1): p. 208-213.

Mancuso S., 1999. Fractal geometry-based image analysis of grapevine leaves using the box counting algorithm. Vitis, 38(3): p. 97-100.

Mancuso S., 2001. The fractal dimension of grapevine leaves as a tool for ampelographic research. HarFA - Harmonic and Fractal Image Analysis, 6-8.

Mandelbrot B.B., 1983. The fractal geometry of nature - Revised and Enlarged Edition, W.H. Freeman and Co., New York, 495 p.

Meier U., Graf H., Hess M., Kennel W., Klose R., Mappes D., Seipp D., Stauss R., Streif J., van den Boom T., 1994. Phänologische Entwick-lungsstadien des Kernobstes (*Malus domestica* Borkh. und *Pyrus communis* L.), des Steinobstes (Prunus-Arten), der Johannisbeere (Ribes-Arten) und der Erdbeere (*Fragaria x ananassa* Duch.). Nachrichtenbl. Deut. Pflanzenschutzd, 46: p. 141-153.

Mokhtarpour H., Teh C.B.S., Saleh G., Selamat A.B., Asadi M.E., Kamkar B., 2010. Non-destructive estimation of maize leaf area, fresh weight, and dry weight using leaf length and leaf width. Communications in Biometry and Crop Science, 5(1): p. 19–26.

Mugnai S., Pandolfi C., Azzarello E., Masi E., Mancuso S., 2008. *Camellia japonica* L. genotypes identified by an artificial neural network based on phyllometric and fractal parameters. Plant Systematics and Evolution, 270: p. 95–108.

Nachtigall G.R., Dechen A.R., 2006. Seasonality of nutrients in leaves and fruits of apple trees. Scientia Agricola, 63(5): p. 493-501.

Nežádal M., Zmeškal O., Buchníček M., 2001. The Box-Counting: Critical Study. 4th Prediction Conference, Institute of Information Technologies, Faculty of Technology, Tomas Bata University in Zlin, October 25-26:18.

Oancea S., 2007. Fractal analysis as an useful method in ampelography. Analele Ştiinţifice ale Universităţii "Al. I. Cuza" Iaşi Tom. III, S. Biofizică, Fizică medicală şi Fizica mediului, p. 53-56.

Pandey S.K., Singh H., 2011. A Simple, Cost-Effective Method for Leaf Area Estimation. Journal of Botany, ID 658240: p. 1-6.

Plotze R.O., Falvo M., Pádua J.G., Bernacci L.C., Vieira M.L.C., Oliveira G.C.X., Bruno O.M., 2005. Leaf shape analysis using the multiscale Minkowski fractal dimension, a new morphometric method: a study with *Passiflora* (Passifloraceae). Canadian Journal of Botany, 83: p. 287–301.

Poblete-Echeverría C., Fuentes S., Ortega-Farias S., Gonzalez-Talice J., Yuri J.A., 2015. Digital cover photography for estimating leaf area index (LAI) in apple trees using a variable light extinction coefficient. Sensors, 15: p. 2860-2872.

Rasband W.S., 1997. Image J. U. S. National Institutes

of Health. Bethesda, Maryland, USA, p. 1997-2014.

Rossatto D.R., Casanova D., Kolb R.M., Bruno O.M., 2011. Fractal analysis of leaf-texture properties as a tool for taxonomic and identification purposes: a case study with species from neotropical Melastomataceae (Miconieae tribe). Plant Systematics and Evolution, 291: p. 103–116.

Rouphael Y., Mouneimne A.H., Ismail A., Mendoza-de Gyves E., Rivera C. M., Colla G., 2010a. Modeling individual leaf area of rose (*Rosa hybrida* L.) based on leaf length and width measurement. Photosynthetica, 48(1): p. 9–15.

Rouphael Y., Mouneimne A.H., Rivera C. M., Cardarelli M., Marucci A., Colla G., 2010b. Allometric models for nondestructive leaf area estimation in grafted and ungrafted watermelon (*Citrullus lanatus* Thunb). Journal of Food, Agriculture and Environment, 8(1): p. 161-165.

Sala F., Boldea M., 2014. Fractal analysis of trunk bark in relation to age of trees: case study in plum. Bulletin UASVM Horticulture 71(2): p. 299-307.

Sala F., Arsene G.G., Iordănescu O., Boldea M., 2015. Leaf area constant model in optimizing foliar area measurement in plants: A case study in apple tree. Scientia Horticulturae, 193: p. 218-224.

Tricot C., 1995. Curves and Fractal Dimension. Springer-Verlag, New York.

Vlcek J., Cheung E., 1986. Fractal analysis of leaf shapes. Canadian Journal of Forest Research, 16(1): p. 124-127.

Wen X.P., Pang X.M., Matsuda N., Kita M., Inoue H., Hao Y.J., Honda C., Moriguchi T., 2008. Over-expression of the apple *spermidine synthase* gene in pear confers multiple abiotic stress tolerance by altering polyamine titers. Transgenic Research, 17(2): p. 251-263.

Voss R., 1985. Random fractal forgeries. In: Earnshaw R. (Ed.) Fundamental algorithms for computer graphics, Sringer Verlag, Berlin, p. 805-835.

Williams L.E., 1887. Growth of "Thompson Seedless" grapevines: I. Leaf area development and dry weight distribution. Journal of American Society and Horticultural Science, 112(2): p. 325-330.

Williams L.E., Ayars J.E., 2005. Grapevine water use and the crop coefficient are linear functions of the shaded area measured beneath the canopy. Agricultural and Forest Meteorology 132: p. 201–211

Wünsche J.N., Lasko A.N., 2000. The Relationship between leaf area and light interception by spur and extension shoot leaves and apple orchard productivity. HortScience, 35(7): p. 1202-1206.

Xu Y., Huang S., Ji H., Fernmüller C., 2012. Scale-space texture description on SIFT-like textons. Computer Vision and Image Understanding, 116(9): p. 999-1013.

Zhao Y., Jia W., Hu R.-X., Min H., 2013. Completed robust local binary pattern for texture classification. Neurocomputing, 106: p. 68-76.

Zhang S., Li B., Liu Y., Zhang L., Wang Z., Han M. 2011. Fractal characteristics of two-dimensional images of 'Fuji' apple trees trained to two tree configurations after their winter pruning. Scientia Horticulturae, 130: p. 102-108.

28

DISSIPATION OF ACETOCHLOR AND RESIDUE ANALYSIS IN MAIZE AND SOIL UNDER FIELD CONDITIONS

Irina Gabriela CARA[1], Florin Daniel LIPŞA[1], Mihai Sorin CARA[1], Lavinia BURTAN[2], Denis ȚOPA[1], Gerard JITĂREANU[1]

[1]Ion Ionescu de la Brad University of Agricultural Sciences and Veterinary Medicine of Iasi, 3 Mihail Sadoveanu Alley, Iaşi, 700490, Romania, Email: coroirina@yahoo.com

[2]National Research and Development Institute for Soil Science, Agrochemistry and Environment Protection – ICPA Bucharest, 61 Mărăşti Blvd., District 1, Bucharest, Romania, Email: lavinia.parvan@yahoo.com

Corresponding author email: topadennis@yahoo.com

Abstract

Acetochlor is a widespread used herbicide in maize crops; however, the environmental risk of its residues in the soil-plant system remains unknown. There was assessed the dissipation dynamics of acetochlor doses and its impact on residue level and microbial activity in soil over a season of vegetation. Since the herbicide was applied to the soil surface, its degradation varies as a dependence of concentration, soil type, pH, organic matter and environmental conditions. The field soil samples extraction in different imposed conditions of depths, time and herbicide application revealed a moving deeper of doses. The increased dose (80%+Rd) affects the persistence of acetochlor in the top layer by increasing its half-life from 14 to 17 days. Dissipation followed a first order kinetics. The diversity of soil microbial community changed after the introduction of acetochlor doses. An evident increase of bacteria and soil microorganisms was observed; however, fungal growth was prone to be inhibited. The higher concentration of herbicide was found to be safe, as well as the residues of acetochlor below maximum residue limits (MRL) at the end of maize crop season.

Key words: adsorption, degradation, microorganisms, residues.

INTRODUCTION

Environmental contamination with pesticides and heavy metals as a result of agricultural and industrial activities represents an ongoing concern (Gavrilescu, 2010). The protection of crops is a top priority for the agricultural productivity improvement to sustain an exponential growing population (Gupta et al., 2012). Sustainability of agricultural technologies, including intensive agriculture, integrated pest management and ecological farming is focused on soil quality, the relation between its use and management and the environment (Székács et al., 2014).

The widespread use of pesticide in agriculture leads to soil contamination, surface water and groundwater. These pesticides are broadly applied to crops in different stages of cultivation (preemergence, postemergence) to provide protection against pests and thus to prevent/reduce agricultural losses and to improve the production yield (Zhang et al., 2011). There are studies that organochlorine compounds are toxic, bioaccumulative and tend to persist due to the lipophilic characteristics (Gavrilescu, 2010; Miclean et al., 2011). Herbicides are the main type of pesticide causing public concern because of their implied short and long term risks for ecosystems and for human health (Pogăcean et al., 2009). However, many of these compounds have been proved a mutagenic and carcinogenic character. Acetochlor (2-chloro-N-ethoxymethyl-6'-ethylaceto-o-toluidide) is a selective herbicide used in conditions of pre-emergence or preplant to control mainly annual grasses and broadleaf weeds; its physical properties are given in Table 1. Acetochlor possessed the capability to be adsorbed by shoots and germinating plants roots and inhibits cell division by blocking protein synthesis (Tomlin et al., 2009). As a member of acetanilide herbicide family, it was widely detected as a pollutant in soil and surface water at concentrations values higher than the European Union accepted limit for drinking water of 0.1 μg L^{-1}. It has become a special concern due to its large quantity of use and potential for transport and accumulation. In 1997, acetochlor was already the fourth most used herbicide in US agriculture, with an annual use of approximately 14800 tons, which increased up to 16400 tons in 2002. The use of it in EU and also in Romania is in a lesser extend; it was ranked as the seventh most used herbicide in the EU in 2003, with an annual use of 2300 tons (Nadin et al., 2009).

Acetochlor is a B-2 carcinogen and may be removed in conditions that exceeds the limits of 0.10 μg L^{-1} in groundwater or 2.00 μg·L^{-1} as an annual average in surface water (Xiaoyin et al., 2011). Also it has been shown that acetochlor could induce metamorphosis of ranid species and accelerated T-3-induced metamorphosis in amphibians (Crump et al., 2002; Li et al., 2009).

Despite of its high ecological risks and wide range applications, there are only a few available data concerning acetochlor in environmental protection and its persistence under field conditions. Previous studies (Chao et al., 2007; Xiao et al., 2005; Zhou et al., 2006,) based on adsorption and degradations experiments have been demonstrated a high/medium risk of soil contamination, especially in phaeozem soil type.

Table 1. Physical chemistry data for acetochlor

Chemical structure	(chemical structure image)
Molecular formula	$C_{14}H_{20}ClNO_2$
Molecular weight	269.77 g mol^{-1}
Vapor pressure	3.4 x 10^{-8} mm Hg at 25°C
K$_{ow}$	300
Water solubility	233 mg L^{-1}

There is no report about its presence in cheornozem soil type from Romanian in temperate climate conditions. In this paper, the research was focused on the investigation of acetochlor evolution in soil and plants of maize fields. Gas chromatographic mass spectrometer was standardized for the quantitative determination of acetochlor from soil and plants, and on the residues level. Persistence studies were also carried out under field conditions to evaluate the impact of acetochlor

on the soil microbial communities in the rhizosphere soil zone.

MATERIALS AND METHODS

Reagents

Acetochlor (95% purity) was supplied by Dr. Ehrenstorfer GmbH, Augsburg, Germany. The commercial formulation of acetochlor (Guardian 820-860 g·L^{-1} active ingredient, Monsanto) was used for soil treatment. Analytical reagents including acetone, n-hexane and dichloromethane were bought from Merck, Germany and used for sample processing and extraction.

Field experiments

The area for field experiments on microbial activity and on the persistence of acetochlor in soil and plants was located at Ezăreni – The Experimental Farm of the Agricultural University Iasi (47°07′ N latitude, 27°30′ E longitude), Romania using a split plot design. The geographic location of the sampling site is depicted on Figure 1.

The soil is a cambic chernozem (SRTS, 2012) (haplic chernozem WRB-SR, 2006), with a clay-loamy texture, 6.96 pH, 3.06% humus content and a medium level of fertilization, without irrigation (Table 2).

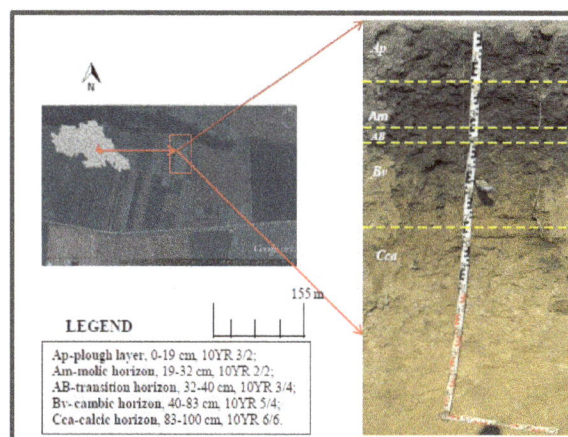

Figure 1. Spatial distribution of sampling and soil profile in Ezăreni area (Iaşi County)

Maize (variety Pioneer PR38V91) was sown in field plots and the size of each plot was 18 m x 7 m. Guardian (820-860 g L^{-1} a.i. acetochlor) was applied at three different dosages, i.e. 2.2 L ha^{-1} (Recommended dose), 3.1 L ha^{-1} (40%+Recommended dose) and 3.96 L ha^{-1}

(80%+Recommended dose) as a pre-emergent spray on maize crop at 3 days after sowing with the help of a knapsack sprayer. The high concentration (80%+Recommended dose) simulates a spill during the filling, while 40%+Recommended dose was to imitate overlap application of the herbicide.

Table 2. Main properties of the soil

Measurements	Amount
Bulk density (g cm^{-3})	1.33
pH (1:2.5)	6.96
Clay (g kg^{-1})	41.8
C_{oc} (%)	1.30
Ca+Mg+Na+K	17.47 meq 100 g^{-1}
Humus	3.06
Texture class	clay-loamy

Data collection

Soil samples for acetochlor persistence were randomly collected from 0-25 cm depth using a tube auger from 7-8 spots in each plot. Approximately 1000 g of soil was collected from each plot. The samples were taken at intervals of 0, 5, 10, 15 and 30 day after the initial herbicide treatment and after the crop harvest time from all the treated plots.

Plant samples from each plot were collected (500 g) at the crop harvest time. The samples were then subjected to different treatments: mixed thoroughly, air dried ground and passed through a 2 mm sieve and stored in sterile glass bottles in the dark at 4°C until analysis.

In order to assess the effect of acetochlor on soil microbiology, soil samples were collected before and after the treatment considering an interval of 7, 14 and 21 days. The procedure of samples collection included samples at a depth of 10 cm. After this step, they were processed by grinding and homogenization in a sterile mortar.

Pesticide extraction and residue analysis

The soil and plant samples were used in a solvent extraction procedure using accelerated solvent extraction (ASE) according with the Environmental Protection Agency (EPA) method 3545 for the analysis of organic compounds in solid matrices. A total quantity of 10 g from each sample was mixed in a mortar with 3 g Diatomaceous earth and the mixture was added directly to the extraction cell containing cellulose extraction filters. The extraction was performed under optimized conditions: extraction solvent acetone-hexane (1:1, v/v); temperature: 140°C; pressure: 1500 psi; heat-up time: 5 min; flush volume: 60%; purge: N2 60 s; number of cycles: 1.

Gas chromatographic analysis of acetochlor was performed on Agilent 7832 GC equipped with a mass spectrometer detector an auto-sampler, a split-splitless injector and a HP-5, fused silica capillary column. The column oven temperature program was used in different steps as follows: initial temperature 50°C, increased to 200°C at a rate of 30°C/min, increased to 280°C at 10°C/min and held for 1 min, and then increased to 310°C and held for 3 min. The injector temperature was set to 250°C in splitless mode (volume injected 1.00 μL) and MS temperature was 280°C. The carrier helium (99.999%) with a flow rate of 0.8 mL·min^{-1} was selected based on the instrument optimization results provided by the manufacturer's identification of peak and compared with the retention time of the compound with the standard solution.

Determination of soil microorganisms

The total numbers of microorganism and colony forming units (CFUs) of fungi and bacteria were determined by serial dilution and plating into selective media methods.

One gram of soil was mixed with 9 mL sterile water (dilution 10-1) and then 1 mL of the dilution 10-1 was poured into 9 mL sterile water (dilution 10-2). After a successive tenfold dilution series, 10-2 to 10-6 dilution were prepared. Aliquots (0.1 mL) of 10-2 to 10-6 dilution were spread on simple PDA (potato-dextrose-agar) medium for the total number of microorganisms. Similarly, aliquots (0.1 mL) of 10-2 to 10-6 dilution were spread on PDA with streptomycin (35 mg·kg^{-1}) medium for the number of bacteria. The numbers of bacteria or fungi were counted using the plate counting method after 24 hours for the bacteria colonies and 5 days for the fungi colonies. The experiments were performed in triplicates.

Recovery assay

A recovery assay was conducted to confirm the validity of the method described above. Known amounts of acetochlor were added to 10 g soil samples to give final spiked concentrations of

0.01 and 0.5 mg·kg^{-1} of dry soil. Extraction procedure and analysis were performed in triplicate as described previously.

RESULTS AND DISCUSSIONS

Evaluation of recovery

The average recoveries of acetochlor from the soil are shown in Table 3. The recoveries of acetochlor from soil ranged from 80.9% to 96.04% with a relative standard deviation (RSD) of less than 1.4%. The limits of detection and quantification were found to be 0.2 ng·g^{-1} and 0.67 ng·g^{-1} of dry soil, respectively. This data indicated that the extraction method is satisfactory for the analysis of residual acetochlor from soil.

Table 3. Average recovery and relative standard deviation of different samples

Fortified level (mg·kg^{-1})	Mean recovery (%)	Relative standard deviation (%)
0.01	81.6	1.4
0.5	94.1	0.4

Persistence and mobility of acetochlor

One application of acetochlor was giving residues to maize crop at all three doses applied. In the case of all three rates of application the highest amounts of acetochlor were always found in the top 0-10 cm soil layer (Figure 2).

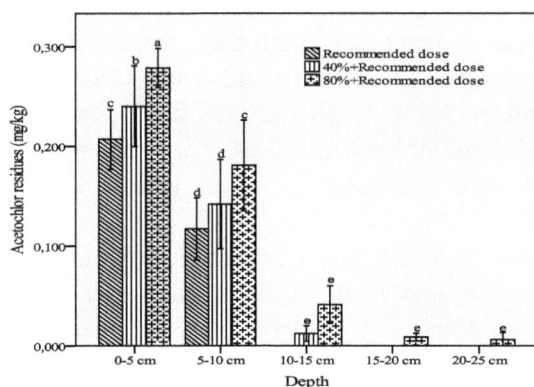

Figure 2. Distribution of acetochlor residues under field conditions. All values represent means ± standard deviation of triplicate samples. Means with different letters are significantly different (p<0.05) by Duncan test

This fact reflected a medium potential leaching of acetochlor herbicide, most like due to concentrations applied and rainfall events. It

should be also noticed a variation for concentration values between 0.301 and 0.304 mg·kg^{-1} acetochlor, after 5 days after application of 40 and 80% + Rd in the top 0-5 cm. This case indicates a lesser adsorption strength in the conditions of increasing of herbicide concentration.

As pointed out by adsorption isotherms (data not shown), the affinity of the acetochlor molecules and soil particles decreases with increases in acetochlor concentration. Several authors have reported L-type isotherms for acetochlor (Giles et al., 1974; Weber et al., 1989; Hiller et al., 2008).

According to these circumstances and taking into account the physicochemical properties, acetochlor seems to be more likely levigated particularly at high concentration. Also, the low adsorption and the rainfall event three hours after pesticide application lead to a dispersion to lower depths, especially at 80%+Rd followed by 40%+Rd. The residual acetochlor detected in depth of 5-10 cm were in the range of 0.169-0.259 mg·kg^{-1} but lower than in the case belonging to the depth of 0-5 cm soil. Baran et al. (2004) reported that the residual acetochlor had been detected at 60-70 cm depth of a Luvisol 7 days after being directly sprayed onto the soil surface.

In the experimental measurements (Figure 3) after 30 days of pesticide application it was shown that only 53.75% of the initially applied pesticide remained in 0-10 cm depth at 80%+Rd against, 49.15% in 40%+Rd and 44.89% in Rd. It can be observed a decreasing of acetochlor concentration value as a dependence of the increasing of time.

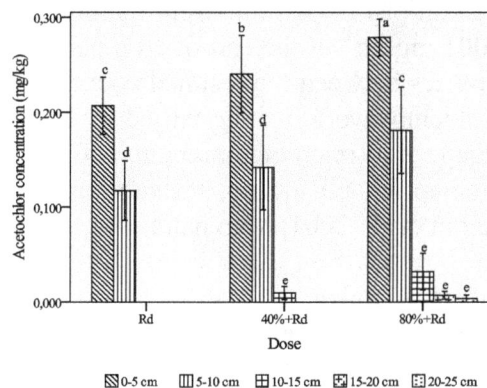

Figure 3. Soil residues after 30 days. All values represent means ± standard deviation of triplicate samples. Means with different letters are significantly different (p<0.05) by Duncan test

This fact is attributed to the combined effect between acetochlor degradation and its dispersion to a lower depth level. As the tested soil had relatively high humus content (3.06%) and proportion of clay-sand (22.9-41.8%), which probably caused the acetochlor molecules to be absorbed and therefore quickly degraded, as the acetochlor in the surface soil was prone to undergo highly chemical and biological loss and volatilization (Jablonkai 2000; Ma et al., 2004; Dictor et al., 2008; Hiller et al., 2009; Zhen et al., 2012).

In the case of recommended dose (Rd) after 30 days, it was observed the acetochlor residues were below calibration curves at the depth of 10-15 cm and no residue below 15 cm depth. While at 40%+Rd, herbicide residues reached up to 20 cm but no residues were after 25 cm. However, at values of 80%+Rd, in 10-15 cm depth, 9% of the initially applied concentration was reached after 30 days of acetochlor application. The traceable herbicide concentration also reached the depth 20-25 cm but no higher than 0.01 mg·kg^{-1}. Possible routes of acetochlor dissipation in the environment include plant species, climatic conditions, photo conversion and biotransformation via soil microorganisms and soil.

The moment corresponding to the end of crop period suggests that traceable concentrations were reached and otherwise completely degraded. In the conditions of the applied dose of 80%+Rd the residual acetochlor remaining at the harvest time were higher in surface soil 0-10 cm and around of 0.0027 mg·kg^{-1}. Similarly, at Rd and 40%+Rd variants, concentrations persist but were mainly limited to 0.001 mg·kg^{-1} at the end of crop period. The differences between the residual concentrations of acetochlor were mainly caused by leaching, the changes of microbial structure and function and the correlated specific metabolic pathways (Baudoin et al., 2001; Marchand et al., 2002).

Residues in maize

The active ingredient acetochlor was below calibration curves at recommended dose and 40%+Rd whereas at higher field rate (80%+Rd) was 0.0011 mg·kg^{-1} at harvest time. These residue levels for acetochlor could be related to the conjugation with GSH and cysteine, which has been observed in some plants as a mechanism of resistance to the herbicidal activity of the compound. The possible ways of acetochlor degradation and dissipation in environment and plants include plant and soil uptake and biotransformation via soil microorganism on soil and conversion to simpler products on plant surfaces.

Taken into account the final residues of acetochlor in maize were below the EPAs MRL (0.01 mg·kg^{-1}), it could be considered as safe for human beings and environment.

Microbial activity

Many of the pesticide used in modern agriculture present a high potential to influence the number and functions of a diverse range of soil microorganisms that contribute to soil microbiological processes and thus to soil fertility (Saha et al., 2012).

The effect of acetochlor doses on the total number of microorganisms is shown in Figure 4. Significant increase of soil biological activity was observed in all the variants where acetochlor was applied.

The number of microorganism g^{-1} in 40%+Rd variant was significantly increased compared to Rd and 80%+Rd variants, on 7 and 14 day and much greater than the level of control (before herbicide application) on day 21. Furthermore, the Rd and 80%+Rd biological activity, was almost equal and some lower compare to first variant. As it was noticed in our research and many other studies, the soil microorganisms generally react to herbicide molecule by increasing their biomass and activity although inhibitory (at 80%+Rd) effects have also been noted. This observation was in agreement with previously published studies (Zhen et al., 2012).

Compared to the number of microorganism g^{-1}, the ratio between the mains group of microorganism not only were significantly higher but also present a difference derived from the doses applied. The log values of fungal biomass at 40%+Rd ranged from 3.9 to 4.6 and were significantly increased from day 7 to day 21; in the case of Rd and 80%+Rd, their log values varied between 3.8 and 4.2 and were decreased to 0.465-0.477 log units from day 14 to day 21. The increase in fungal diversity may have resulted from the release of additional organic carbon as the acetochlor degraded, because the degradation half-life of acetochlor in soil is between 3 and 6 days (Zheng et al., 2001). Guo et al. (2009) reported that the soil fungi population increased at day 7 after application of 25 and 75 mg·kg^{-1} acetochlor, decreased at day 14, and recovered thereafter. Le et al., (2010) reported that the diversity indices changed rapidly after application of 250 mg·kg^{-1} and were lower than those from 50 and 150 mg·kg^{-1} acetochlor treatment. Comparing with these results, our research is more consistent.

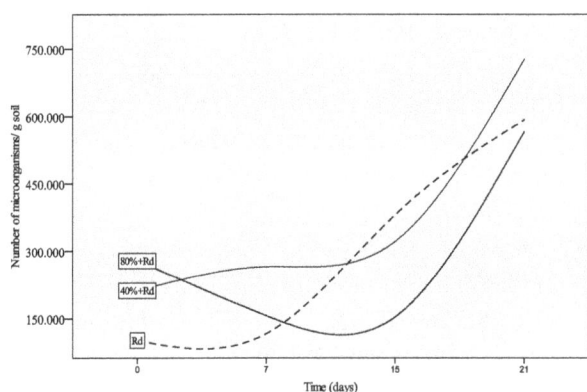

Figure 4. The effect of acetochlor doses on the total number of microorganisms

For the bacterial biomass completely reverse trends responding to acetochlor doses were observed: at 80%+Rd their log values were significantly decreased by 0.270 units on day 7 and 0.424 on day 14 respectively; thereafter, increased in day 21 and were almost equal to Rd variant. An increase in bacterial CFUs was observed in Rd and 40%+Rd from 7 to 21 day. Feng et al. (2008) studied the effects of acetochlor on soil microbiology and showed that acetochlor decreased the biomass of bacteria soon after application, but the biomass of bacteria recovered to a level similar to that of the control over time.

The increase in fungal diversity after acetochlor application may be explained due to the negative effect of acetochlor on bacterial population. This fact could result in a decrease in competition for nutrients among the remaining soil microbes (Le et al., 2010).

The soil fungal composition was altered after increased acetochlor treatment. As a consequence of the control soil (before acetochlor application) displayed no changes in the number of fungus, all changes in soil on the frequency were in response to acetochlor application. The results showed that acetochlor application led to an increase in the proportion of some common fungi such Penicillium, Fusarium and Aspergillus. In spite of a high dose, only one genus proliferated, while identical species in all three variants were obtained. Our results showed that the most significant shift appeared in day 7, whereas the drastic change in the Penicillium abundance occurred in day 14 and 21.

The temporal alteration of soil microbial might be due to different responses of microbial groups to the applied acetochlor: bacteria, affect in early stage the metabolism of soluble compounds whiles fungi degraded resistant complexes in the later decomposition phase.

Degradation

Concerning the chloroacetanilide herbicide effect on soil microbial activity to its persistence in soil it was necessary to establish the degradation kinetics of this molecule. The dissipation patterns of acetochlor in chernozem soil are presented in Figure 5.

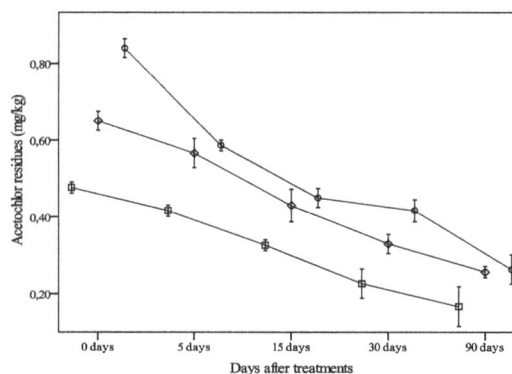

Figure 5. Degradation curves of acetochlor in soil. All values represent mean ± standard deviation of triplicate samples

The degradation followed a first order kinetics for all three doses. In soil, dissipation was quick during the first week and then slowed down from third week onwards.

The dissipation rate was lowest at recommended dose and 40%+Rd and highest in 80%+Rd. The half-lives of acetochlor were calculated as 13.86 and 17.32 days respectively with r2 values of 0.967 and 0.978 respectively.

Our results are in agreement with the reports of Xiao et al. (2006), where it was reported that the degradation rates may be influenced by the initial concentrations, because in soil the degradation rate of acetochlor is faster at the lowest concentration (5 mg·kg^{-1}) than at a highest one (80 mg·kg^{-1}).

Mills et al. (2001) reported that half-life of acetochlor in surface soils was 18 days, whilst in subsurface soils down to 4.6 mbs, range from 2 to 88 days. These data shows a relatively fast decline of residues in field soil due to environmental conditions that control soil temperature and moisture content.

CONCLUSIONS

Acetochlor was found in the Am horizon of the examined soil and it dissipated faster with concentration that not exceeds the EU contamination limits.

The residue trends of all three doses are in the same pattern but at different concentration levels.

Studies have shown that a fraction of acetochlor was retaining by cation exchange mechanism and therefore, the compound was available for plants and microorganisms.

All three doses applied caused changes in soil microbial diversity with enhancement of bacteria and microorganisms at a higher concentration.

ACKNOWLEDGEMENTS

This paper was accomplished with the support of European Social Fund, Human Resources Development Operational Programme 2007-2013, project no. POSDRU/159/1.5/S/132765.

REFERENCES

Andras Szekacs, Maria Mortl, Gábor Fekete, Ágnes Fejes, Béla Darvas, Miklós Dombos, Orsolya Szecsy, Attila Anton, 2014. Monitoring and biological evaluation of surface water and soil micropollutants in Hungary. Carpathian Journal of Earth and Environmental Sciences, 9(3), p. 47-60.

Chao L., Zhou Q.X., Chen S., Cui S., Wang M., 2007. Single and joint stress of acetochlor and Pb on three agricultural crops in northeast China. Journal of Environmental Sciences, 19(6), p. 719-724.

Crump D., Werry K., Veldhoen N., Van Aggelen G., Helbing C., 2002. Exposure to the herbicide acetochlor alters thyroid hormone dependent gene expression and methamorphosis in Xenopus Laevis. Environmental Health Perspectives, 110(12), p. 1199-1205.

Gavrilescu M., 2010. Behaviour of persistent pollutants and risks associated with their presence in the environment - integrated studies. Environmental Engineering and Management Jou., 9, p. 157-1531.

Gupta Manika, Garg N.K., Himanshu Joshi, Sharma M.P., 2012. Persistence and mobility of 2.4-D in unsaturated soil zone under winter wheat crop in subtropical region of India. Agriculture, Ecosystems and Environment, 146, p. 60-72.

Li W., Zha J.M., Li Z.L., Yang L.H., Wang Z.L., 2009. Effects of exposure to acetochlor on the expression of thyroid hormone related genes in larval and adult rare minnow (Gobioscypris rarus). Aquatic Toxicology, 94, p. 87-93.

Miclean Mirela, Oana Cadar, Cecilia Roman, Claudiu Tănăselia, Lucrina Ştefănescu, Codruţa Ileana Stezar, Ioan Ştefan Groza, 2011. The influence of environmental contamination of heavy metals and organochlorine compounds levels in milk. Environmental Engineering and Management Journal, 10(1), p. 37-42.

Mills M.S., Hill I.R., Newcombe A.C., Simmoms N.D., Vaughan P.C., Verity A.A., 2001. Quantification of acetochlor degradation in the unsaturated zone using two novel in situ field techniques: comparisons with laboratory generated data and implications for groundwater risk assessments, Pest Management Science, 57, p. 351-359.

Nadin P., 2009. The use of Plant Protection Products in the European Union, European Comission, Luxembourg.

Pogăcean Manuela Olga, Maria Gavrilescu, 2009. Plant protection products and their sustainable and environmentally friendly use, Environmental Engineering and Management Jou., 8(3), p. 607-627.

Supradip Saha, Deashis Dutta, Rajib Karmakar, Deb Prasad Ray, 2012. Structure toxicity relationship of chlororacetanilide herbicide: Relative impact on soil microorganisms. Environmental Toxicology and Pharmacology, 34, p. 307-314.

Tomlin C.D.S., 2009. The Pesticide Manual, 15th edition, British Crop Protection Council, Alton, Hants, UK.

Xiao H., Zhou Q.X., Ma L.Q., 2005. Joint effects of acetochlor and urea on germinating characteristics of crops seeds. Science in China Series C-Life Sciences, 48(S1), p. 1-6.

Xiao N., Jing B., Ge F., Liu X., 2006. The fate of herbicide acetochlor and its toxicity to Eisenia Fetida under laboratory conditions. Chemosphere, 62, p. 1366-1373.

Xiaoyin Sun, Qixing Zhou, Wenjie Ren, Xuhui Li, Liping Ren, 2011. Spatial and temporal dsitribution of acetochlor in sediments and riparian soils of the Songhua River Basin in northeastern China. Journal of Environmental Sciences, 23(10), p. 1684-1690.

Yaping Zang, Jun Yang, Ronghua Shi, Qingde Su, Li Yao, Panpan Li, 2011. Determination of acetanilide herbicides in cereals crops using accelerate solvent extraction, solid-phase extraction and gas chromatography-electron capture detector. Journal of Separation Science, 34, p. 1675-1682.

Zhou Q.X., Zhang Q.R., Liang J.D., 2006. Toxic effects of acetochlor and methamidophos on earthworms Eisenia fetida in Phaeozem, Northeast China. Journal of Environmental Scinces, 18(4), p. 741-745.

***SRTS, 2012. Romanian System of Soil Taxonomy, Sitech, Craiova, 157.

***WRB-SR, 2006. World Reference Base for Soil Resources, World Soil Report No. 103 FAO, Rome, 60 p.

29

RESPONSE OF *In vitro* CULTIVATED EGGPLANT (*Solanum melongena* L.) TO SALT AND DROUGHT STRESS

Ely ZAYOVA, Philip PHILIPOV, Trendafil NEDEV, Daniela STOEVA

Institute of Plant Physiology and Genetics, 21, Acad. G. Bonchev Street, 1113 Sofia, Bulgaria

Corresponding author email: e_l_y@abv.bg

Abstract

Salinity and drought are some of the major abiotic stresses that adversely affect eggplant productivity and quality. The response of in vitro cultivated plants of three eggplant cultivars - Lagra Negra (LN), China-A2 (CH) and Black Beauty (BB) to sodium chloride (NaCl) and polyethylene glycol (PEG$_{8000}$) induced stress was studied. The Murashige and Skoog, 1962 nutrient medium (MS + 1% sucrose and 0.8% agar) supplemented with NaCl (0; 50; 100; 150 and 200 mM) or PEG (0; 3; 8 and 10%) was used. The survival rate, percentage of rooted plants, plant height and root length were recorded after three weeks of cultivation. The addition of 50 mM NaCl to the MS culture medium influenced positively the plants growth, development and rooting. The survival percentage significantly decreased while increasing salt concentration in the culture medium, especialy at 100 mM NaCl. Concentration of 150 mM NaCl was toxic for the plants of all studied cultivars. The in vitro response of eggplant to drought stress was tested as well. The percentage of surviving plants also decreased while increasing PEG concentration in the culture medium. The appropriate concentrations of the abiotic agent (100 mM NaCl for salinity and 8% PEG for drought) for in vitro selection of tolerant eggplant plants were established. Salinity and drought due to NaCl and PEG respectively resulted in higher and more pronounced levels of proline in the cultivar LN plants, probably due to their better adaptation to abiotic stress.

Key words: *eggplant, in vitro selection, NaCl and PEG induced stress, survival rate, proline.*

INTRODUCTION

Eggplant (*Solanum melongena* L.) is a vegetable crop species belonging to the Solanaceae family. Eggplant possesses insufficient levels of resistance to biotic (diseases and pests) and abiotic (high salinity and drought) stress that cause serious crop losses. This problem has been addressed by hybridizing *S. melongena* L. with wild resistant *Solanum* species, which present a wide genetic diversity and are a source of useful agronomic traits (Anisuzzaman et al., 1993; Blestos et al., 1998). However, this approach is limited by sexual incompatibility (Collonier et al., 2001a) and difficulties in obtaining fertile progenies (Gleddie et al., 1986). The application of the biotechnological aspects of egg plant improvement has been used as a powerful tool for solving many target tasks in vegetable breeding. *In vitro* methods are successfully applied to improve both the resistance to diseases and pests and the tolerance to abiotic stress of eggplant (Collonier et al. 2001b; Kashyap et al., 2002; Kantharajah and

Golegaonkar, 2004). Eggplant tissues present high morphogenetic potential that makes this plant a suitable model for studies in different areas of plant science (Magioli and Mansur, 2005; Lakshman and Ravali, 2016).
Eggplant is moderately sensitive to salinity. The increasing of NaCl negatively affects almost all aspects of plant development, including germination and vegetative growth (Akinci et al., 2004; Krusteva et al., 2007). It affects as well the early growth characteristics of this species. Growth parameters in seedlings and young plants are also influenced by increasing the NaCl (Akinci et al., 2004). The plant response to increased salinity varies among species and cultivars. Drought is one of the most important abiotic stresses reducing crop growth and yield of eggplant. Tissue and cell cultures may be used to obtain plants tolerant to drought (Mohamed et al., 2000). Polyethylene glycol (PEG), sucrose, mannitol or sorbitol are osmotic stress agents used in the *in vitro* selection (Hassan et al., 2004). However, the effect of proline accumulation in osmotic adjustment is still controversial and

varies according to the type (Silveira et al., 2003; Meloni et al., 2004). Proline accumulation is a part of the adaptation to unfavorable environmental conditions, and a response to disturbance of the water balance of the cell at salt stress, low temperatures, heavy metals, and high acidity. Proline is synthesized in large quantities only when growth is sharply reduced due to stress. The key enzyme responsible for the accumulation of proline in the cell is pyrroline-5-carboxylate synthetase; at stress conditions, this enzyme is accumulated in large quantities (Kavi Kishor et al., 1995). The accumulation of proline can be used as a biochemical marker for salt tolerance in eggplant.

The aim of the present study was to investigate the influence of salinity (with NaCl) and drought (with PEG) on the plant growth, development and survival of *in vitro* cultured plants of different eggplant cultivars, and to select forms with enhanced tolerance to the studied abiotic factors.

MATERIALS AND METHODS

Initial plant materials and culture medium

The three eggplant commercial cultivars, Larga Negra (LN); China-A2 (CH) and Black Beauty (BB) used in the experiment were obtained from Institute of Plant and Genetic Resources, Agricultural Academy, Sadovo, Bulgaria. To study the effect of salinity and drought of the medium on the plant growth and development of the three eggplant cultivars, the basic Murashige and Skoog, 1962 medium (MS) containing 1% sucrose and 0.8% agar was used. This medium served as a control. To induce salinity and drought, the control medium was supplemented with 0; 50; 100; 150 and 200 mM NaCl and PEG at concentrations of 0; 3; 8, and 10%, respectively, with 40 plants. After application of the salt and drought stress to the three cultivars, the number of survived and rooted plants was counted after 21 days of cultivation. The survival rate was defined as the ratio of the number of survived plants in each count and the number of the initially transferred plants. Morphological parameters, such as the height of the aerial plant part and the root length, were measured. The plants showed normal morphological development, including minimal depression

and development of the vegetation tip, which was estimated by the normal growth of true leaves, at least 0.5 cm in length. After this period of cultivation, the roots and the damaged leaves were removed and the plants were transferred to the same medium free of NaCl or PEG.

Culture conditions

All media were adjusted to pH 5.8 with 1N NaOH or HCl, prior to gelling with agar-agar and were sterilized by autoclaving (121 °C for 20 min.). Cultures were maintained in growth room at $24\pm1°C$ temperature, under 16/8 h light/dark regime, 40 µmol m^{-2} s^{-1} light intensity provided by white fluorescent light tube 36 W (Phillips).

Content of free proline

The content of free proline was determined according to the method of Bates et al., 1973 in the control and experimental variants on day 21 after induction of stress conditions. The extinction of the top layer, colored in pink, was measured spectrophotometrically at a wave length of 514 nm. The amount of proline was calculated by a standard curve (γ proline/ ml), and the resulting value was defined by the formula:

$$[\mu mol/gFW] = \frac{E.\ V\ (ml\ toluene)}{FW\ (g).115.1}$$

Statistical analysis

The data were subjected to one-way ANOVA analysis of variance for comparison of means, and significant differences were calculated according to Fisher's least significance difference (LSD) test at the 5% significance level using a statistical software package (Statgraphics Plus, version 5.1 for Windows). Data were presented as means ± standard error.

RESULTS AND DISCUSSIONS

The effect of NaCl stress on the three eggplant cultivars

The effect of salt and drought stress on the growth of *in vitro* cultures, derived from seeds of three eggplant cultivars (LN, CH and BB) was examined. To assess the salt stress of the medium, the *in vitro* plants were transferred to a MS rooting medium containing 0, 50, 100, 150 and 200 mM NaCl for 21 days of cultivation. The plant growth was significantly reduced with the increase of NaCl concentration. The data about the influence of NaCl on survival rate, rooting and some morphological parameters of the *in vitro* cultivated plants of the three eggplant cultivars are presented of Tables 1 and 2. In our studies, we searched for this selective concentration of NaCl, which killed 70-80% of the plants, and allowed 20-30% of them to be viable for further development. The results show that the concentration of 50 mM NaCl in the culture medium has a positive influence on the growth, development and rooting of the plants from all three tested origins. After 21 days of cultivation on culture medium with 50 mM NaCl, 100% survival rate of all origins was registered. The plants were characterized by well-developed leaves and roots. Upon addition of 1% NaCl in the culture medium, the number of survived plants reduced by almost half, since part of the leaves turned yellow during the first week of culture (Figure 1). During cultivation of the plants in the culture medium with 150 mM NaCl, the plant number that survived salt stress significantly decreased. Only for cv. LN 15% survival was reported; however, the leaves turned yellow and rooting was not observed after three weeks of cultivation (Figure 1). The few survived plants were characterized with a very slow growth and by the end of the culture period, the plant growth was terminated. These results showed that at this stage of *in vitro* development the cv. LN was more tolerant to salinity compared to the cv. CH and cv. BB.

The concentration of 200 mM NaCl was toxic to the plants of the three studied cultivars. They were not able to overcome the salt stress and died - a fact that indicates that they are very sensitive to high concentration of NaCl. Inhibition of the development of the vegetative tip and scorching of leaves were also observed. Plant survival is often chosen as the main criterion for salt tolerance in eggplant (Akinci et al., 2004). This study has shown that the survival of *in vitro* cultivated plants depends on the cultivar and the stress intensity. Thus, the cv. LN maintained a survival rate of 50% with 100 mM NaCl and was classified as salt tolerant, and cv. BB - as salt sensitive. Depending on the cultivars, high NaCl concentration affected significantly the root growth. The positive effect of 50 mM NaCl concentration was demonstrated by the increased plant height and root length of the three cultivars, whose dimensions were larger compared to the control plants (Table 2). Well-developed plants with dark green leaves were observed; the height of the aerial plant part was 4.5; 3.7 and 4.1 cm for the three cultivars, and the roots length - 4.3, 4.0 and 3.7 cm, respectively. At 100 mM NaCl concentration the plant height was significantly reduced (2.0, 1.6 and 1.8 cm, respectively for the three origins) and the root length decreased to 1.1; 1.0 and 1.5 cm, due to necrosis observed in some of them. It was found that the root growth is more adversely affected by the application of salt stress than the aerial plant part.

Table 1. Influence of NaCl on the survival of *in vitro* cultivated *S. melongena* plants

Content of NaCl in the culture medium	Survived plants (%)			Rooted plants (%)		
	cv. LN	cv. CH	cv. BB	cv. LN	cv. CH	cv. BB
Control	100	100	100	80	55	70
50 mM NaCl	100	80	100	100	60	100
100 mM NaCl	50	40	30	35	20	25
150 mM NaCl	15	0	0	0	0	0

n = 40 plants

In fact, the slower growth is adaptive function for the plants survival under stress and the degree of salinity tolerance is often inversely proportional to the growth rate (Queiros et al., 2007). The reduction in growth not only helps the plant to conserve energy for tolerance but also limits the risk of inherited damage

(Hossain et al., 2007). During the further cultivation of the plants in the culture medium free of NaCl, the characteristic morphological type was recovered. The lack/presence of reparation capacity after stress may be used as one of the criteria for abiotic stress tolerance (Yasar et al., 2006).

Table 2. Influence of NaCl on the plant height and root length of *in vitro* cultivated *S. melongena* plants

Content of NaCl in the culture medium	Plant height (cm)			Root length (cm)		
	cv. LN	cv. CH	cv. BB	cv. LN	cv. CH	cv. BB
Control	3.8 ± 0.94^c	3.3 ± 0.86^b	3.4 ± 0.89^b	3.2 ± 0.81^b	2.5 ± 0.71^b	2.6 ± 0.78^b
50 mM NaCl	4.5 ± 1.32^d	3.7 ± 0.92^c	4.1 ± 1.15^c	4.3 ± 1.18^c	4.0 ± 1.24^c	3.6 ± 0.95^c
100 mM NaCl	2.0 ± 0.65^b	1.6 ± 0.28^a	1.8 ± 0.46^a	1.1 ± 0.38^a	1.0 ± 0.29^a	1.5 ± 0.52^a
150 mM NaCl	1.4 ± 0.57^a	0	0	0	0	0

Data are presented as means of 40 plants per treatment ± standard error. Different letters indicate significant differences assessed by the Fisher LSD test (5%) after performing ANOVA multifactor analysis.

K   50   100   150   200   mM NaCl

Figure 1. Effect of NaCl on the growth and rooting of *in vitro* cultivated plants of *S. melongena*, cv. LN

There was a progressive accumulation of proline while increasing the amount of NaCl (50, 100 and 150 mM) in the culture medium compared to a control, in which the levels of proline were significantly lower (Figure 2).
At 150 mM NaCl this amino acid peaked. At the studied concentrations of NaCl, the lowest value of the biochemical markers was reported for the plants in cv. BB, and the highest one - in cv. LN. Analysis of the plants indicated that the level of free proline has increased in response to the increasing concentrations of NaCl in the medium. A large increase in response to salt stress (100 and 150 mM) was observed, although the growth of the plants at these concentrations was minimal. It has been found that the level of proline in the plants tolerant to saline stress is higher than that in the sensitive ones (Ashraf and Harris, 2004).

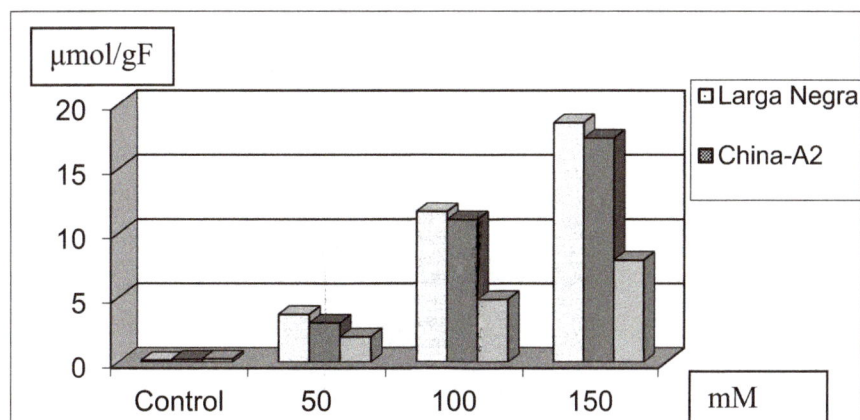

Figure 2. Content of proline in *S. melongena*, treated to salt stress

The effect of PEG stress on the three eggplant cultivars

Table 3 present details about the effect of PEG on the studied parameters in *in vitro* eggplant cultures. When the nutrient medium was supplemented with 3% PEG, the percentage of survived plants of all origins decreased. However, a significant difference with the control had not been proven. The percentage of rooting was almost equal to that in the control medium for the three cultivars, *i.e.* this concentration has a positive impact on root growth and plants rooting (Table 3, Figure 3). Concentration of 8% PEG proved to be suitable for selecting plants that survived the drought stress of the medium. Survived plants were 60, 45 and 40% for the three cultivars respectively, and the rooted ones were 50, 40 and 30%. The treatment of the plants with 8% PEG for a period of 21 days has a strong reducing effect on the plants height and roots length. The plants with withered leaves, with significantly reduced dimensions of the aerial plant part (1.9, 1.6 and 2.0 cm, respectively for the three cultivars) and reduced roots (1.5, 1.3 and 1.5 cm, respectively) were observed on this nutrient medium (Table 4). The slower growth and development of the plants under drought

conditions of the medium may be regarded as genetically conditioned adaptive response for their survival, while passing the stress effects (Yasar et al., 2006). Using various concentrations of PEG, different behaviors depending on the cultivar were observed. The results showed that 8% PEG significantly slowed down the growth and decreased the plants height and the roots length of cv. BB, while these morphological parameters of cv. LN were less affected by this concentration. The depressive effect of PEG takes place during root growth along the vegetative stage of development. The poor development of these parts is due to the increase of the osmotic pressure in the medium and to nutritional imbalance of the plant caused by a lack of absorption and/or transport of nutrients to the stem (Evelin et al., 2009).

Table 3. Influence of PEG on the survival of *in vitro* cultivated *S. melongena* plants

Content of PEG in the culture medium	Survived plants (%)			Rooted plants (%)		
	cv. LN	cv. CH	cv. BB	cv. LN	cv. CH	cv. BB
Control	100	100	100	80	75	70
3% PEG	90	75	70	80	70	70
8% PEG	60	45	40	50	40	30
10% PEG	10	0	0	0	0	0

n = 40 plants

Table 4. Influence of PEG on the plant height and root length of *in vitro* cultivated *S. melongena* plants

Content of PEG in the culture medium	Plant height (cm)			Root length (cm)		
	cv. LN	cv. CH	cv. BB	cv. LN	cv. CH	cv. BB
Control	4.2±1.14[c]	4.0±0.96[c]	3.8±0.89[b]	3.2±0.91[c]	2.7±0.85[b]	2.6±0.86[b]
3% PEG	2.5±0.83[b]	2.2±0.74[b]	2.0±0.67[a]	2.4±0.74[b]	2.0±0.61[b]	2.1±0.71[b]
8% PEG	1.9±0.51[b]	1.6±0.48[a]	2.0±0.58[a]	1.5±0.44[a]	1.3±0.37[a]	1.5±0.47[a]
10% PEG	0.8±0.23[a]	0	0	0	0	0

Data are presented as means of 40 plants per treatment ± standard error. Different letters indicate significant differences assessed by the Fisher LSD test (5%) after performing ANOVA multifactor analysis

Figure 3. *In vitro* rooted plants of *S. melongena*, cv. LN on MS medium a) control and b) with 3% PEG

The impact of drought on free proline content was directly dependent on the concentration of PEG as the rise in its level increased the content of proline compared to the control. The differences in the response of the three cultivars of eggplant to PEG induced stress was noticeable, since cv. LN accumulated more proline than the other cultivars. From Figure 4 it is seen that the increased concentration of the acting agent in the culture medium led to an increase of proline accumulation in the plants compared to the controls, in which the levels of proline were significantly lower. While higher levels of proline were estimated in cv. LN plants, the lowest levels of proline were found in cv. BB.

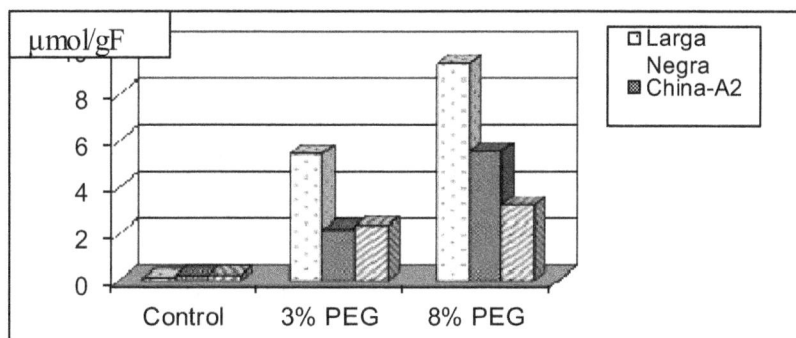

Figure 4. Content of proline in *S. melongena*, treated with PEG

The experimental results were similar for both types of abiotic stress - salinity and drought, as observed by the varietal differences. The higher content of proline in cv. LN showed some plasticity in its response to abiotic stress applications. The results obtained are in accordance with those reported by several other authors: the drought conditions inhibit the growth processes of the plants, leading to a number of metabolic changes.

CONCLUSIONS

The selective concentrations of 100 mM NaCl for salinity and 8% PEG for drought of the medium were identified. These *in vitro* stress conditions allow rapid screening of eggplant cultivars with increased tolerance to salinity and drought. Cv. LN overcomes in a greater extent both stresses of the culture medium in comparison with the other two origins – cv. CH and cv. BB. The presented *in vitro* model can be used for preliminary evaluation of different eggplant genotypes tolerance to salinity and drought of the culture medium. The accumulation of proline is considered a marker for abiotic stress, protecting tissues from damage.

REFERENCES

Akinci I., Akinci S., Yilmaz K., Dikici H., 2004. Response of eggplant varieties (*Solanum melongena* L.) to salinity in germination and seedling stages. New Zealand Journal of Crop and Horticultural Science, 32, p. 193-200.

Anisuzzaman M., Kamal H.M., Islam R., Hossain M., Joarder O.I., 1993. Genotypic differences in somatic embryogenesis from hypocotyls explants in *Solanum melongena* L., Plant Tissue Culture, 3, p. 35-40.

Ashraf M., Harris J.C., 2004. Potential biochemical indicators of salinity tolerance in plants. Plant Science, 166, p. 3-16.

Bates L.S., Waldren R.B., Teare E.D., 1973. Rapid determination of free proline for water stress studies. Plant and Soil, 39, p. 205-207.

Blestsos F.A., Roupakias D.G., Tsaktsira M.L., Scaltsojannes A.B., Thanassoulopoulos C.C., 1998. Interspecific hybrids between three eggplant (*Solanum melongena* L.) cultivars and two wild species (*Solanum torvum* Sw. and *Solanum sisymbriifolium* Lam.). Plant Breeding, 117, p. 159–164.

Collonnier C., Mulya K., Fock I., Mariska I., Servaes A., Vedel F., Souvannavong V., Ducreux G., Sihachakr D., 2001a. Source of resistance against Ralstonia solanaceraum in fertile somatic hybrids of eggplant (*Solanum melongena* L.) with *Solanum aethiopicum* L., Plant Science, 160, p. 301-313.

Collonnier C., Fock I., Kashyap V., Rotino G.L., Daunay M.C., Lian Y., Mariska I.K., Rajam M.V., Servaes A., Ducreux G., Sihachakr D., 2001b. Applications of biotechnology in eggplant, Plant Cell, Tissue and Organ Culture, 65, p. 91-107.

Evelin H., Kapoor R., Giri B., 2009. Arbuscular mycorrhizal fungi in alleviation of salt stress: a review. Annals of Botany, 104, p. 1263-1280.

Gleddie S., Keller W., Setterfield G., 1986. Production and characterization of somatic hybrids between *Solanum melongena* L. and *S. sisymbriifolium* Lam., Theoretical and Applied Genetics, 71, p. 613-621.

Hassan N.M., Serag M.S., El-Feky F.M., 2004. Changes in nitrogen content and protein profiles following *in vitro* selection of NaCl resistant mung bean and tomato. Acta Physiologiae Plantarum, 26, p. 165-175.

Hossain Z., Mandal A.K.A., Datta S.K., Biswas A.K., 2007. Development of NaCl tolerant line in *Chrysanthemum morifolium* Ramat through shoot organogenesis of selected callus line, Journal of Biotechnology, 129, p. 658-667.

Kantharajah A.S. and Golegaonkar P.G., 2004. Somatic embryogenesis in eggplant. Scientia Horticulturae, 99, 2, p. 107-117.

Kashyap V., Kumar V., Collonier C., Fusari F., Haicour R., Rotino G., Sihachkr D., Rajam M., 2002. Biotechnology of eggplant. Scientia Horticulturae, 1846, p. 1-25.

Kavi Kishor P.B., Hong Z., Miao C.H., Hu Ch.A. and Verma D.P.S., 1995. Overexpression Of Δ^1-Pyrroline-5-Carboxylate Synthetase increases proline production and confers osmotolerance in transgenic plants. Plant Physiology, 108, p. 1387-1394.

Krusteva L., Nikova V., Philipov Ph., Zayova E., Petkova A., 2007. Response of certain eggplant lines (*Solanum melongena* L.) to abiotic stress during germination. International Scientific Confereence, Plant Genetic Stoks-The Basis of Agriculture of Today , 13-14 June 2007, Sadovo, Bulgaria, 1, p. 215-217.

Lakshman N.M. and Ravali B., 2016. Plant regeneration in eggplant (*Solanum melongena* L.) a review. International Journal of Plant, Animal and Environmental Sciences, 6, p. 121-126

Magioli C. and Mansur E., 2005. Eggplant (*Solanum melongena* L.): tissue culture, genetic transformation and us as an alternative model plant. Acta Botanica Brasilica, 19, p. 139-148.

Meloni D.A., Gullota M.R., Martinez C.A., Oliva M.A., 2004. The effects of salt stress on growth, nitrate reduction and proline and glycine betaine accumulation in *Prosopis alba*. Brazilian Journal of Plant Physiology, 16, p. 39-46.

Mohamed M.A.H., Harris P.J.C., Henderson J., 2000. *In vitro* selection and characterisation of a drought tolerant clone of *Tagetes minuta*. Plant Science, 159, p. 213-222.

Murashige T. and Skoog F., 1962. A revised medium for rapid growth and bioassays with tobacco cultures. Physiologia Plantarum, 15, p. 473-497.

Silveira J.A.G., Viegas R.A., Rocha I.M.A., Moreira A.C.O.M., Moreira R.A., Oliveira J.T.A., 2003. Proline accumulation and glutamine synthetase activity are increased by salt-induced proteolysis in cashew leaves. Journal of Plant Physiology, 160, p. 115-123.

Queiros F., Santos F., Salema I., 2007. *In vitro* selection of salt tolerant cell lines in *Solanum tuberosum* L. Biologia Plantarum, 51, p. 728-734.

Yasar F., Ellialtioglu S., Kusvuran S., 2006. Ion and Lipid Peroxide Content in Sensitive and Tolerant Eggplant Callus Cultured under Salt Stress, European Journal of Horticultural Science, 71, 4, p. 169-172.

PERMISSIONS

LIST OF CONTRIBUTORS

Theodora Borugă, Costică Ciontu, Iulian Borugă and Dumitru-Ilie Săndoiu
University of Agronomic Sciences and Veterinary Medicine of Bucharest, 59 Mărăşti Blvd.,District 1, Bucharest, Romania

Ortansa Csutak, Emilia Sabău, Diana Pelinescu, Viorica Corbu, Ioana Cîrpici and Tatiana Vassu
University of Bucharest, Faculty of Biology, Department of Genetics, 1-3 Aleea Portocalelor,060101 Bucharest, Romania

Floricel Cercel, Mariana Stroiu and Petru Alexe
"Dunarea de Jos" University of Galati, Faculty of Food Science and Engineering, 111 Domneasca Street, 800201, Galati, Romania

Daniela Ianiţchi
University of Agronomic Sciences and Veterinary Medicine of Bucharest, 59 Mărăşti Blvd., District 1, 011464, Bucharest, Romania

Georgeta Dicu, Daniela Horhocea and Daniel State
SC Procera Agrochemicals Romania SRL, 47 Muncii Street, 915200, Fundulea, Calarasi, Romania

Viorel Ion
University of Agronomic Sciences and Veterinary Medicine of Bucharest, 59 Marasti Blvd,District 1, 011464, Bucharest, Romania

Nicoleta Ion
Apiculture Research and Development Institute of Bucharest, 42 Ficusului Blvd, District 1, 013975, Bucharest, Romania

Daniel Cristina
University of Agronomic Sciences and Veterinary Medicine of Bucharest,59 Mărăşti Blvd., District 1, Bucharest, Romania
National Agricultural Research and Development Institute Fundulea,1 Nicolae Titulescu Street, 915200, Fundulea, Călăraşi, Romania

Matilda Ciucă
National Agricultural Research and Development Institute Fundulea,1 Nicolae Titulescu Street, 915200, Fundulea, Călăraşi, Romania

Călina-Petruţa Cornea
University of Agronomic Sciences and Veterinary Medicine of Bucharest,59 Mărăşti Blvd., District 1, Bucharest, Romania

Nicolae Ionescu
Agricultural Research and Development Station Pitesti, Pitesti-Slatina Road no. 5, 117030, Pitesti, Romania

Pompiliu Chirilă
National Agricultural Research and Development Institute Fundulea, N. Titulescu Road no. 1,915200 Fundulea, Romania

Doru-Gabriel Epure, Marius Becheritu and Cristian-Florinel Cioineag
Probstdorfer Saatzucht Romania SRL, 20 Siriului Street, District 1, Bucharest,Romania

Georgi Komitov and Dimitar Kehajov
Agricultural University - Plovdiv, 12 Mendeleev Avenue, Plovdiv, Bulgaria

Mustafa Germec
Akdeniz University, Department of Food Engineering, 07058, Antalya, Turkey
Cankiri Karatekin University, Department of Food Engineering, 18100, Cankiri, Turkey

Ali Ozcan, Cansu Yilmazer, Nurullah Tas, Zeynep Onuk, Fadime Demirel and Irfan Turhan
Akdeniz University, Department of Food Engineering, 07058, Antalya, Turkey

Mihaela Ghiduruş and Mioara Varga
University of Agronomic Sciences and Veterinary Medicine of Bucharest,59 Mărăşti Blvd., 011464, Bucharest, Romania

Mevlüt Gül and Halil Parlak
University of Süleyman Demirel, Agriculture Faculty, Department of Agricultural Economics,32260 Isparta, Turkey

Muhammad Muddassir, Muhammad Mubushar and Muhammad Abubakar Zia
Department of Agricultural Extension and Rural Society, College of Food and Agricultural Sciences, King Saud University, Riyadh 11451, Kingdom of Saudi Arabia

Muhammad Shahid and Syed Muhammad Waqar Ahsan
Institute of Agricultural Extension and Rural Development, University of Agriculture, Faisalabad-38040, Pakistan

Ahmed Awad Talb Altalb
Department of Agricultural Extension and Technology Transfer, Faculty of Agriculture and Forestry, University of Mosul, Mosul, Iraq

Mehmood Ali Noor
Institute of Crop Science, Chinese Academy of Agricultural Sciences, Key Laboratory of Crop Physiology and Ecology, Ministry of Agriculture, Beijing 100081, China

Süleyman Kizil
Dicle University, Faculty of Agriculture, Department of Field Crops, 21280 Diyarbakir, Turkey

Khalid Mahmood Khawar
Ankara University, Faculty of Agriculture, Department of Field Crops, 06100 Ankara, Turkey

Veronica Tanasa
University of Agronomic Sciences and Veterinary Medicine of Bucharest,59 Mărăşti Blvd., District 1, Bucharest, 011464, Romania
Institute of Research and Development for Industrialization and Marketing of Horticultural Products - HORTING, 1A Intrarea Binelui, District 4, Bucharest,042159, Romania

Radu I. Tanasa
National Institute of Research "Cantacuzino", 103 Splaiul Independentei,District 5, 050096, Bucharest, Romania

Madalina Doltu
Institute of Research and Development for Industrialization and Marketing of Horticultural Products - HORTING, 1A Intrarea Binelui, District 4, Bucharest,042159, Romania

Gabriela Hristea
National Institute of Research and Development in Electrical Engineering ICPE-CA,
313 Splaiul Unirii, District 3, 030138, Bucharest, Romania

Andra Poruţiu, Iulia Mureşan, Felix Arion, Tudor Sălăgean and Teodor RUSU
University of Agricultural Sciences and Veterinary Medicine, 3-5 Mănăştur Street, 400372,Cluj-Napoca, Romania

Raluca Fărcaş
Technical University, 128-130 21 Decembrie 1989 Avenue, 400604 Cluj-Napoca, Romania

Galia Panayotova and Lubov Pleskuta
Trakia University, Faculty of Agriculture, 6000 Stara Zagora, Bulgaria

Svetla Kostadinova
Agricultural University, Faculty of Agronomy, Mendeleev 12, 4000 Plovdiv, Bulgaria

Neli Valkova
Field Crops Institute, 6200 Chirpan, G. Dimitrov 2, Bulgaria

Cristina Andreea Oprea, Doru Ioan Marin, Ciprian Bolohan, Alexandra Trif, Leonard Ilie, Mihai Gîdea and Aurelian Penescu
University of Agronomic Sciences and Veterinary Medicine of Bucharest, 59 Marasti Blvd, District 1, Bucharest, Romania

Adrian Trulea, Teodor Vintilă, Nicolae Popa and Georgeta Pop
University of Agricultural Sciences and Veterinary Medicine of Banat "Regele Mihai I al României", 119 Calea Aradului, Timişoara, Romania

Minodora Tudorache, Ioan Custura, Ilie Van and Andrei Marmandiu
University of Agronomic Sciences and Veterinary Medicine of Bucharest, 59 Mărăşti Blvd, District 1, Bucharest, Romania

Paul Anton
Aviagen Romania

Gheorghe Valentin Roman, Lenuta Iuliana Epure and Maria Toader
University of Agronomic Sciences and Veterinary Medicine of Bucharest, 59 Mărăşti Blvd,Distric t 1, 011464, Bucharest, Romania

Antonio-Romeo Lombardi
SC Lombardi Agro SRL, Macin, Tulcea County

Ely Zayova, Ludmila Dimitrova and Maria Petrova
Institute of Plant Physiology and Genetics, 21, Acad. G. Bonchev, Str., 1113 Sofia, Bulgaria

Milena Nikolova
Institute of Biodiversity and Ecosystem Research 23, Acad. G. Bonchev, Str., 1113 Sofia, Bulgaria

Florin Sala, Ciprian Rujescu and Cristian Constantinescu
Banat University of Agricultural Sciences and Veterinary Medicine "King Michael I of Romania", Timisoara, Calea Aradului 119, 300645, Timişoara, Romania

Mihaela Popa and Narcisa Băbeanu
University of Agronomic Sciences and Veterinary Medicine of Bucharest, 59 Mărăşti Blvd.,District 1, Bucharest, Romania

Gabriel Florin Anton and Elena Petcu
National Agricultural Research and Development Institute Fundulea, 1 Nicolae Titulescu Street, Romania

Luxița Rîşnoveanu
Agricultural Research and Development Station from Brăila, Romania

Ioannis Roussis, Ilias Travlos, Dimitrios Bilalis and Ioanna Kakabouki
Agricultural University of Athens, School of Agriculture, Engineering and Environmental Sciences,75 Iera Odos Street, 118 55 Athens, Greece

Florin Sala, Olimpia Iordănescu and Alin Dobrei
Banat University of Agricultural Sciences and Veterinary Medicine, "King Michael I of Romania",from Timisoara, Str. Calea Aradului 119, 300645, Timişoara, Romania

Irina Gabriela Cara, Florin Daniel Lipşa, Mihai Sorin Cara, Denis Țopa and Gerard Jităreanu
Ion Ionescu de la Brad University of Agricultural Sciences and Veterinary Medicine of Iasi,3 Mihail Sadoveanu Alley, Iaşi, 700490, Romania

Lavinia Burtan
National Research and Development Institute for Soil Science, Agrochemistry and Environment Protection – ICPA Bucharest, 61 Mărăşti Blvd., District 1, Bucharest, Romania

Ely Zayova, Philip Philipov, Trendafil Nedev and Daniela Stoeva
Institute of Plant Physiology and Genetics, 21, Acad. G. Bonchev Street, 1113 Sofia, Bulgaria

Index